ELECTROCHEMICAL POWER SOURCES

THE ELECTROCHEMICAL SOCIETY SERIES

ECS-The Electrochemical Society
65 South Main Street
Pennington, NJ 08534-2839
http://www.electrochem.org

A complete list of the titles in this series appears at the end of this volume.

ELECTROCHEMICAL POWER SOURCES

Batteries, Fuel Cells, and Supercapacitors

VLADIMIR S. BAGOTSKY
ALEXANDER M. SKUNDIN
YURIJ M. VOLFKOVICH

Published by John Wiley & Sons, Inc., Hoboken, New Jersey
Published simultaneously in Canada

For general information on our other products and services or for technical support, please contact our
Customer Care Department within the United States at (800) 762-2974, outside the United States at
(317) 572-3993 or fax (317) 572-4002.

Wiley also publishes its books in a variety of electronic formats. Some content that appears in print may
not be available in electronic formats. For more information about Wiley products, visit our web site at
www.wiley.com.

Library of Congress Cataloging-in-Publication Data:

Bagotskii, V. S. (Vladimir Sergeevich)
  Electrochemical power sources : batteries, fuel cells, and supercapacitors / Vladimir S. Bagotsky,
Alexander M. Skundin, Yurij VM. Volfkovich.
     1 online resource.
  Includes bibliographical references and index.
  Description based on print version record and CIP data provided by publisher; resource not viewed.
    ISBN 978-1-118-94253-6 (epub) – ISBN 978-1-118-94251-2 (pdf) – ISBN 978-1-118-46023-8 (cloth)
1. Electric batteries. 2. Fuel cells. 3. Supercapacitors. I. Skundin, A. M. II. Volfkovich, Yurij V., 1940-
III. Title.
  TK2896
  621.31′242–dc23
                                        2014023307

Printed in the United States of America

10 9 8 7 6 5 4 3 2 1

Dedicated to Professor V. S. Bagotsky.

CONTENTS

FOREWORD

When the major part of this book was written, Vladimir Bagotsky, its initiator and the first author passed away in his home in Boulder, Colorado, at the age of 92. It was a shock for us as all our scientific life was connected with Bagotsky who was our teacher and friend. We hope this book will be some kind of a memorial to this outstanding scientist, recognized authority in electrochemistry and battery science.

A. SKUNDIN
YU. M. VOLFKOVICH

ACKNOWLEDGEMENTS

We are very much obliged to V. Bagotsky's daughter Natalia Bagotskaya and to his granddaughter Katya Lysova for their invaluable assistance at the manuscript preparing.

We are grateful also to Dr. Marie Ehrenburg, who translated several chapters from Russian.

A. SKUNDIN
YU. M. VOLFKOVICH

PREFACE

In 1980 Academic Press London & New York published the book *Chemical Power Sources* written by two of the authors of this book (V.S. Bagotsky and A.M. Skundin). Now, almost 35 years later, this book is outdated and has become an obsolete rarity.

During this period in the field of batteries two entirely new directions emerged, which are now mass-produced (i) *nickel–metal hydride batteries* that practically replaced same-sized nickel–cadmium batteries (not a word about these batteries can be found in our 1980 book) and (ii) *lithium ion batteries* that are the only possible power source for cell phones and other small-size electronic equipment, and thus, they substantially helped to change our everyday life. In the chapter about lithium batteries of the 1980 book containing 10 of the 370 pages, the not yet developed lithium ion batteries could not be mentioned. The same is true for *supercapacitors* and *photogalvanic devices.*

The aim of the authors of the present book is to update in a concise manner the information contained in the 1980 book and to add all new relevant information published up to 2012.

With the kind permission of Elsevier (which is the legal successor of Academic Press), in this book, excerpts of the 1980 book that do not need renewal are extensively used.

Chapters 1–9, 14–25, and 32 have been written by V.S. Bagotsky, Chapters 10–13 have been written by A.M. Skundin, and Chapters 26–31 have been written by Yu. M. Volfkovich.

SYMBOLS

Symbol	Meaning (values)	Dimensions
Roman symbols		
c_j	Concentration	mol/dm^3
D_j	Diffusion coefficient	cm^2/s
E	Electrode potential	V
E	Equilibrium electrode potential	V
ϵ	Electromotive force	V
F	Faraday constant	9485 C/mol
G	Gibbs energy	kJ/mol
H	Enthalpy	kJ/mol
i	Current density	mA/cm^2
i^0	Exchange current density	mA/cm^2
I	Current	A, mA
M	(1) Mass	kg
	(2) Molar concentration	mol/dm^3
n	Number of electrons in the reaction's elementary act	none
p	Power density	W/kg, W/L
	Power	W, kW
Q, q	Heat	kJ, eV
R	(1) Resistance	Ω
	(2) Molar gas constant	8.314 J/mol K
S	(1) Entropy	kJ/K
	(2) Surface area	cm^2
T	Absolute temperature	K
U	Cell voltage	V
w	Energy density	kWh/kg, kWh/L
W	Work, useful energy	Wh, kWh

Symbol	Meaning (values)	Dimensions
Greek symbols		
γ	Roughness factor	None
δ	Thickness	cm
λ_e	Amount of coulombs	None
η	Efficiency	None, %
σ	Conductivity	S/cm
Subscripts		
ads	Adsorbed	
app	Apparent	
e	Electrical	
exh	Exhaust	
ext	External	
h.e.	Hydrogen electrode	
i	Under current	
loss	Energy loss	
o.e.	Oxygen electrode	
ox	Oxidizer	
S	Per unit area	
red	Reducer	
V	Per unit volume	
j	Any ion, substance	
0	Without current	
+	Cation	
−	Anion	

ABBREVATIONS

AC	alternating current
Ah	ampere-hour
AFC	alkaline fuel cell
APU	auxiliary power unit
CD	current density
CHP	combined heat and power
CNT	carbon nanotube
CTE	coefficient of thermal expansion
DBHFC	duirect borohydride fuel cell
DCs	direct current
DCFC	direct carbon fuel cell
DEFC	direct ethanol fuel cell
DFAFC	direct formic acid fuel cell
DHFC	direct hydrazine fuel cell
DLFC	direct liquid fuel cell
DMFC	direct methanol fuel cell
DSA®	dimensionally stable anode
DOC	depth of charge
DOD	depth of discharge
EM	electron microscopy
EMF	electromotive force
eV	electron-volt
EPS	electrochemical power source
ET-PEMFC	elevated temperature PEMFC
FCV	fuel cell vehicle

GDL	gas diffusion layer
GLDL	gas liquid diffusion layer
ICE	internal combustion engine
ICV	internal combustion vehicle
IT-SOFC	interim temperature SOFC
IRFC	internal reforming fuel cell
LHV	lower heat value
LT-SOFC	low temperature SOFC
LPG	liquefied petroleum gas
MCFC	molten carbonate fuel cell
MEA	membrane-electrode assembly
OCP	open-circuit potential
OCV	open-circuit voltage
PVC	polyvynil chloride
SLI	starting, lighting, ignition
VRLA	valf-regulated lead acid (battery)
Wh	watt-hour

INTRODUCTION

Electrochemical Power Sources (EPS) are autonomous devices based on electrochemical phenomena that produce electrical current and power and can be used in conditions when the connection to electrical grid power is not possible (e.g., for mobile and portable devices or in case of a grid failure). The most representative and widely used EPS type are batteries.

A *battery* is a device destined for the electrochemical conversion of the energy of a chemical reaction between two solid reactants to electrical energy. It is impossible to imagine human activity without batteries. Hundreds of millions of batteries are used worldwide for personal and domestic needs (wrist watches, cell phones, cameras, personal computers, audio and video players, hearing aids and other different electronic and medical devices), and also in different means of transport (ICE and hybrid cars, passenger carriages, planes, ferries, liners). They are also used in different municipal buildings (telephone stations, power backup in hospitals). A huge number of batteries are used for military purposes (soldiers' personal equipment, guided missiles, drones, rockets).

The definition of *fuel cells* is similar to the definition of batteries, but an important distinction is that in fuel cells the chemical reaction takes place between gaseous liquid and/or liquid reactants. The definition of compound batteries is also similar to that of batteries, the only difference being that in *compound batteries* the chemical reaction takes place between a solid reactant on one of the electrodes and a gaseous and/or liquid reactant on the other electrodes.

Two varieties of EPSs are now in the state of wide development and are beginning to be used in different fields: viz. (i) fuel cells for electric cars, small power plants for individual cottages, and for large grid power plants and (ii) supercapacitors (high-capacitance electrochemical capacitors) as rechargeable power sources with higher energy values and power capabilities, and with much better cycleability properties than those in existing storage batteries that are used in parallel with batteries for starting purposes of ICE cars, delivering the necessary initial peak power (especially at low temperatures) and thus increasing the battery lifetime.

Improvements of existing EPSs and the development of new kinds of EPS are the results of intense R&D work performed in industrial and academic institutions of many countries. The limited size of this book prevents the possibility to single out the contributions of the numerous researchers in these institutions. Therefore, in the book the number of references in most chapters is limited only to some important work in the corresponding field, particularly to achievements of historical significance. The main emphasis in the references is given to monographs and review papers, containing more detailed information about the contributions of different authors.

PART I

BATTERIES WITH AQUEOUS ELECTROLYTES

1

GENERAL ASPECTS

1.1 DEFINITION

Batteries are a variety of *galvanic cells*, that is, devices containing two (identical or different) electron-conducting *electrodes*, which contact an ion-conducting *electrolyte*. Batteries are destined to convert the energy of a chemical reaction between solid electrode components into electrical energy providing an electric current (when the circuit is closed) between two not-identical electrodes having different values of the *electrode potential* (positive and negative terminals). A battery comprises one or several single galvanic cells. In each such cell a comparatively low voltage is generated, typically 0.5–4 V for different classes of cells. Where higher voltages are required, the necessary number of cells is connected in series to form a galvanic battery. Colloquially, the term "battery" is often used to denote single galvanic cells acting as electrochemical power sources as well as groups of single cells. This is retained in this book. Some battery types retain the term "cell" even for groups of single cells (e.g., fuel cell, not fuel battery). The term "cell" is also used when it is necessary to compare different aspects of single-cell and multicell batteries.

1.2 CURRENT-PRODUCING CHEMICAL REACTION

Reactions in batteries are chemical reactions between an oxidizer and a reducer. In reactions of this type, the reducer being oxidized releases electrons while the oxidizer

Electrochemical Power Sources: Batteries, Fuel Cells, and Supercapacitors, First Edition.
Vladimir S. Bagotsky, Alexander M. Skundin, and Yurij M. Volfkovich
© 2015 John Wiley & Sons, Inc. Published 2015 by John Wiley & Sons, Inc.

being reduced accepts electrons. An example of such a redox reaction is the reaction between silver oxide (the oxidizer) and metallic zinc (the reducer):

$$Ag_2O + Zn \rightarrow 2Ag + ZnO \tag{1.1}$$

in which electrons are transferred from zinc atoms of metallic zinc to silver ions in the crystal lattice of silver oxide. When reaction (1.1) is allowed to proceed in a jar in which silver oxide is thoroughly mixed with fine zinc powder, no electrical energy is produced in spite of all the electron transfers at grain boundaries. This is because these transfers occur randomly in space and the reaction energy is liberated as heat that can raise the temperature of the reaction mixture to dangerous levels. The same reaction does occur in batteries, but in an ordered manner in two partial reactions separated in space and accompanied by electric current flow (Fig. 1.1).

In the simple case a battery (cell) consists of two *electrodes* made of different materials immersed in an electrolyte. The electrodes are conducting metal plates or grids covered by reactants (*active mass*); the oxidizer is present on one electrode, the reducer on the other. In silver–zinc cells the electrodes are metal grids, one covered with silver oxide and the other with zinc. An aqueous solution of KOH serves as electrolyte. Schematically, this system can be written as

$$Zn \ |KOHaq| \ Ag_2O \tag{1.2}$$

When these electrodes are placed into the common electrolyte enabling electrolytic contact between them, an open circuit voltage (OCV) ϵ develops between them (here $\epsilon = 1.6$ V), zinc being the negative electrode. When they are additionally connected by an electronically conducting external circuit, the OCV causes electrons

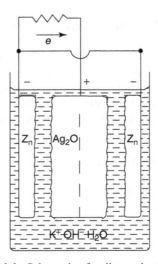

Figure 1.1. Schematic of a silver–zinc battery.

to flow through it from the negative to the positive electrode. This is equivalent to an electric current I in the opposite direction. This current is the result of reactions occurring at the surfaces of the electrodes immersed into the electrolyte: zinc being oxidized at the negative electrode (anode)*

$$Zn + 2OH^- \rightarrow ZnO + H_2O + 2e^- \tag{1.3}$$

and silver oxide being reduced at the positive electrode (cathode)

$$Ag_2O + H_2O + 2e^- \rightarrow 2Ag + 2OH^- \tag{1.4}$$

These electrode reactions sustain a continuous flow of electrons in the external circuit. The OH^- ions produced by reaction (1.4) in the vicinity of the positive electrode are transported through the electrolyte toward the negative electrode to replace OH^- ions consumed in reaction (1.3). Thus, the electric circuit as a whole is closed. Apart from the OCV, the current depends on the cell's internal resistance and the ohmic resistance present in the external circuit. Current flow will stop as soon as at least one of the reactants is consumed.

In contrast to what occurred in the jar, in the batteries, the overall chemical reaction occurs in the form of two spatially separated partial electrochemical reactions. Electric current is generated because the random transfer of electrons is replaced by a spatially ordered overall process (*current-producing reaction*).

1.3 CLASSIFICATION

By their principles of functioning, batteries can be classified as follows:

1. **Primary (single-discharge) batteries**. A primary battery contains a finite quantity of the reactants participating in the reaction; once this quantity is consumed (on completion of discharge), a primary battery cannot be used again ("throw-away batteries").

2. **Storage (multiple-cycle) batteries** (also called secondary or rechargeable batteries). On the completion of discharge, a storage battery can be recharged by forcing an electric current through it in the opposite direction; this will regenerate the original reactants from the reaction (or discharge) products. Therefore, electric energy supplied by an external power source (such as the grid) is stored

---

*This definition of the negative electrode as anode (at which an oxidation reaction take place) and the positive electrode as cathode (with a reduction reaction) is valid only for the discharge process of batteries. For the charging process of storage batteries (as well as for electrolyzers) when the current flow and the electrode reactions are in the direction opposite to that during discharge, the opposite definition is encountered, that is, the negative electrode works as cathode and the positive electrode as anode. For this reason in the case of storage batteries preferably only the terms of positive or negative electrode instead of anode or cathode should be used.

in the battery in the form of chemical energy. During the discharge phase this energy is delivered to a consumer independent of the grid. During the charging phase the electrode reactions and the overall current-producing reaction occur in the direction opposite to that during discharge. Thus, these reactions must be chemically reversible (the notion of chemical reversibility must not be confused with that of thermodynamic reversibility). Good rechargeable batteries will sustain a large number of such charge–discharge cycles (hundreds or even thousands). The classification into primary and storage batteries is not rigorous because under certain conditions some primary battery may be recharged and storage batteries after a single use are sometimes discarded.

The silver–zinc battery is a storage battery: after discharge, it can be recharged by forcing through it an electric current in the reverse direction. In this process the two electrode reactions (1.3) and (1.4) as well as the overall reaction (1.2) go from right to left.

3. **Fuel cells.** In the fuel-cell mode of operation, reactants are continuously fed into the cell (or battery) while reaction products are continuously removed. Hence, fuel cells (the more appropriate term of fuel battery is not commonly used) can deliver current continuously for a considerable length of time, which largely depends on external reactant storage.

Batteries are also classified according to their chemistry (their system), that is, the chemical nature of reactants. The above-mentioned battery with silver oxide as an oxidant at the positive electrode and metallic zinc as negative electrode is called "silver–zinc battery."

Sometimes other methods of classification are also used, for example, on the basis of the application (stationary or mobile batteries), shape (cylindrical, prismatic, disk-shape batteries), size (miniature, small-sized, medium-sized, or large-sized batteries), electrolyte type (alkaline, acidic, or neutral electrolyte, with liquid or solid (solidified), or molten salt electrolyte), voltage (low voltage or high voltage batteries), electric power generation (low power or high power batteries), and so on.

1.4 THERMODYNAMIC ASPECTS

Each electrode j of a battery brought into contact with the electrolyte develops a certain *electrode potential* E_j. The concept of "potential" is an experimental, undefined parameter, that is, it has a real physical meaning and reflects a real physical phenomenon, but cannot be determined from experimental data (even from thought experiments). Only potential differences between the given electrode and another electrode (reference electrode) are measurable. (Similarly, the height of a certain geographic point is defined and can be measured only when referred to the height of another point, e.g., sea level). Values of electrode potentials are commonly referred to as the potential of the standard hydrogen electrode (SHE). Potentials of different electrodes can be either negative (i.e., more negative than the potential of the SHE) or positive. The OCV of a battery U is the potential difference between the positive electrode and the negative electrode:

$$U = E_+ - E_- \tag{1.5}$$

According to this definition, the OCV is always positive (provided the potentials of both electrodes are referred to the same reference electrode).

Thermodynamically, electrode reactions can be either reversible or irreversible. In case of a reversible reaction, the electrode potential is called reversible (thermodynamic electrode potential). The corresponding OCV is traditionally called "electromotive force" (EMF) and is denoted as ϵ.

The EMF of a battery with reversible electrodes can be defined by the thermodynamic relation

$$\epsilon = \frac{\Delta G}{nF} \tag{1.6}$$

where ΔG is the difference of the Gibbs energy G during the current-producing reaction—the difference of the Gibbs energies of all reactants and all reaction products—n is the number of electrons taking part in one elementary act of the electrode reaction, $F = 96485$ C/mol is the Faraday constant. The reversible potential of an electrode contacting an electrolyte, all ions of which have a thermodynamic activity $a_j = 1$ mol/l, is called standard electrode potential and is denoted by E_j^0.

Values of standard electrode potentials for some reactions are shown in Table 1.1. Values for other reactions as well as of the Gibbs energy G for different reaction components can be found in special reference books.

TABLE 1.1. Standard Electrode Potentials (25 °C)

Reaction	E^0, V (SHE)	Reaction	E^0, V (SHE)
$Li^+ + e^- \rightleftarrows Li$	−3.045	$HgO + H_2O + 2e^- \rightleftarrows Hg + 2OH^-$	0.098
$K^+ + e^- \rightleftarrows K$	−2.935	$Sn^{4+} + 2e^- \rightleftarrows Sn^{2+}$	0.154
$Ca^{2+} + 2e^- \rightleftarrows Ca$	−2.866	$Cu^{2+} + e^- \rightleftarrows Cu^+$	0.153
$Na^+ + e^- \rightleftarrows Na$	−2.714	$AgCl + e^- \rightleftarrows Ag + Cl^-$	0.2224
$Mg^{2+} + 2e^- \rightleftarrows Mg$	−2.363	$Hg_2Cl_2 + 2e^- \rightleftarrows 2Hg + 2Cl^-$	0.2676
$Al^{3+} + 3e^- \rightleftarrows Al$	−1.662	$Cu^{2+} + 2e^- \rightleftarrows Cu$	0.337
$Ti^{2+} + e^- \rightleftarrows 2e^- \rightleftarrows Ti$	−1.628	$Fe(CN)_6^{3-} + e^- \rightleftarrows e(CN)_6^{4-}$	0.36
$Zn(OH)_2 + 2e^- \rightleftarrows Zn + 2OH^-$	−1.245	$Cu^+ + e^- \rightleftarrows Cu$	0.521
$Mn^{2+} + 2e^- \rightleftarrows Mn$	−1.180	$I_2 + 2e^- \rightleftarrows 2I^-$	0.536
$2H_2O + 2e^- \rightleftarrows H_2 + 2OH^-$	−0.822	$O_2 + 2H^+ + 2e^- \rightleftarrows H_2O_2$	0.682
$Zn^{2+} + 2e^- \rightleftarrows Zn$	−0.764	$Fe^{3+} + e^- \rightleftarrows Fe^{2+}$	0.771
$S + 2e^- \rightleftarrows S^{2-}$	−0.48	$Br_2 + 2e^- \rightleftarrows 2Br^-$	1.065
$Fe^{2+} + 2e^- \rightleftarrows Fe$	−0.441	$O_2 + 4H^+ + 4e^- \rightleftarrows 2H_2O$	1.229
$Cd^{2+} + 2e^- \rightleftarrows Cd$	−0.403	$Cl_2 + 2e^- \rightleftarrows 2Cl^-$	1.358
$Ni^{2+} + 2e^- \rightleftarrows Ni$	−0.250	$PbO_2 + 4H^+ + e^- \rightleftarrows Pb^{2+} + 2H_2O$	1.455
$Sn^{2+} + 2e^- \rightleftarrows Sn$	−0.136	$Ce^{4+} + e^- \rightleftarrows Ce^{3+}$	1.61
$2H^+ + 2e^- \rightleftarrows H_2$	0.0000	$F_2 + 2e^- \rightleftarrows 2F^-$	1.87

1.5   HISTORICAL DEVELOPMENT

In 1791 the Italian physiologist Luigi Galvani (1737–1798) demonstrated in remarkable experiments that muscle contraction similar to that produced by the discharge of a Leyden jar, will occur when two different metals touch the exposed nerve of a frog. This phenomenon was in part correctly interpreted in 1792 by the Italian physicist Alessandro Volta (1745–1827), who showed that this galvanic effect originates from the contacts established between these metals and between them and the muscle tissue. In March 1800, Volta reported a device designed on the basis of this same phenomenon, which could produce "inexhaustible electric charge." Now known as the *Volta pile*, this was the first example of an electrochemical device: an electrochemical power source (*a battery*). No special oxidizer was used in the Volta pile and this role was played by water molecules that were reduced at the silver cathode to gaseous hydrogen. As a result of such a weak oxidizer, the OCV of a single cell in this pile was only about 0.4 V. If high voltages and high discharge current were needed, very large batteries had to be built. One pile manufactured in 1803 comprised 2100 individual cells.

The Volta pile (certainly not inexhaustible!) was of extraordinary significance for the developments both in the science of electricity and of electrochemistry, because a new phenomenon, a continuous flow of charges (an electric current), hitherto not known, could for the first time be realized. Soon various properties and effects of the electric current were discovered, including many electrochemical processes. In May 1801, William Nicholson and Sir Anthony Carlisle in London electrolyzed water-producing hydrogen and oxygen. The Volta pile was also of significance for the development of many new fields of science of technology in the nineteenth century.

In order to circumvent the limited possibilities of the original Volta pile, the following period saw the development of other battery systems in which special oxidizers were introduced. In 1836, J. F. Daniell (1796–1845) developed a cell with an oxidizer in the form of copper ions in a copper sulfate solution. Cells with the use of nitric acid as oxidizer were developed in 1838 by W. R. Grove (1811–1896) and in 1841 by R. Bunsen (1811–1899). Cells containing sodium bichromate dissolved in sulfuric acid were developed in 1843 by Ch. Poggendorff (1824–1876) and in 1856 by Grenet.

A considerable improvement of electrical batteries was achieved after the replacement of liquid oxidizers (mainly in aqueous solutions) by solid oxidizers, in the form of different metal oxides. In 1865, the French engineer G.L. Leclanché (1839–1882) made a battery containing manganese dioxide as an oxidizer (positive electrode) and zinc as a reducer (negative electrode) and an aqueous solution of ammonium chloride as an electrolyte. In the following, the liquid electrolyte in this battery was replaced by an electrolyte solidified by different gelling agents. These "dry" Leclanché batteries proved to be very simple with regard to manufacture and reliable in usage. As early as 1868, more than twenty thousands of such cells were being manufactured. A further advance in battery technology was the development of rechargeable batteries.

In 1859, the French scientist Gaston Planté (1834–1889) made the first prototype of a lead acid rechargeable battery. An alkaline nickel–cadmium rechargeable battery was developed in 1899 by the Swedish engineer W. Jungner (1869–1924) and an alkaline nickel–iron battery was developed two years later by the well-known American inventor Thomas A. Edison (1847–1931). Up to the seventh decade of the nineteenth century, electrochemical batteries remained the only sources of electrical current and power.

After the appearance of the Volta pile and other improved versions of batteries, extended experiments with the new phenomenon of a continuous electrical current became possible and soon different properties of this current could be established: in 1820 Ampère's law of interaction between electrical currents; in 1827 Ohm's law of proportionality between voltage and current; in 1831 Joule's law of the thermal effect of electrical current; in 1831 Faraday's law of electromagnetic induction, and many others. These achievements led to the development of the *theory of electrodynamics* and practice of electrical engineering and, as a result, to the appearance of a revolutionary new power source: the electromagnetic generator invented in 1866 by Werner von Siemens (1816–1872), which soon surpassed their predecessors both in electrical and economic parameters.

After the development of the electromagnetic generator, a large-scale production of electric power became possible ("grid electricity"). Nevertheless, despite the worldwide expansion of electric grids, batteries retained their significance as autonomous power sources up to now. According to a 2001 Report (cited from the online encyclopedia Wikipedia), the worldwide battery industry generates US$ 48 billion in sales every year, with 6% annual growth. One of the reasons for the widespread acceptance of batteries is the tremendous large range of power that can be delivered. Wrist watches are powered by miniature batteries with a power of about 10^{-5} W. Huge storage batteries with power up to 10^9 W are used in submarines. The mass of a single power unit can vary from 0.1 g to a hundred tons. It is striking that both miniature and huge batteries operate with the same high efficiency. No other type of electric power source could be said to be as flexible or as versatile.

1.6 NOMENCLATURE

As yet a complete worldwide nomenclature for batteries fully specifying all their characteristics including shape and size (which determine battery interchangeability) is not established. Different countries and different battery manufacturers use different systems for designating and labeling batteries (see in the online encyclopedia Wikipedia the entry "List of battery sizes"). Two battery types widely used for household purposes have well-established designations that facilitate interchangeability: (i) cylindrical "dry" batteries—AA, A, B, C, and so on, with dimensions (height + diameter in millimeters), for example, for AA (44 + 10), for A (50 + 13.5), and for D (49 + 24), and (ii) disc cells—for example, A2325, where A is the battery chemistry type, 23 is the diameter in millimeters, and 25 is the height in 0.1 mm (i.e., 2.5 mm).

REVIEWS AND MONOGRAPHS

Bagotsky VS, Skundin AM. *Chemical Power Sources*. London: Academic Press; 1980.

Daniel C, Besenhard JO, editors. *Handbook of Battery Materials*. 2nd ed. Chichester, Weinheim: Wiley-VCH; 2011.

Heise GW, Cahoon NC, editors. *Primary Batteries*. Vol. 1. New York: Wiley; 1971.

Liebhafsky HA, Cairns EJ. *Fuel Cells and Batteries*. New York: Wiley; 1968.

Linden D, Reddy TB, editors. *Handbook of Batteries*. McGraw-Hill; 2002.

Vincent CA, Scrosati B. *Modern Batteries. An Introductory to Electrochemical Power Sources*. London: L. Edward Arnold Ltd.; 1997.

2

MAIN BATTERY TYPES

2.1 ELECTROCHEMICAL SYSTEMS

The numerous existing battery types vary in their size, structural features, and nature of the chemical reactions. They vary accordingly in their performance and parameters. This variety reflects the diverse conditions under which cells operate, each field of application imposing its specific requirements.

All batteries are based on a specific electrochemical system, that is, a specific set of oxidizer, reducer, and electrolyte. Conditionally, an electrochemical system is written as

$$\text{oxidizer}|\text{electrolyte}|\text{reducer} \qquad (2.1)$$

Often, the oxides of certain metals are used as the oxidizer. In the names of systems and batteries, though, often only the metal is stated, so that the example reported above is called a silver–zinc, rather than silver oxide–zinc battery (or system).

Batteries are known for about 100 electrochemical systems. Today, many of them are of mere historical interest. Commercially, batteries of less than two dozens of systems are currently produced. The largest production volumes are found in just three systems: primary zinc–manganese batteries (today with an alkaline electrolyte, in the past with a salt electrolyte), rechargeable lead acid batteries, and rechargeable alkaline (nickel–cadmium, nickel–iron) batteries. Batteries of these systems have been manufactured for more than a century, and until today are widely used.

Electrochemical Power Sources: Batteries, Fuel Cells, and Supercapacitors, First Edition.
Vladimir S. Bagotsky, Alexander M. Skundin, and Yurij M. Volfkovich
© 2015 John Wiley & Sons, Inc. Published 2015 by John Wiley & Sons, Inc.

Two more types have gained increasing importance during the second half of the twentieth century: nickel hydride storage batteries and a variety of lithium batteries. Other battery systems are of relatively limited use, mainly to supply power needs in military devices.

2.2 LECLANCHÉ (ZINC–CARBON) BATTERIES

$$MnO_2 \ (C)|NH_4Cl \ aq|Zn \qquad\qquad (2.2)$$

For over a 100 years now primary *manganese–zinc batteries* (in the past called zinc–carbon cells) have been produced and used as the major primary battery. Their popularity is because of a favorable combination of properties: they are relatively cheap, have satisfactory electrical parameters and a convenient storage life, and offer convenient utilization. Their major disadvantage is a strong voltage decrease during progressive discharge; depending on the load, the final voltage is just 50–70% of the initial value.

The manganese–zinc batteries are manufactured as leak-proof "dry" batteries having the electrolyte soaked up by a matrix.

The first zinc–carbon cell made in 1865 by the French engineer G.-L. Leclanché was a glass jar containing an aqueous solution of ammonium chloride into which were immersed an amalgamated zinc rod (the negative electrode) and a porous earthenware pot packed with a mixture of manganese dioxide and powdered coke and containing a carbon-rod current collector at the center (positive electrode). Quite soon a zinc can served as the anode and cell container replaced the zinc rod.

The discharge reaction at the positive electrode

$$MnO_2 + H_2O + e^- \rightarrow MnOOH + OH^- \qquad\qquad (2.3)$$

can be regarded as a process of cathodic intercalation of hydrogen atoms into the lattice of MnO_2. This causes the electrolyte near the cathode to become alkaline, and as a result ammonium ions decompose forming free ammonia.

The anodic oxidation of zinc in salt solutions produces Zn^{2+} ions, and in practice is accompanied by various secondary reactions resulting in the formation of barely soluble complex compounds. Zinc ions diffuse to zones with higher pH where, after hydrolysis, they precipitate as oxychlorides $ZnCl_2 \cdot xZn(OH)_2$ or hydroxide $Zn(OH)_2$. Crystals of $Zn(NH_3)_2Cl_2$ formed by interaction with free ammonia also precipitate. These products all shield the active materials of both electrodes, increase the internal resistance and the pH gradient, and produce deterioration of the cell parameters. The zinc ions can also react with the product of discharge of the positive electrode to hetaerolite $ZnO \cdot Mn_2O_3$ forming a new solid phase.

Thus, the electrode processes occurring in manganese–zinc batteries with salt electrolytes are complicated, and their thermodynamic analysis is difficult. In a rough approximation disregarding secondary processes, the current-producing reaction can be described by the following equation:

$$Zn + 2MnO_2 + 2H_2O \rightarrow Zn(OH)_2 + 2MnOOH \tag{2.4}$$

Often the equation

$$Zn + 2MnO_2 + 2NH_4Cl \rightarrow Zn(NH_3)_2Cl_2 + 2MnOOH \tag{2.5}$$

is used, but it also fails to supply an exhaustive description of the process, inasmuch as the actual ampere-hour capacity of a battery can be higher than that corresponding to the amount of ammonium chloride in Equation (2.5). The battery's open circuit voltage (OCV) decreases during discharge and formation of the variable composition mass. On prolonged storage of undischarged batteries, their OCV also decreases.

The most widespread production and use has a cylindrical-shaped version of these batteries shown in Figure 2.1. A cylindrical zinc can (1) serve simultaneously as cell container and as anode (negative electrode). It is lined with a paper separator (2) carrying a layer of the electrolyte paste on the outside. The zinc can is enclosed in a jacket (3) of thin steel. The cathode (4) pressed with the cathode mixture (MnO_2 ores and carbonaceous materials) with a central disposed carbon current collecting rod (5) is inserted into the can and pressed from above decreasing the electrolyte gap down to $1.13–1.20$ mm. The lid (7) is retained in position by swaging the steel jacket. The ring (6) insulates the can from the lid and seals the cell. Cells not intended for use in batteries are enclosed in a cardboard jacket bearing the manufacturers' label. The main advantage of metal-clad cells is efficient sealing.

From the 1960s onward, alkaline manganese–zinc batteries started to be produced. They have appreciably better electrical performance parameters but do not differ in their operating features from the Leclanché batteries, are produced in identical sizes, and can be used interchangeably with them. Thus, a gradual changeover occurred and the phase-out of the older system is now almost complete.

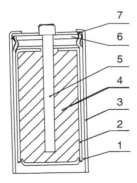

Figure 2.1. Schematic of a paper-lined cylindrical manganese–zinc Leclanché battery.

2.3   THE ZINC ELECTRODE IN ALKALINE SOLUTIONS

Metallic zinc was used as material for the negative electrode in the earliest electrical cell, Volta's pile, and is still employed in a variety of batteries including batteries with alkaline electrolytes.

The operation of zinc anodes in alkaline solutions (mainly 20–40% KOH) involves specific features. In the anodic dissolution of zinc

$$Zn + 4OH^- \rightarrow ZnO_2^{2-} + 2H_2O + 2e^- \tag{2.6}$$

the consumption of alkali is high, because two OH^- ions are needed for each electron liberated, and zincate ions are formed as a soluble product (this is the so-called primary process of zinc electrode dissolution). The solubility of zincate ions in alkaline solutions having the above concentration is $1-2$ mol/l. When saturation has been reached, zinc hydroxide starts to sediment on the zinc surface and the primary process practically stops. Here the capacity of the zinc electrode is limited by the available volume of alkali solution, rather than by the amount of zinc; about 10 ml of the solution are needed for each ampere-hour. When the current density is very low the zinc electrode continues to function in the saturated zincate solution, its dissolution now producing insoluble zinc oxide (the secondary process of zinc oxidation):

$$Zn + 2OH^- \rightarrow ZnO + H_2O + 2e^- \tag{2.7}$$

Because, under these conditions, discharge of the battery as a rule results in the production of one OH^- ion for each electron at the positive electrode (Eq. 2.7), the secondary process overall occurs without the consumption of alkali, and a solution volume of $1-2$ ml/Ah is practically sufficient for the operation of the cell.

Thus, there are two possible modes of utilizing zinc anodes in alkaline solutions. In the first, and older, mode only the primary process is used, with monolithic zinc anodes and a large volume of electrolyte. In the second mode, the secondary process is employed, with powdered zinc anodes at which the true current densities are much lower than at smooth electrodes.

2.4   ALKALINE MANGANESE–ZINC BATTERIES

$$MnO_2 | 27 - 40\% KOH \text{ aq} | Zn \tag{2.8}$$

2.4.1   Primary Alkaline Manganese–Zinc Batteries

Compared with the Leclanché batteries, the alkaline manganese–zinc batteries (often labeled simply as "alkaline batteries") offer better performance at high discharge currents and lower temperatures and a better shelf life. They are more expensive than the Leclanché batteries, but their cost per unit of energy is competitive while sufficient raw materials for a mass production of these batteries are available. Their capacity at low current drains is 50% higher, and at high current drains where Leclanché batteries have a much lower capacity, alkaline batteries have a capacity that is higher by factors of 3–6.

Rather than natural ores as in Leclanché batteries, electrolytic manganese dioxide (EMD) which is produced by anodic oxidation of Mn^{2+} ions at graphite electrodes in solutions of manganese salts is used as the active material for the positive electrode. Owing to a higher conductivity of the alkaline solution and lack of precipitation of solid $Zn(NH_3)_2Cl_2$, a smaller volume of electrolyte solution than in the Leclanché batteries is needed in the pores of the active mass. Hence, alkaline batteries contain more MnO_2 than Leclanché batteries of the same size. The zinc can of the Leclanché cells with a smooth surface is ineffective as anode. To provide high performance, powdered zinc electrodes with a highly extended surface are used. Therefore alkaline manganese–zinc batteries have a so-called "inside-out" construction as illustrated in Figure 2.2.

The active material of the cathode (6) is pressed into the inner surface of a steel can (2). A separator (3) of unwoven plastic fabric and/or cellophane is inserted into the can, which contains the electrolyte and prevents internal shortings. A petal-shaped brass current collector is in the central part of the cell. The space between the separator and the current collector is filled with the anode paste (7), which consists of the alkaline solution gelled with carboxymethyl cellulose (CMC) and zinc powder. An additional amount of pure electrolyte (9) is inside the current collector. To provide exchangeability with conventional cylindrical cells, the upper side of the cell has a bulge (1) that serves as the positive terminal. The bottom (13) serves as the negative terminal. To improve internal contact a pressure spring (12) is often used. The can is inserted into a metal jacket (4) with the insulator (5).

Performance

At high-drain discharges alkaline manganese–zinc batteries have an ampere-hour capacity that is higher than that of Leclanché batteries. The OCV of not discharged alkaline manganese–zinc batteries is 1.5–1.7 V. These batteries have a good storability and low-temperature performance. They retain 90% of their ampere-hour capacity after 1 year's storage at room temperature and 7–10% at a discharge temperature of −40 °C.

Alkaline batteries are used in many household items. This includes MP3 players, CD players, digital cameras, pagers, toys, lights, and radios, to name a few. The most widespread production and application have cylindrical alkaline batteries which now can be found literally in every household. To provide interchangeability their size is standardized with that of Leclanché batteries. Most commonly the following sizes are used:

Size	Diameter (mm)	Height (mm)
AAA	10	44
AA	13.5	50
D	24	49

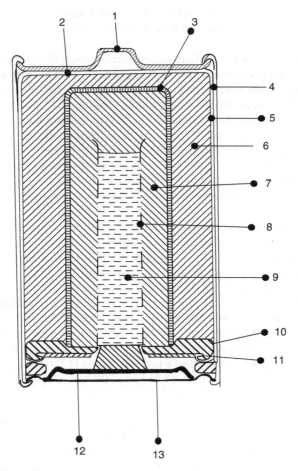

Figure 2.2. Schematic of an alkaline manganese–zinc battery.

2.4.2 Rechargeable Alkaline Manganese–Zinc Batteries

In the years after World War II transistor radios came into wide use, along with tape recorders and numerous other appliances, requiring high-capacity small-size power sources. A "dry cell" of this type was suggested in 1912. But these batteries became available only in the beginning of the 1950. The first generation of the rechargeable alkaline battery's technology was developed by Battery Technologies Inc. in Canada and licensed to Pure Energy, EnviroCell, Rayovac, and Grandcell. Subsequent patent and advancements in technology were introduced eventually. The types produced include AAA, AA, C, D, and snap-on 9-Volt batteries.

Rechargeable alkaline manganese–zinc batteries have a chemistry and general design principle analogous to those in primary manganese–zinc batteries. A thorough sealing (for safe recharge) resists leakage that a recharge would cause, provided a proper charging unit is used. Their cycle life depends on the depth of discharge (DOD). During cycling at a DOD of 50% they can be almost-fully recharged after about 12 cycles with an end-of-recharge voltage of 1.42 V. After deep discharges, they can be brought to their original ampere-hour capacity only after a few deep charge–discharge cycles.

2.5   LEAD ACID BATTERIES

$$PbO_2|30 - 40\%H_2SO_4aq|Pb \tag{2.9}$$

Lead acid batteries are the storage batteries most widely used at present. This is readily explained by their low price, high reliability, and good performance. Their cycle life is a few hundred charge–discharge cycles, though for some battery types, it extends to even more than a thousand cycles.

The first working lead cell manufactured in 1859 by the French scientist, Gaston Planté, consisted of two lead plates separated by a strip of cloth coiled and inserted into a jar with sulfuric acid. A surface layer of lead dioxide was produced by electrochemical reactions in the first charge cycle. Later developments led to electrodes made by pasting a mass of lead oxides and sulfuric acid into grids of lead–antimony alloy (for lead acid batteries the electrodes are often called plates).

2.5.1   Current-Producing Reactions

In the charged lead battery, the negative electrode contains sponge lead; the positive electrode contains lead dioxide PbO_2. The current-producing reactions during charging (ch) and discharge (disch) are described by the following equations:

$$(+)\ PbO_2 + 3H^+ + HSO_4^- + 2e^- \underset{ch}{\overset{disch}{\rightleftarrows}} PbSO_4 + 2H_2O \tag{2.10}$$

$$(-)\ Pb + HSO_4^- \underset{ch}{\overset{disch}{\rightleftarrows}} PbSO_4 + H^+ + 2e^- \tag{2.11}$$

$$(cell)\ PbO_2 + Pb + 2HSO_4^- \underset{ch}{\overset{disch}{\rightleftarrows}} 2PbSO_4 + 2H_2O \tag{2.12}$$

(at the concentrations used in the batteries, sulfuric acid is practically dissociated into H^+ and HSO_4^- ions). Thus, discharge of the battery consumes sulfuric acid and produces barely soluble lead sulfate on both electrodes. This reaction mechanism

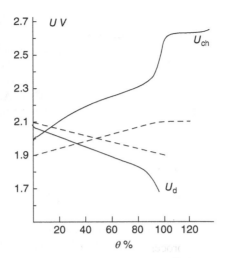

Figure 2.3. Typical charging–discharge curves of a lead acid battery. The broken lines indicate the EMF values measured after disconnecting the current.

was suggested as early as 1883 by Gladstone and Tribe in their theory of "*double sulphation.*" The concentration of sulfuric acid drops from 30 to 40% (depending on battery type) in a charged lead acid battery to 12–24% at the end of discharge.

2.5.2   Charging–Discharging Curves

Typical charging–discharging curves for lead acid storage batteries are shown in Figure 2.3

2.5.3   Battery Design

Most single lead acid cells or parts of multicell batteries have a similar design. The electrode assembly is placed into the box of an insulating material. The end electrodes are always negative. The plates are welded to a current collector with a vertical bar. Separators are placed between positive and negative plates. The lower side of the plates rest on prismatic supports on the bottom of the box, providing a mud space for the particles shed from the plates. The distance between the upper edges of the plates and the cover is not less than 2–3 cm. The materials used for battery manufacture should be resistant to prolonged action of concentrated sulfuric acid (e.g., lead alloys and some plastics).

2.5.4   Passivation

During discharge the active materials are not fully utilized: at low current drains, the degree of utilization is 40–60%. At high drains, it drops to 5–10%, which is because of the concentration polarization, that is, a sharp decrease to almost zero in sulfuric acid concentration in the pores of the positive electrode. At small drains, a premature drop in discharge voltage is caused by the passivation of the electrodes coming about by shielding of the active materials (both lead and lead dioxide) by the formation of a dense, fine-grained layer of lead sulfate.

Special additives such as barium sulfate, potassium lignosulfonate, and tanning agents are introduced into the active material in order to reduce passivation of the negative electrodes. The organic additives are adsorbed at the surface of lead and lead sulfate where they hinder the formation of new nuclei of lead sulfate and promote the growth of larger crystal grains and formation of looser layers. At the same time, these additives prevent the sintering of the spongy lead during cycling and storage (hence the term "expander").

2.5.5 Sulfation

If a lead acid storage battery is stored in a discharged state or is regularly undercharged a highly undesirable process, the so-called *sulfation* occurs on the electrodes, (particularly on the negative electrode). This process consists of gradual transformation of fine-grain lead sulfate into a hard dense layer of large grain sulfate. A cell with sulfated electrodes is difficult to charge because passage of a charging current produces on the negative electrode only hydrogen evolution instead of lead sulfate reduction. To prevent sulfation regular recharging of cells is recommended. To restore cell capacity, cells with sulfated electrodes are filled with diluted sulfuric acid or even with deionized water and charged at very low currents.

2.5.6 Forming Lead Acid Battery Electrodes

New (and also activated dry-charged) lead acid storage batteries need to be formed prior to normal use in order to increase their performance. Forming is achieved by carrying out from 2 to 4 preliminary charge/discharge cycles. During this process in the discharge phase lead sulfate is formed on both plates whereas in the charging phase lead sulfate is converted to lead dioxide on the positive electrode and to pure sponge metallic lead at the negative plate. Thus, during forming the electrode's surface areas are substantially increased.

An increase of cycle life under high-rate working conditions, and also of high-power capability of lead acid batteries can be achieved by adding expanded graphite and some other materials to the negative active mass formulation (Pavlov et al., 2005; Muneret et al., 2005).

2.5.7 Valve-Regulated Lead Acid Batteries (VRLA)

A valve-regulated lead acid battery is an almost hermetic battery in which the quantity of electrolyte is limited and gases evolving during operation are not released to the atmosphere, but recombined to water. Particularly, oxygen evolved during charging at the positive electrode reacts at the negative electrode also forming water (oxygen recombination cycle). To protect against intense gas evolution during overcharge the battery is equipped with valves that open when the internal gas pressure reaches a value of about 20–45 kPa. There is no need to periodically add water to this type of battery because during normal operation the concentration of the sulfuric acid does not change. For this reason, VRLAs are also called maintenance-free or low-maintenance storage batteries.

The first VRLA batteries were developed in the 1970s. In 1978 automobiles started using the VRLA battery for starting, lighting and ignition (SLI) purposes. Because they required little maintenance from the 1990s they were used in prototypes of electric vehicles. VRLA batteries have contributed to the reduction costs of electric vehicles and subsequently their sales.

A VRLA battery uses an electrolyte solution immobilized either by adding gel-forming substances (gel version) or being absorbed in a porous glass-fiber mat (absorbed glass mat (AGM) version). In the AGM version, the glass mat serves simultaneously as separator between positive and negative electrodes and as absorbent for the electrolyte. The amount of the liquid electrolyte is limited so that about 10% of the electrode space remains void. This allows the oxygen evolved during charging at the positive electrode to diffuse to the negative electrode and react there. In the gel version the dissipation of heat generated at the electrodes during the charge/discharge reaction to the outside is somewhat better than in the AGM version because in this version the gel completely contacts the plates and the battery container. Therefore, it is necessary to review and clarify the key parameters for the use of VRLA batteries with respect to the optimum conditions. Several series of chemical and electrochemical reactions occur in VRLA batteries particularly when in a charge or float charge condition. These reactions give specific properties such as minimal water loss (low maintenance) and create specific precautions for use. VRLA batteries with AGM separators have been in use for over 20 years even in different standby applications. These applications are increasingly varied, especially with regard to environmental conditions.

The operation of VRLA batteries is connected with the following problems: (i) corrosion of the positive electrode's grid because of persistent overcharge and to high values of temperature during operation, (ii) irreversible sulfation of the negative electrode because of lack of charge and (iii) degradation of the active mass because of loss of water during frequent overcharges, and depending on the frequency and depth of battery discharge, determining how the active mass is utilized.

2.6   ALKALINE NICKEL STORAGE BATTERIES

Alkaline storage batteries with a nickel hydroxide positive electrode were developed with three different negative electrodes, containing cadmium, iron, or a metal hydride

$$NiOOH|20 - 22\%KOH \; aq|M \qquad\qquad (2.13)$$

or

$$NiOOH|20 - 22\%KOH \; aq|MeH \qquad (2.13)$$

(M stands for cadmium or iron, MeH for metal hydride).

2.6.1 Nickel–Cadmium Storage Batteries

The Ni–Cd storage batteries have a long cycle life (a few thousand charge–discharge cycles), are compact in size, and easy to operate. The first patent for a Ni–Cd battery was granted in 1899 to the Swedish engineer Waldemar Jungner. Pocket-type positive electrodes were used in the first alkaline storage batteries. In 1928, first experiments were made with a new electrode type based on sintered porous nickel ("sintered electrodes"). In the early 1930s, the development of sealed alkaline batteries was reported (an important point in view of unpleasant and potentially hazardous leakage of alkali from a battery). Both battery types were developed further after World War II, and by about 1950 mass production of these new battery types started.

2.6.2 Nickel–Iron Storage Batteries

The first alkaline nickel–iron storage batteries were proposed in 1901 by Thomas Alva Edison. They have designed features and properties similar to those of nickel–cadmium batteries, although some of their parameters are different. The main difference resides in the potential of the negative electrodes: the equilibrium potential of the cadmium electrode is 20 mV more positive, the equilibrium potential of the iron electrode is 50 mV more negative than the equilibrium hydrogen potential in the same solution. This difference is small, but very significant inasmuch as the iron electrode can corrode with concomitant hydrogen evolution, leading to an appreciable self-discharge of the battery. For cadmium this process is thermodynamically unfeasible, hence cadmium is corrosion-resistant and the self-discharge of nickel–cadmium batteries very low. Nickel–iron batteries are very rugged much cheaper than nickel–cadmium batteries, and are mainly used in industrial transport applications.

2.6.3 Nickel Hydride Storage Batteries

An important advance in alkaline nickel storage batteries has been the introduction of the so-called metal hydrides as the active material in the negative electrode. Around 1950 it had already been evident that certain metal alloys can reversibly take up considerable quantities of hydrogen into their crystal lattice, that is, form metal hydrides. The equilibrium potential of such alloys after hydrogen uptake in a solution is close to that of the hydrogen electrode, and on anodic polarization, the hydrides yield electrical charge equivalent to the amount of absorbed hydrogen. Nowadays such hydride electrodes are widely employed to make alkaline storage batteries, which in their design are similar to Ni–Cd batteries but exhibit a considerably higher capacity than these. These two types of storage battery are interchangeable, because the potential of the hydride electrode is similar to that of the cadmium electrode. The metal alloys used to prepare the hydride electrodes are multicomponent alloys, usually with a high content of rare-earth elements. These cadmium-free batteries are regarded as environmentally preferable.

2.6.4   Electrochemical Processes

The main current-producing reactions on the negative electrodes of alkaline nickel storage batteries can be written as

$$\text{for Cd and Fe} : M + 2OH^- \underset{\text{ch}}{\overset{\text{disch}}{\rightleftarrows}} M(OH)_2 + 2e^- \tag{2.14}$$

$$\text{for MeH} : MeH + OH^- \underset{\text{ch}}{\overset{\text{disch}}{\rightleftarrows}} Me + H_2O + e^- \tag{2.15}$$

and the reaction on the positive electrode as

$$(+)2NiOOH + 2H_2O + 2e^- \underset{\text{ch}}{\overset{\text{disch}}{\rightleftarrows}} 2Ni(OH)_2 + 2OH^- \tag{2.16}$$

The processes taking place on the positive electrode actually are more complicated. Several modifications of nickel oxides exist which in particular differ in their degrees of hydration, so the above equations do not correctly describe the water balance in the reaction. The hydroxide of divalent nickel is formed as β-$Ni(OH)_2$ and has a lamellar structure with disordered crystal lattice. This disorder has a beneficial effect on the electrochemical activity. Charging typically produces an oxide hydroxide of trivalent nickel in the form of β-NiOOH. However, at high alkali concentrations or high charge currents, γ-NiOOH having a large specific volume may form. The resultant swelling leads to a deterioration of contact and deformation of the electrode.

The conductivity of pure $Ni(OH)_2$ is very low, but increases markedly, already at slight degrees of oxidation. After discharge, the battery's active mass contains a residual 20–40% of nonreduced NiOOH and thus remains sufficiently highly conductive. On charging, higher nickel oxides of the type of $NiO_x \cdot yH_2O$ (where x varies between 1.6 and 1.8) as well as NiOOH are formed. The higher oxides are unstable and spontaneously liberate excess oxygen by decomposition. Owing to the temporary existence of higher oxides on charging, the OCV of freshly charged batteries is higher and amounts to 1.45–1.7 V. As these oxides decompose it gradually decreases to the steady-state value of 1.30–1.34 V.

2.6.5   Design Varieties of Alkaline Nickel Storage Cells

Pocket Electrodes
In the older versions of Ni–Cd storage cells and in the Ni–Fe cells pocket-type electrodes were used. In this version the pressed active mass of the electrodes are

placed in small flat boxes (pockets) manufactured from perforated strips of 0.1 mm thick mild steel. The width of the pockets is about 13 mm and the length is requested by the required width of the electrode. In the electrode, the pockets are arranged in horizontal rows with neighboring pockets locked together by jointly beaded edges.

Sintered Electrodes

In these electrodes, the active materials are present in pores of a sintered nickel support plate. This plate is manufactured by sintering highly disperse nickel powder produced by thermal decomposition of nickel pentacarbonyl $Ni(CO)_5$. The plates are filled by impregnating them in alternation with concentrated solutions of salts of the corresponding metals (Ni or Cd) and with an alkali solution serving to precipitate insoluble oxides or hydroxides. As the sintered plates provide a better contact with the active components, batteries with sintered electrodes allow using much higher discharge current densities than batteries with pocket electrodes but at the same time are more expensive.

Sealed Batteries

During charging of the batteries hydrogen and/or oxygen may evolve at the electrodes. Complete sealing will be admissible only when pressure buildup by evolved gases is avoided. For sealed Ni–Cd batteries this problem was solved by an appropriate balance of reactants in the battery. This must be such that during charging, reduced nickel is exhausted and anodic oxygen evolution starts at the positive electrode, long before cathodic hydrogen evolution starts at the negative electrode. To this end an excess of the nonreduced cadmium oxide, CdO, is provided in the negative electrode, and battery design is such that oxygen evolved at the positive electrode easily reaches the surface of the negative electrode and reacts with the metallic cadmium formed during charging. The oxygen thus is caught in a closed cycle of anodic evolution and cathodic reaction and cannot accumulate.

2.7   SILVER–ZINC BATTERIES

$$AgO|40\% \text{ KOH aq}|Zn \tag{2.17}$$

2.7.1   Silver–Zinc Storage Batteries

Attempts to develop a rechargeable battery with zinc negative electrodes were made repeatedly in the 1890s. The first successful silver–zinc battery was developed in 1941 by a French scientist, H. André, whose cell design had two essentially new features: the use of a swelling cellophane-type material for separators and a strictly limited volume of electrolyte solution, so that the zinc electrode operated only via the secondary process. Despite their high cost, these batteries immediately attracted great attention, owing to their excellent performance, with specific energies as high as 130 Wh/kg or 300 Wh/dm^3 (three to four times more than for other storage batteries), and their slight load dependence.

Rechargeable zinc electrodes have two major problems. During charging of the battery and electrolytic zinc deposition (reaction 2.6 from right to left), zinc crystallizes in the form of thin, branching crystals (dendrites) grow into the solution and rapidly reach the opposite electrode leading to an internal short circuit and breakdown of the battery. During cycling of the battery, a gradual migration of active materials of the zinc electrode can be observed (i.e., shape change of the electrode). There are many possible causes for such a migration (nonuniform current distribution over the electrode surface, gravitational forces, etc.). Using a swelling separator and operating with only the secondary zinc process, it was possible to mitigate the consequences of these adverse effects, but so far their causes are not well understood, and no means are known to completely prevent them. For these reasons the cycle life of silver–zinc storage batteries is rather limited; depending on the cycling mode it amounts to no more than 30–100 cycles.

Maintenance
The silver–zinc storage battery can or even should be stored in the state of complete discharge because then the silver electrode does not contain silver oxide which could be dissolved. Overcharge is dangerous owing to the increased possibility of formation and growth of zinc dendrites and oxidation of the separator with the evolved oxygen. Therefore, charging is discontinued at the final voltage 2.0–2.5 V.

2.7.2 Primary Silver–Zinc Batteries

There are two construction types of batteries: (i) miniature batteries, similar to the mercury–zinc button batteries, which are now produced and used in large quantities for hearing aids and for wrist watches and (ii) comparatively large automatic activated reserve-type batteries, which are activated by a forced injection of the alkaline electrolyte into the electrode compartment and which are mainly employed for aerospace applications.

Economic Parameters
The batteries with silver electrodes contain about 4 g/Ah of metallic silver; 75–90% of it is in the active material of the positive electrode. Therefore, large silver–zinc batteries can be used only for special applications when other battery types are unsuitable and high costs are acceptable.

REFERENCES

André H. Bull Soc Fr Electr 1941;60:1132.
Edison Th. A. US patent #678,722. 1901
Gladston JH, Tribe A. Nature 1881;25:221, 461.
Jungner W. Swedisch patent # 1556722. 1901.
Leclanché G-L. Compt Rend 1876;83:54.
Muneret X, Gobé V, Lemoine C. J Power Sources 2005;144:322.
Pavlov D, Kirchev A, Monahov B. J Power Sources 2005;144:521.
Planté G. Compt Rend 1859;49:402; 50, 640 (1860).

MONOGRAPHS AND REVIEWS

Bode H. *Lead-Acid Batteries*. NY: Wiley; 1971.

Fleisher A, Landers JJ. *Zinc Silver Oxide Batteries*. NY: Wiley; 1971.

Heise G.W. and N.C. Cahoon. *Primary Batteries*. 1. *Wiley*: NY; (1971).

McBreen JJ, Cairns EJ. *Advances of Electrochemistry and Electrochemical Engineering*. Vol. 1. NY: Wiley; 1978. p 375.

Milner PCC, Thomas UB. In: Delahay P, Tobias W, editors. *Advances of Electrochemistry and Electrochemical Engineering*. Vol. 5. NY: Interscience; 1967.

Rand DAJ, Moseley PT, Garche J, Parker CD, editors. *Valve-Regulated Lead Acid Batteries*. Amsterdam: Elsevier; 2004.

Rüetschi P. J Power Sources 1977;2:3.

3

PERFORMANCE

3.1 ELECTRICAL CHARACTERISTICS OF BATTERIES

3.1.1 OCV and Discharge Voltage

The open circuit voltage (OCV), U_0, of a battery depends on the electrochemical system selected for it and is somewhat affected by the electrolyte concentration, degree of discharge of the battery, temperature, and other factors. Once these parameters have been fixed, the OCV then is a fairly reproducible quantity.

Because of electrode polarization and ohmic voltage drops, the discharge voltage of a battery, U_d, is lower than the OCV and depends on the value of the discharge current, I_d. The battery voltage diminishes with increasing discharge current. The electrical power delivered by the battery ($P = I_d \cdot U_d$) at first increases with the increasing current and then, because of the voltage drop, passes through a maximum value. A typical dependence of discharge voltage and power on load current (current–voltage curve) is plotted in Figure 3.1.

The discharge voltage strongly depends on the structural and technological features of the battery, on temperature, and on numerous other factors. The spread in values of the discharge voltage is greater than that of OCV values.

The functional dependence of the discharge voltage, U_d, on discharge current I_d is sometimes represented by a simplified linear equation

$$U_d = U_0 - I_d . R_{app} \tag{3.1}$$

Electrochemical Power Sources: Batteries, Fuel Cells, and Supercapacitors, First Edition.
Vladimir S. Bagotsky, Alexander M. Skundin, and Yurij M. Volfkovich.
© 2015 John Wiley & Sons, Inc. Published 2015 by John Wiley & Sons, Inc.

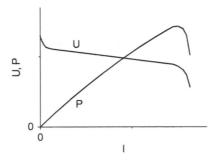

Figure 3.1. Typical current–voltage curve and battery power as function of load current.

where it is assumed that the apparent internal resistance, R_{app}, is constant. This is a rather crude approximation, especially when the current–voltage curve is S-shaped. The formal parameter, R_{app}, includes not only the internal ohmic resistance of the battery but also terms arising from the polarization of the electrodes.

3.1.2   Discharge Curves

When a battery undergoes discharge, a gradual decrease of the voltage is normally observed. Typical plots of U_d against the time of discharge, τ, or the amount of electric charge, Q_d, delivered are reported in Figure 3.2 for different battery types. The degree of voltage falloff differs between battery types, varying from 5 to 10% from its initial level, $(U_d)_{init}$, in some systems, to 50% in other systems. This decrease in voltage may be caused by higher polarization of the electrodes arising from an altered ratio of

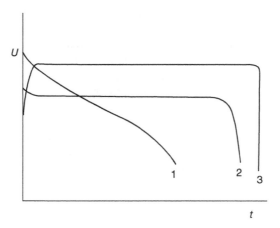

Figure 3.2. Typical battery discharge curves. 1—"Steep" discharge curve typical, for example, for Leclanché cells; 2—"flat" discharge curve typical, for example, for mercury–zinc cells; 3—discharge curve with initial "dip" typical, for example, for thionyl chloride–lithium cells.

reactants and reaction products at the electrodes and/or by increased ohmic resistance. The terminal decrease of voltage at the end of discharge may be steep or gradual. For this reason, it is recommended to terminate discharge at a certain cutoff voltage, $(U_d)_{fin}$, even though the reactants are not yet completely consumed.

For energy estimates, it is convenient to use the parameter of mean discharge voltage under given discharge conditions, $\bar{U}_d$, which is defined as the mean integral or mean arithmetic value of discharge voltage over a given discharge period τ.

3.1.3   Discharge Current and Discharge Power

The discharge current of a battery depends on the external circuit resistance, R_{ext}, and is given by Ohm's law as $I_d = U_d/R_{ext}$ or, when using the simplified expression (3.1), as

$$I_d = \frac{U_0}{(R_{app} + R_{ext})} \tag{3.2}$$

The power delivered by a battery during discharge is found as $P = U_d \cdot I_d$ or, with Equations (3.1) and (3.2), as

$$P = \frac{U_0{}^2 R_{ext}}{(R_{app} + R_{ext})^2} \tag{3.3}$$

Neither the discharge current nor the power output are the sole characteristics of a battery, as both are determined by the external resistance, R_{ext}, selected by the user. However, the maximum admissible discharge current, I_{adm}, and associated maximum admissible power, P_{adm}, constitute important characteristics of all battery types. These performance characteristics place a critical lower bound value, U_{crit}, on battery voltage; certain considerations (such as overheating) make it undesirable to operate at discharge currents above I_{adm} or battery voltages below U_{crit}. To a certain extent the choice of values for I_{adm} and U_{crit} is arbitrary. Thus, in short-duration (pulse) discharge, significantly higher currents can be sustained than in long-term discharge.

3.1.4   Battery Capacity and Stored Energy

The electric charge, Q_d, which has passed through the external circuit over a discharge period, τ, is given by $Q_d = \int_0^\tau I_d dt$ or, for $I_d = const$, simply by $Q_d = I_d \cdot \tau$. In the literature on batteries, this charge is expressed in ampere-hours (Ah). The total charge that can be delivered by the battery during a full discharge is called its ampere-hour capacity, C. As a rule, this parameter is stated in the battery specifications for defined discharge modes (in terms of discharge currents and temperatures).

The energy delivered during discharging a battery for τ hours is

$$W_e = \int_0^\tau U_d I_d dt \approx \bar{U}_d Q_d \tag{3.4}$$

This energy is measured not only in Joules (1 J = 1 W s) but also in watt-hours (1 Wh = 3600 J.). Both the maximum ampere-hour capacity and the watt-hour energy depend on the amount of reactants in a battery at the start of discharge. As the discharge current increases, both the discharge voltage and the reactant utilization coefficient diminish. A number of attempts to find a relationship between the actual battery capacity and the discharge current were reported. An approximation found in 1897 by Peukert for lead acid batteries is sometimes used:

$$Q_d = \frac{k}{I_d^\alpha} \tag{3.5}$$

The parameter ampere-hour capacity of a battery is more common (and mostly used in battery specifications) than the parameter watt-hour energy as it is easier to measure.

After a battery has been partially discharged its state can be characterized by the parameter depth of discharge (d.o.d) $\theta_d = Q_u/Q_r$(Q_u–used capacity, Q_r–rated capacity)

3.2 ELECTRICAL CHARACTERISTICS OF STORAGE BATTERIES

Both the ampere-hour capacity and the watt-hour energy of a storage battery are referred to one complete discharge of the battery after it has been charged to full capacity. While a storage battery undergoes recharging, the charging voltage, U_{ch}, will be higher the higher the charging current, I_{ch}. The charging voltage will also increase with time as the amount of charge stored increases. A battery is charged either to a certain final charge voltage, $(U_{ch})_{fin}$, or to the point where the battery has accepted an amount of charge defined as its charging capacity. The mean discharge voltage U_d of a storage battery is always lower than the mean charging voltage U_c. The discharge ampere-hour capacity of an ideal storage battery is equal to its charging ampere-hour capacity. Charging, however, is frequently accompanied by side reactions. For instance, in aqueous solutions, hydrogen and/or oxygen often start to be generated before charging is completed ("boiling" of the electrolyte). For this reason, the ampere-hour efficiency of capacity (or current) utilization, $\mu_Q = Q_d/Q_{ch}$, often is less than unity.

3.3 COMPARATIVE CHARACTERISTICS

Often the electric and other characteristics of batteries differing in their size, design, or electrochemical system need to be compared. The easiest way is by using normalized (reduced) parameters. Thus, current density serves as a measure of the relative reaction rate. Therefore, plots of voltage against current density provide a useful characterization of a battery reflecting its specific properties independent of its size. Not only the current, but also battery characteristics such as power, capacity, or energy

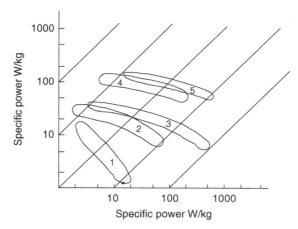

Figure 3.3. Dependence of specific energy on specific power for different storage batteries (Ragone plots).

can be referred to unity surface area (W/cm^2, Ah/cm^2, or Wh/cm^2). Likewise, the apparent internal resistance can be referred to unit surface area, but then the relevant unit will be ohm $\cdot$ cm^2 (with a rising surface area, current, power, and energy increase, but the internal resistance decreases).

Widely used parameters that are important to the consumer, are the specific energy or power per unit mass or per unit volume ($w = W/M$ or W/V) in watt-hour per kilogram or watt-hour per liter and the specific power ($p = P/M$, or P/V) in watts per kilogram and watts per liter. In each battery type, the specific energy is a falling function of specific power. Plots of w against p (Fig. 3.3) often called Ragone (1968) plots yield a clear illustration of the electrical performance parameters of given types of batteries, and are very convenient for their comparison.

Sometimes the discharge (or charging) current is characterized by the time needed for a complete discharge of the battery (e.g., 8-h discharge with a current of 5 A for a battery with a capacity of 40 Ah). Another parameter convenient for comparing different discharge modes is the inverse value—the normalized current referred to the battery's rated capacity $j = A/Ah$. In the abovementioned example the normalized current is 0.125 (commonly used without mentioning the units /h).

3.4 OPERATIONAL CHARACTERISTICS

Battery characteristics strongly depend on the operating temperature. As a rule, both the discharge voltage and the reactant utilization coefficient are lower at lower temperatures. However, increased temperatures are conducive to side reactions (such as corrosion processes) and thus reduce battery efficiency. Therefore, each battery type is designed for a specific temperature range within which its characteristics will be within the prescribed limits.

Another group of important battery characteristics are the lifetime parameters. For primary batteries and for charged storage batteries, a factor of paramount importance is the rate of self-discharge. *Self-discharge* may be the result of processes occurring at one of the electrodes (e.g., corrosion of zinc in batteries with zinc anodes or the decomposition of higher metal oxides in batteries with oxide cathodes), or it may be the result of processes taking place in the battery as a whole (internal short circuits following the growth of metallic dendrites between the electrodes or to a "shuttle action" of impurities in the electrolyte, which are oxidized at the positive electrode and reduced at the negative electrode, or external short circuits because of insufficient insulation between the electrodes). The rate of self-discharge determines shelf life (the maximum admissible interval between manufacturing and utilization in discharge) and service life (including the time of discharge) of primary batteries. The service life of storage batteries is usually understood as being the number of charging/discharge cycles.

For storage batteries produced and sold in a dry state (without the electrolyte) and for reserve-type batteries that need activation prior to use, it is necessary to distinguish the shelf life in the dry (not activated) state and the service time after adding the electrolyte or activating the battery.

Normally, the discharge capacity (in ampere-hour and watt-hour) is reduced in the course of cycling. Sometimes, the capacity increases over the first several cycles, passes through a maximum, and then diminishes. The practical cycle life of a battery (the admissible number of cycles not affected by a capacity or voltage decreases below a certain limit) depends on the cycling condition (temperature, depth of discharge, etc.). From the economic standpoint, it is also of interest to know the total energy delivered by a storage battery during its whole cycle life.

The last group of parameters relates to the handling of batteries. These include the mechanical strength (e.g., with respect to vibrations and shock accompanying transportation), maintenance, and "fool-proofness" (a term used by Thomas A. Edison as one of the characteristics of storage batteries).

REFERENCES

Peukert W. Elektrotechn Zeitschr 1897;18:287.

Ragone D.V. Proc. Soc. Automotive Engs. Conf., 1968; Warrendale, PA.

4

MISCELLANEOUS BATTERIES

4.1 MERCURY–ZINC BATTERIES

$$HgO|\ KOH\ aq|Zn \qquad\qquad (4.1)$$

Alkaline mercury–zinc batteries were manufactured as sealed cells of low capacity (0.05–15 Ah). They contain mercury oxide HgO and a limited amount of electrolyte (about 1 ml/Ah) absorbed in a porous matrix, so they operate only according to the secondary process of the zinc electrode. Modern mercury–zinc batteries were developed by S. Ruben in the beginning of the 1940s. His "button" construction was so effective that large-scale production started in the United States as early as World War II and after the war in other countries. A schematics of the button construction is shown in Figure 4.1

The active material of the positive electrode (1) consisting of mercury oxide HgO and 5–15% fine purified graphite is pressed into the nickel-plated steel case (6). Zinc powder (2) is pressed into the steel cover (4) and amalgamated. The separator (3) consists of several layers of a special alkali-resistant filter paper. The separator and the powdered zinc electrode are impregnated with a 40% KOH solution saturated with zincate. After assembly the battery is sealed by bending the edges of the case (6). The terminals (case and cover) are insulated by a rubber or plastic spacer (5).

On the positive electrode, discharge results in direct reduction of mercuric oxide to metallic mercury (without formation of intermediate products or

Electrochemical Power Sources: Batteries, Fuel Cells, and Supercapacitors, First Edition.
Vladimir S. Bagotsky, Alexander M. Skundin, and Yurij M. Volfkovich
© 2015 John Wiley & Sons, Inc. Published 2015 by John Wiley & Sons, Inc.

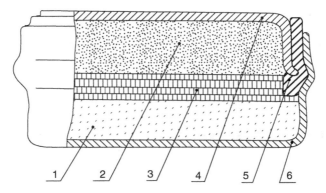

Figure 4.1. Alkaline mercury–zinc button battery.

variable—composition phases):

$$HgO + H_2O + 2e^- \rightarrow Hg + 2OH^- \tag{4.2}$$

Mercury–zinc batteries have a very stable open-circuit voltage (OCV) $(1.352 \pm 0.002 \text{ V})$.

Miniaturized button batteries (all three dimensions below 8–10 mm) were used for hearing aids and similar small-sized appliances. For some time hearing aid batteries were based on the mercury–zinc system. Because of the toxicity of mercury compounds, the production and use of mercury–zinc batteries in many countries were held up and they are now replaced with silver–zinc or zinc–air batteries.

4.2 COMPOUND BATTERIES

In some batteries two different types of electrodes are used: on one side a conventional battery electrode with solid reactant and reaction products, and on the other side an electrode with gaseous and/or liquid reactant and products of the type used in fuel cells. Such batteries are called *compound batteries* or sometimes *semi fuel cells* (this term is rather poor).

4.2.1 Metal–Air Batteries

The most important group of compound batteries are metal–air batteries in which readily oxidized metals are used as anodes and so-called air-breathing electrodes are employed as cathodes. The cathode reactant—air oxygen—is readily available everywhere and need not to be stored in the battery, thus, providing a considerable economy in the battery mass and volume and a corresponding increase of the specific ampere-hour capacity

1. Zinc–air batteries

The best-known example of a metal–air battery is the miniature button-type zinc–air battery widely used in hearing aids and other electronic devices of small size. Metallic zinc is the negative electrode in these batteries (usually in the form of highly disperse powder). When current is drawn, the zinc dissolves anodically in a concentrated alkaline solution according to the equation:

$$(-)\, Zn + 2OH^- \rightarrow ZnO + H_2O + 2e^- \quad E^0 = -1.254\,V \qquad (4.3)$$

Electrodes built according to the principles of the oxygen (air) electrodes of alkaline fuel cells are used as the positive electrode in these batteries:

$$(+)\, {}^1\!/_2 O_2 + H_2O + 2e^- \rightarrow 2\,OH^- \quad E^0 = 0.40\,V \qquad (4.4)$$

$$\text{Overall}\ \ Zn + {}^1\!/_2 O_2 \rightarrow ZnO \quad \varepsilon^0 = 1.654\,V \qquad (4.5)$$

As the currents drawn from the small zinc–air batteries in hearing aids are very low, it is not necessary to use expensive catalysts in the air electrodes of these batteries. The catalytic activity of activated carbon is quite sufficient for sustaining these loads. These batteries are designed for a moderate service life (several weeks), so the risk of carbonation of the alkaline electrolyte solution by traces of CO_2 from the air is not large. The hole providing access of air to the electrode is closed off with adhesive tape or paper that is removed prior to using the battery, in order to provide protection against carbonation during prolonged storage.

When the supply of metallic zinc is used up, the zinc–air battery stops working and cannot be reanimated. Repeated attempts have been made to build rechargeable zinc–air storage batteries. The reactions reported above are basically reversible, that is, can also occur in the opposite (charging) direction. For a number of reasons associated, both with the zinc and with the air electrodes, such rechargeable zinc–air batteries are as yet very unreliable and have so far not found any practical application, but research work in this direction continues.

2. Iron–air batteries

Another metal–air power source is the *iron–air* battery. The reactions occurring in these batteries are analogous to those occurring in zinc–air batteries, the only difference being that iron oxides Fe_2O_3 and/or Fe_3O_4 are the final product of anodic iron oxidation. This means that this battery is based on the well-known iron corrosion reaction. It may be assumed to be the cheapest kind of electrochemical power source. Earlier on, iron–air batteries with rather massive iron and carbon–air electrodes had been built. The electrical capacity (and size) of these batteries was relatively large (tens or hundreds of ampere-hours). These batteries were very simple to maintain. In the former Soviet Union, such batteries were used for power supply to unmanned installations such as railroad signaling equipment.

During the past decades many groups in different countries tried to "revive" the old Fe–air system. Hang et al., 2006 used Fe_2O_3 deposited on carbonaceous materials. The Fe_2O_3-loaded carbon material was prepared by chemically depositing Fe_2O_3 on carbon. $Fe(NO_3)_3$ was impregnated on carbon with different weight ratios of iron-to-carbon in an aqueous solution, and the mixture was dried and then calcinated for 1 h at 400°C in flowing argon. Comparison of Fe/C-mixed and Fe_2O_3-loaded carbon electrodes indicates that higher capacities are obtained in the latter case following the highly uniform distribution of iron on the carbon surface. K_2S and FeS were employed as additives for electrolyte and electrode, respectively, to suppress hydrogen evolution and improve the cycleability of the Fe/C composite air battery anode. Among the carbons used, nanocarbons such as tubular carbon nanofibers (CNF), platelet, vapor-grown carbon fibers, and acetylene black improved the discharge capacity of the Fe/C electrode. The FeS additive showed a larger beneficial effect for the Fe/C composite electrode than K_2S with regard to cycleability and capacity. The roles of the K_2S and FeS additives in determining the behavior of the Fe/C electrode were interpreted based on the presence of adsorbed sulfide ion. Iron sulfide interacts with Fe(I), Fe(II), and Fe(III) in the oxide film to promote the dissolution of iron and enhance the bulk conductivity of the electrode, leading to improvements in the cycleability. Iron corrosion with hydrogen evolution—the most important drawback of nickel–iron storage batteries—was suppressed significantly by using both FeS and K_2S additives.

4.2.2 Nickel–Hydrogen Storage Batteries

The nickel–hydrogen storage batteries were first developed in Russia in 1964. In Russia and in a number of other countries, a moderate production output of such storage batteries was started to satisfy the needs of space technology. In them the nickel oxide electrodes used in the conventional alkaline nickel–cadmium storage batteries served as positive electrodes. Hydrogen electrodes from alkaline fuel cells batteries, which at that time had already found some application, were used as negative electrodes. When current is drawn (during discharge), the following reactions occur in these storage batteries:

$$(+)NiOOH + H_2O + e^- \rightarrow Ni(OH)_2 + OH^- \qquad E^0 = 0.49 \text{ V} \qquad (4.6)$$

$$(-)\tfrac{1}{2}H_2 + OH^- \rightarrow H_2O + e^- \qquad E^0 = -0.828 \text{ V} \qquad (4.7)$$

$$\text{Overall } NiOOH + \tfrac{1}{2}H_2 \rightarrow Ni(OH)_2 \qquad \varepsilon^0 = 1.22 \text{ V} \qquad (4.8)$$

During charging, the reactions occur in the opposite direction. Therefore, hydrogen gas is evolved at the negative electrode during charging, is recovered into tanks, and kept there (usually under higher pressure) until needed for discharge.

Such storage batteries became feasible when it was realized that gaseous hydrogen is almost completely inert (unreactive) toward the trivalent nickel hydroxide. It was

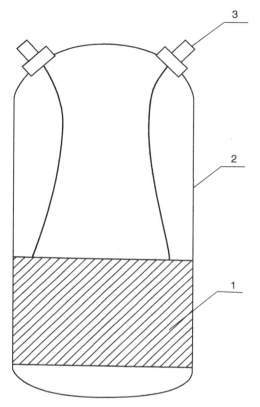

Figure 4.2. Schematic of a nickel–hydrogen storage battery: (1) electrode block, (2) container, (3) terminal.

possible, therefore, to place the block of charged positive electrodes into the hydrogen tank, greatly simplifying battery designs. A very convenient aspect in the use of these batteries was the fact that from the hydrogen pressure in the tank, the battery's depth of discharge could be determined. The number of discharge–recharge cycles sustained by nickel–hydrogen storage batteries is exceptionally high. Thus, in various long-lived communication satellites and in NASA's Mars Global Surveyor, batteries of this type were used. For general applications, these batteries were later replaced by nickel hydride storage batteries (Section 2.6.6), which are much easier to handle. Figure 4.2 shows the schematics of a nickel–hydrogen storage battery

4.3 BATTERIES WITH WATER AS REACTANT

The Volta pile (the historical first battery) did not use specially added oxidants. Water molecules served in it as oxidant (donors of protons for cathodic hydrogen evolution).

The efficiency of this battery was low as the rate of hydrogen evolution on a silver electrode is low. However, the possibility to use hydrogen evolution from water in the cathodic current-producing reaction is very attractive. A battery built according to this concept has only two metal electrodes (plates): one serving as dissolving anode and the other (catalytically active, such as steel or nickel) as current collector for hydrogen evolution. This simple construction does not make use of powdered materials, separators, and so on. The basic reactant (water) is freely available anywhere, especially in marine applications. The current-producing reaction in these batteries can be written as

$$M \rightarrow M^{n+} + ne^- \tag{4.9}$$

where M is the dissolving metal (Zn, Mg, Al, Li, etc.). The use of lithium anodes in such batteries became practically feasible when it was found that at certain concentrations of LiOH in water, lithium is partially passivated and its self-dissolution is sharply inhibited though a certain activity for the anodic dissolution remains. When the battery starts operating the concentration of LiOH in the vicinity of the lithium electrode gradually increases until after some time (depending on the rate of water circulation) an optimal concentration is reached. Batteries of this type are used mainly for marine applications (e.g., in floating navigation and communication buoys).

4.4 STANDARD CELLS

The unit of voltage or difference of potentials (volt) is defined as the voltage produced in a circuit a current of 1 A generating a power of 1 W. This definition is not convenient for practical measurements. Therefore secondary voltage standards are used—that is, the OCV values of standard cells. The standard cells serve not for delivering current but for maintaining constant voltage during various measurements, in compensation circuits and for testing calibrating instruments. Ordinarily for measurements of a battery OCV and of electrode potentials, an accuracy of ± 1.0 mV is necessary. When determining thermodynamic parameters, the accuracy must be much higher (± 10 μV). The device used for measuring these parameters should not cause any current flow in the cell. Currents of 10^{-5} to 10^{-3} A would arise when connecting an ordinary permanent magnet moving coil voltmeter to the cell. Therefore now electronic voltmeters with a very low-current drain (less than 10^{-8} A) are used. In the past, before the availability of electronic voltmeters, compensating potentiometers based on the Wheatstone bridge principle were used. These potentiometers were calibrated with a cell for which the OCV was stable and reproducible. The best-known standard cell was proposed in 1892 by Edward Weston and in 1908 officially adopted for metrological purposes. The cell is sealed into H-shaped glass vessels (Fig. 4.3). The positive electrode is mercury in contact with a paste of crystalline Hg_2SO_4 and $CdSO_4$ $8/3H_2O$, and the negative electrode is a 6–12% cadmium amalgam in contact with crystals of $CdSO_4$ $8/3H_2O$. A saturated $CdSO_4$ solution is used as electrolyte. These cells have an OCV of 1.01864 ± 0.00002 V at 20°C.

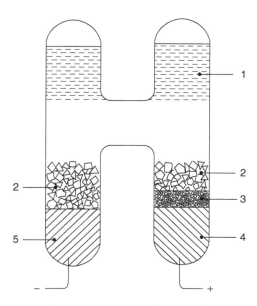

Figure 4.3. Standard Weston cell.

Less critical measurements in everyday laboratory or industrial work are made with the unsaturated Weston cell in which no crystals of $CdSO_4 8/3H_2O$ are used and the saturated $CdSO_4$ solution is replaced by a lower-concentration solution (saturated at $4°C$). The OCV of this cell is 1.0192 ± 0.0002 V, that is, its reproducibility is lower. As the solution concentration does not vary with temperature, in contrast to the saturated Weston cell, the temperature coefficient of the unsaturated cell is very small (about 1 $\mu V/°C$) and the cell does not require careful thermostating. The unsaturated cells are placed into plastic cases with copper shields to provide for temperature equilibrium at both electrodes.

4.5 RESERVE BATTERIES

The term *Reserve battery* is applied to single-use primary batteries with comparatively active reactants in which in order to increase shelf life the electrodes are brought into contact with the liquid electrolyte (activated) only immediately prior to their (usually short-time) discharge. This concept does not apply to dry-charged storage batteries, which for the same purpose are manufactured and sold without electrolyte, and when being filled with the electrolyte solution allow a longtime cycling. Several versions of reserve batteries are distinguished depending on the method of activation.

4.5.1   Manually Activated Reserve Batteries

These batteries are manufactured without electrolyte and are filled before operating with a liquid electrolyte solution either manually or by an auxiliary feeding device. In some batteries of this type an anhydrous solid alkali or salt is placed in special bags into the interelectrode gaps to be dissolved when water is added. In particular this principle is realized in water-activated reserve batteries with magnesium anodes and metal chlorides (univalent MgCl, $PbCl_2$, poorly soluble AgCl) as cathode material. The first MgCl and AgCl batteries were developed in 1943 in the United States and in 1949 production of CuCl batteries was started. As water-activated magnesium reserve batteries can be activated by sea water, they are used mainly for maritime applications, such as various alarm devices and survival aids for plane passengers.

4.5.2   Reserve Batteries with Automatic Electrolyte Injection of Electrolyte (Ampoule Batteries)

In these types of batteries the electrolyte solution is kept in special containers (ampoules) inside the battery. The battery can be activated either pneumatically (Fig. 4.4) by rupturing a special membrane with an exploding cartridge (2). Gas released from a high-pressure gas bottle (1) compresses an elastic ampoule (3) containing the electrolyte breaks through a membrane at the ampoule neck and fills the electrode chamber (5). Reverse valve (4) prevents the electrolyte reflux.

4.5.3   Reserve Batteries with Molten Electrolyte (Thermal Batteries)

In normal condition the electrolyte (a salt mixture) is solid and has no conductivity so that the batteries can be stored for a long time. Batteries of this type contain two parts: (i) the battery proper and (ii) a device for activating the battery by exploding pyrotechnic heaters ignited by fuses and thus instantly raising the temperature above

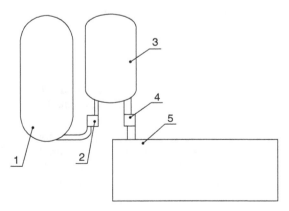

Figure 4.4. Schematic of an ampoule-type reserve battery.

the melting temperature of the electrolyte. Thermal batteries are described more in detail in Section 15.2.

Automatically activated reserve batteries that have an activation time below 1 s are very expensive. They are used for military purposes in devices that require a combination of long shelf life with constant readiness for a speedy application.

REFERENCE

Hang BT, Watanabe T, Egashira M, Watanabe I, Okada SH, Jun-ichi Yamaki J. J Power Sources 2006;155:461.

REVIEWS AND MONOGRAPHS

Hamer WJ. In: Cahoon NC, Heise GW, editors. *Primary Batteries*. Vol. 1. NY: Wiley; 1971. p 433–477.

Hamlen R, Atwater T. In: Linden D, Reddy TB, editors. *Handbook of Batteries*. McGraw-Hill; 2002.

Jennings CW. In: Cahoon NC, Heise GW, editors. *Primary Batteries*. Vol. 2. NY: Wiley; 1976. p 263–293.

Ruben S. In: Cahoon NC, Heise GW, editors. *Primary Batteries*. Vol. 1. NY: Wiley; 1971. p 207–223.

5

DESIGN AND TECHNOLOGY

The battery design must ensure the conditions required for the current-producing reaction and for practical utilization of the electrical energy released. Among these conditions are:

1. separation of reactants (e.g., by separators),
2. good and reliable contact between the electrodes and the electrolyte,
3. minimum ohmic losses in current collection from the reaction zone to the battery terminals,
4. elimination of possible current leakage,
5. uniform loading of each of the electrodes (in multielectrode batteries),
6. mechanical strength and reliable sealing of the battery.

5.1 BALANCE IN BATTERIES

A problem that appears immediately when a specific single-cell battery is designed is the required ratio of the reactants for the positive and negative electrodes and for the electrolyte. An unnecessary excess in one of the components is undesirable. It is therefore typical for the reactants to be loaded in the stoichiometric ratio given by the current-producing reaction with the actual utilization coefficient of each of

Electrochemical Power Sources: Batteries, Fuel Cells, and Supercapacitors, First Edition.
Vladimir S. Bagotsky, Alexander M. Skundin, and Yurij M. Volfkovich
© 2015 John Wiley & Sons, Inc. Published 2015 by John Wiley & Sons, Inc.

the reactant taking into account. Sometimes difficulties may be encountered as the utilization coefficients depend on the conditions of battery operation (temperature, storage duration, etc.). Sometimes a surplus of one of the reactants is provided to avoid undesirable side reactions. For instance, a surplus of mercury oxide is used in sealed mercury–zinc batteries in order to consume the whole amount of zinc stored and to prevent hydrogen evolution at the end of discharge. Obviously the discharge ampere-hour capacity is then determined by the electrode with a shortage of reactant. In some battery types the electrolyte is an active component of the current-producing reaction (e.g., sulfuric acid in lead acid storage batteries). In the other case the electrolyte composition is constant during the battery life and its volume can be minimized.

Another problem is that of balance between single cells in a multicell battery. Such batteries are balanced when all cells (preferably of the same type) have the same capacity and the same state of charge (SOC). In case of a mismatch between cells connected in series the battery's ampere-hour capacity is determined by the cell with the least available ampere-hour capacity. In case of cells connected in parallel, a failure (e.g., short circuit) of one of the cells can lead to the failure of the whole battery.

5.2 SCALE FACTORS

Each battery type is usually manufactured as a series of models differing in size and hence in the ampere-hour and watt-hour capacities. But despite analogous structure and production processes, not all parameters of batteries are proportional to their size, and sometimes a so-called scale-factor must be introduced. Miniature batteries are characterized by an appreciable higher fraction of structural elements, as it is not always possible to reduce their size proportionally. In large-scale batteries (e.g., storage batteries for submarines and for some stationary applications) another problem is involved—that of uniform current distribution. Normally a current-conducting bus is fixed to the upper part of the electrode. In electrodes with a considerable height, owing to ohmic losses in the electrodes' current collecting grids, the process will concentrate on the upper part of the plate. Moreover the electrodes located near the container walls may be in different thermal conditions compared with those in the middle of the electrode group.

5.3 SEPARATORS

Separators are installed in the interelectrode gaps of practically all batteries with liquid electrolytes in order to prevent accidental electron-conducting contacts between electrodes of opposite sign (e.g., because of mechanical shocks or vibration). At the same time, they must provide free electrolyte access to both electrodes forming an ion-conducting bridge between them. Separators are manufactured of dielectrics in the shape of large-hole meshes (simple separators), porous sheets, or ultra porous (swelling) membranes. In batteries with a matrix electrolyte, porous separators often function as electrolyte carrier retaining the liquid electrolyte close to the electrode surface with capillary forces.

Separators must have a longtime chemical and mechanical stability in the battery environment. They must be sufficiently elastic so as not to break down in the course of battery assembling and be shockproof. In addition, they must be inexpensive, simple in manufacture, with reproducible properties in large-scale production. An ideal separator must introduce only a minimum resistance to ionic current. The conductance attenuation coefficient varies from 1.1 to 1.6 for simple spacers and from 2 to 8 for porous and ultra porous varieties, reaching 15 only in exceptional cases. Depending on the battery type and function, separators either fill the whole electrode gap or only a part of it. In the latter case electrode surface is in free contact with the free liquid electrolyte, which is sometimes essential for sheet-shaped separators to have several rips in order to ensure a gap between them and the electrodes.

5.3.1 Simple Separators

Simple separators (spacers without filtering properties) are made of plastic or ebonite rods or of plastic cords (2 mm in diameter), perforated or corrugated plastic sheets, or fiberglass felts. These separators have large holes or pores (from 0.1 to 4 mm).

5.3.2 Porous Separators

Porous clay ceramic cups were used as separators in batteries with liquid electrolyte in the first half of the nineteenth century. However, the materials employed nowadays as separators have finer pores. Separators of this type are now the most widely used in batteries. Porous separators are sometimes classified as porous having the mean pore radius from 0.1 to 100 μm, and micro porous with smaller radii. Porous separator sheets support mechanically the electrode active mass, preventing its crumbling. In the course of storage battery charging, they suppress dendritic growth of metal deposits (leading to internal short circuits). Micro porous rubber (Mipore) is a high-quality material with small pores and high total porosity, but has the shortcomings of brittleness and high cost. Sintered PVC (Miplast in Russia or Porvic in the United Kingdom) has comparatively large pores. Its production process is simple and the final product is rather cheap.

5.3.3 Membranes

Membranes (swelling separators) are made of polymer materials absorbing either aqueous or nonaqueous solutions. The solvent penetrating the polymer structure increases the distance between macromolecules (the membrane becomes swollen), which make it possible for ions in the solution to move (migrate) across the membrane. In contrast to porous separators, specific interaction forces between individual ions and macromolecules are highly pronounced. As a result, these membranes are mostly selective, that is, the attenuation of transport is different for different ions. Cellophane (hydrated cellulose), polyethylene films with radiation-grafted acrylic acid, and ion-exchange membranes are examples of swelling separators.

5.4   SEALING

Batteries most widely used in the nineteenth century were mostly manufactured in open glass vessels containing the electrodes and a liquid electrolyte solution. Obviously, such batteries could operate only in stationary conditions being in an upright position. The development and wide use of portable devices necessitated the appearance of more convenient, sealed designs that allow operation in any spatial orientation and are reasonably shockproof. Evolution of gases, mostly hydrogen (during battery self-discharge) and both of hydrogen and oxygen (during charging of storage batteries), is a factor making battery sealing and hermetization a difficult problem. In addition to gas evolution, some batteries must have the facility to be opened for filling with the electrolyte solution or for periodic addition of water to adjust the electrolyte concentration. Sealing of storage batteries is also necessary to avoid during charging the release to the ambient medium of a fog containing harmful tiny electrolyte drops transported by the evolving gas. Batteries may be partially sealed with the electrolyte rendered spillproof and the damaging effect of gas evolution reduced, or may be completely sealed.

5.4.1   Partial Sealing

Primary batteries are normally spillproofed by using quasi-solid immobilized gel-type or matrix-type electrolytes. In small-sized primary batteries, a minor accumulation of evolved hydrogen can be avoided by sealing them with materials permeable to hydrogen. In sealed large-sized primary batteries and particularly in storage batteries in which intense gas evolution is possible, safety one-way valves must be fitted. The most popular are rubber valves that in normal conditions are closed and lock out the liquid. When pressure builds up in the battery the valve opens for a short time. More reliable are analogous spring valves.

5.4.2   Complete Sealing

Complete sealing is possible for battery types in which no gases are evolved. An example is the Weston cell that is manufactured in a sealed glass vessel and is not used for current drain and in which self-discharge is thermodynamically impossible. Gassing can also be considered as practically absent in a number of batteries with nonaqueous electrolytes (including storage batteries and some varieties of lithium batteries). If in a battery gas evolution is considerable, complete sealing is possible only when the gas encounters some chemical reaction. Two cases are possible:

1. When, for example, at the end of charging or during overcharging as the result of electrolytic water decomposition, hydrogen and oxygen are evolved in stoichiometric proportion and it is possible to initiate their mutual interaction with the regeneration of the decomposed water. For this purpose often a special catalyst is inserted into the battery.
2. The amount of reactants in the battery is calculated such that during charging, gas evolution starts at one of the electrodes long before at the other one.

For example, in a nickel–cadmium storage battery an excess of cadmium oxide (CdO) is added to the negative electrode, so that during charging oxygen begins to evolve at the positive NiOOH electrode earlier than oxygen at the negative cadmium electrode. The battery design is such that oxygen formed at the positive electrode can easily reach the surface of the negative electrode and react there with metallic cadmium regenerating the amount of cadmium oxide. This situation is called oxygen regeneration cycle. Another opposite variety with a hydrogen regeneration cycle is also possible. With the charging current below a certain value, completely sealed storage batteries permit unlimited overcharging, which is especially important for battery operation in nonserviced installations with automatic recharging units.

A battery is sealed reliably if the joints of its individual parts (container, cover, terminals, valves, etc.) are tight with respect to liquids and gases. Depending on the materials used, the joints are sealed by gluing, soldering, welding, or filling with sealing compounds. The joints must be chemically and thermally resistant (for instance, must not crack because of nonuniform thermal expansion). For ambient temperature batteries, these problems are usually solved without too much effort. But in the case of high-temperature batteries (especially with chemically very aggressive molten salt electrolytes) they are very serious and hinder the development of new battery types.

In batteries with metallic container one terminal must be electrically insulated from it. In low-temperature batteries for this purpose rubber or plastic gaskets are used, which provide a sufficient effective sealing of joints. But in sealed batteries with an alkaline electrolyte solution (e.g., in cylindrical alkaline manganese–zinc batteries), after some time the alkaline solution begins to "creep" around most sealing gaskets and then forms on the outside white patches of alkali carbonates.

In high-temperature battery cells, elastic gaskets cannot be used for sealing, and insulation and must be replaced with ceramic materials.

5.5 OHMIC LOSSES

A single battery cell is a low-voltage power source, so that even slight internal voltage losses (e.g., 0.1 V) produce an appreciable effect on the electrical characteristics of the whole battery. Internal voltage losses in a cell can be distributed over the whole current path from one terminal to the other. In multicell batteries with cells connected in a series, losses in the interconnections must be added to the total internal losses of all cells.

Two types of losses can be distinguished in battery cells: (i) ohmic losses proportional to the current, that is, with a constant value of ohmic resistance R_{Ohm} and (ii) polarization losses because of electrode polarization (change of electrode potential during current drain) with a complex dependence on current, that is, with a variable value of polarization resistance R_{pol}. The total of both ohmic and polarization resistances $R_{app} = R_{Ohm} + R_{pol}$ is called the cell's apparent resistance and allows to use a simplified equation for the dependence of cell voltage on current (Eq. 3.1).

In contrast to other components of ohmic resistance, the electrolyte resistance depends on temperature. As a rule (though not always) this electrolyte resistance is higher than the electrodes' ohmic resistance. When in the cell porous electrodes are used, the resistance of the liquid electrolyte inside pores (from reaction zone to interface with the electrolyte) must also be added to the electrolyte's ohmic resistance. The use of separators in the interelectrode gap or the use of a porous electrolyte-soaked matrix increases the electrolyte resistance. For different separators the corresponding conductance attenuation factors are quoted in Section 5.3. The electrolyte resistance is also increased when gas bubbles accumulate in the interelectrode gap. If the bubble volume comes to 30% of the total gap volume, the resistance increment will be 60–80%.

Electronic conductors in batteries are often connected by clamping, for example, in a clamp with a nut fixing the external cable. Dry clamped-on contacts are usually quite reliable and their contact resistance is low providing measures are taken against loosening the clamp by locking the nut. If, however, the contact is loosened or wetted by electrolyte, the contact resistance can rise rapidly because of surface oxidation or salt accumulation. Furthermore, in this case, an increased contact resistance increases the local heat generation, thereby accelerating the deterioration of the contact.

5.6 THERMAL PROCESSES IN BATTERIES

When a battery is discharged, together with electrical energy, heat is released. The intensity (thermal power) of heat formation depends on the difference between the thermodynamic EMF ϵ of the current-producing reaction and the real discharge voltage U_d: $(\epsilon - U_d)$. Consequently, heat evolution rises with a decreasing discharge voltage. The same is true for the charging process of storage batteries. When sealed storage batteries are overcharged and gas is chemically recycled, no overall current-producing reaction occurs and hence ϵ is equal to zero and the overcharging electrical energy is completely converted into heat. The steady-state temperature difference of the battery relative to the ambient depends on the intensity of heat generation and the conditions of heat removal. For low-size and low-power batteries the raise in temperature is hardly observable. For large-sized batteries, particularly those discharged with high-current values (with low values of discharge voltage) the raise in temperature is considerable. Each battery is characterized by a certain critical maximal admissible temperature T_{crit}, above which all destructive processes (corrosion, aging, etc.) are accelerated. Accordingly, these batteries can also be characterized by a critical maximal current value I_{crit}. To increase the value of I_{crit}, some methods can be used:

1. Gaps between cells in multicell batteries for free access and circulation of air,
2. Air cooling of the battery with a fan,

3. Forced circulation of the liquid electrolyte in an external circuit with a cooling unit,
4. Cell design with external heat exchange,
5. Built-in coolers with liquid coolants.

Specific problems are encountered when the ambient temperature is below the minimum operation temperature of the battery (e.g., in winter conditions and for high-temperature batteries with working temperatures above 50°C.) In this case it is necessary to heat up the battery prior to use. This can be done by internal or external heaters plugged into the electric grid network. If no heating means are available, so-called self-heating is employed, that is, a limited-time preliminary discharge with a low current leading to a gradual raise of the battery's temperature. Both these methods require a certain time interval to reach the desired battery temperature, depending on circumstances, from 15 min to 2–3 h. Starter batteries for ICE cars use for starting the engine a short-time (5 – 15 s) current–pulse discharge with a current value gradually decreasing from about 300–500 A to zero. In winter time at ambient temperatures not below −20°C, some (2–4) preliminary idle starting attempts lead as the result of self-heating, to the raise of the battery's temperature sufficient for the successful starting of the engine. In some cases (still lower temperatures and/or starting powerful Diesel engines) it is very useful to connect a supercapacitor in parallel to the battery. When starting the engine, this supercapacitor, being always charged, takes up the main part of the initial high-current load speeding up the starting procedure and increasing the battery's lifetime.

6

APPLICATIONS OF BATTERIES

6.1 AUTOMOTIVE EQUIPMENT STARTER AND AUXILIARY BATTERIES

Storage batteries are employed in all vehicles with internal combustion engines (ICEs) (cars, motorcycles, airplanes, diesel locomotives, etc.) to start engines and for auxiliary needs. Three functions are assigned to the battery in a car with a carburetor engine: energizing the starter, energizing the ignition system while the generator is not yet operative, and supplying energy to auxiliary circuits like lighting, radio, and so on, when the engine is off. A convenient abbreviation is SLI: starting, lighting, ignition. In addition, aircraft batteries stand by for an emergency, that is, they supply current to all essential systems on board the aircraft when the generator breaks down. Such emergency operation may last till the emergency landing, that is, it may last for more than half an hour.

Certain specific features are typical for starter batteries. First of all starter currents are very high: the normalized current j_d reaches $3-5$ in car batteries and may be even higher in aircraft batteries. A battery must guarantee the starting of the engine in winter, that is, at low temperatures. Starter discharge is usually of short duration, no more than 20 s in automobiles and 60 s in airplanes. Consequently, the starter discharge is normally not deep, with no more than $3-10\%$ of capacity drained by a single starting procedure, immediately followed by recharging. A battery designed

Electrochemical Power Sources: Batteries, Fuel Cells, and Supercapacitors, First Edition.
Vladimir S. Bagotsky, Alexander M. Skundin, and Yurij M. Volfkovich.
© 2015 John Wiley & Sons, Inc. Published 2015 by John Wiley & Sons, Inc.

for repeated starting must be sufficiently quickly recharged. This is carried out by a generator whose output voltage is kept constant by special control facilities with an accuracy of $\pm 3.5 - 8\%$. Starter batteries are required to have a long service life and low self-discharge. Starter batteries may be idle for long periods of time but must be in constant readiness for operation. Finally, vehicle-borne batteries have to withstand substantial mechanical loads (vibration and shock loads).

Battery types used nowadays for starter batteries are mostly lead acid batteries, sometimes nickel–cadmium and silver–zinc ones, the last of these is used only in aircraft applications. The rated voltage of automobile starter batteries is 12 V with the ampere-hour capacity from 40 to 200 Ah. The rated voltage of aircraft batteries is 24 V. A considerable number of vehicle-borne batteries are not meant for engine starting, but power auxiliary systems. Such types are railway carriage-borne, ship-borne batteries. These are used for lighting and for supplying current to air conditioners, transceivers, signaling equipment, and so on, during stops when the generator is off. Batteries used for above applications are lead acid and nickel–iron types. They operate in extreme conditions: wide large temperature range (practically from −50 to +50°C), high humidity elevated contamination levels. Railway carriage batteries carry substantial mechanical loads. In contrast to starter batteries, ship- and carriage-borne batteries undergo rather deep discharges (during long stops) and long overcharging (during long run between stops). The typical ampere-hour capacity for carriage-borne batteries is from 70 to 500 Ah.

6.2 TRACTION BATTERIES

Performance characteristics of internal combustion engines are higher than those of a combination of a battery and an electric motor. Consequently autonomously powered vehicles are mostly powered by ICEs. There are fields, however, where application of these engines is inconvenient and unacceptable. Such are submarines (in submerged condition) and submersible craft in which ICEs cannot be used owing to lack of oxygen and difficulties involved in the disposal of exhaust gases; vehicles in mines where ICEs would constitute an explosion hazard and pollute the atmosphere; transportation systems in factories and in other closed spaces where a vehicle must have high maneuverability, no toxic exhaust, and so on. In all these cases traction electric motors and special traction batteries are employed. The operation mode of traction batteries is very different from that of starter batteries. As a rule traction batteries are discharged with a wide range of loads and to a considerable depth. Moreover they must provide high power in the first several seconds of motion. As with all vehicle-borne batteries traction batteries operate in conditions of high mechanical load. Lead acid and nickel–iron storage batteries are used nowadays as traction batteries. Their capacity ranges from 40 to 1200 Ah and reaches several thousand ampere-hours in the case of submarines.

6.3 STATIONARY BATTERIES

Conditions of operation of stationary batteries are much less severe than those found in transportation systems. They are invariably in a stable position so that careful sealing becomes unnecessary. The requirements on battery and electrode design are greatly simplified by the absence of vibrations and other mechanical loads. Stationary units are often located in special buildings and operate at optimal temperatures. In some applications, however, such as automatic weather stations, stationary batteries work under severe climatic conditions and operate at temperatures from -50 to $+50°C$. The installed capacity often reaches into thousands of ampere-hours. Usually neither dimension nor mass are critical parameters: in other words, high-specific characteristics are not required. Many stationary units are designed for long-term operation, and thus must have long service life, high reliability, and low level of self-discharge. As a rule servicing of such batteries is carried out by skilled personnel. Long-term operation with no servicing is typical for automatic weather stations, relay stations, and the like. Stationary batteries are still in use as high-stability DC sources in telephone exchanges, since high-quality rectification of AC voltage remains a problem without a satisfactory solution. High-capacity stationary units usually based on lead acid batteries are employed in power stations to supply power to auxiliary equipment. Emergency (standby) storage batteries, providing power for lighting and uninterrupted functioning of important systems when the main supply fails, are now used more and more widely. Such batteries are kept in constant readiness; they are designed to operate for several hours until the emergency is over or a diesel generator is assigned to the job. In the last case the stationary battery is used to start the diesel unit. Emergency batteries are installed in hospitals to ensure uninterrupted functioning of surgery rooms. In normal conditions emergency batteries are hardly ever discharged. Self-discharge losses are continuously or intermittently compensated by recharging.

6.4 DOMESTIC AND PORTABLE SYSTEMS

The widest field of battery application both in the number of batteries and in their diversity is that of portable units and domestic appliances. Typically this involves batteries of low and medium capacity from 0.01 to 100 Ah. Battery-powered electrical units include various radio sets, tape recorders, cell phones, TV sets, audio and video players, radio transmitters for civil and military use, portable flashlights for underground and other use, miniaturized devices electronic watches, hearing aids, pocket calculators, toys, some household appliances (electric shavers, measuring instruments including testers, radiation monitoring instruments, etc.), and various medical devices (cardiac pacemakers).

The diversity of battery-powered devices entail a great diversity of operational conditions. Thus, electric torch batteries operate in a wide temperature range while those of cardiac pacemakers are used in strictly isothermal conditions. Batteries in electronic watches discharge uninterruptedly by low-current pulses, while those in transistor radios operate with randomly variable load.

Batteries for military and medical applications must be highly reliable, which would be redundant in electrified toys. Cost is often an important characteristic of a battery: performance characteristics and shelf life may have to be somewhat lowered for the sake of cost reduction. In other cases one of the battery characteristics may be the most important of all. With very few exceptions batteries for portable and domestic appliances must be spillproof and can be transported in any position.

6.5 SPECIAL APPLICATIONS

Specialized applications of batteries cover military and space technology and some scientific instruments. Examples of such battery-powered devices are satellites, space probes, rockets, bathyscaphes and other submersible craft, electrically driven torpedos, meteorological balloons, and so forth. Specialized batteries may be divided into two subdivisions: power sources for short-term loads (typically for a single discharge) and those for long-term low-drain discharge.

Power systems used in space ships typically combine semiconductor solar panels and buffer storage batteries. Storage batteries installed on earth satellites energize the systems while the satellite is in the earth's shadow. For the rest of the time of flight, solar panels recharge the battery. Typically the operation schedule of such batteries may be 20–40 min of discharge and 50–70 min of recharging, which means about 16 charging/discharge cycles (partial as a rule) in 24 h.

As with every item of equipment for critical use, batteries must satisfy exact requirements concerning specific parameters and reliability. They must be designed for prolonged warehouse storage. Throughout their shelf life these batteries must be permanently ready for discharge. As a rule these batteries are expected to operate in a wide range of climatic conditions from arctic to tropical, and at high mechanical loads. Cost considerations normally play only a secondary role when batteries for space applications are selected.

7

OPERATIONAL PROBLEMS

7.1 DISCHARGE AND MAINTENANCE OF PRIMARY BATTERIES

One of the important advantages of primary batteries consists of very simple maintenance procedures. Most such batteries require no servicing. Before a battery is switched on, its appearance and the remaining service life are checked; sometimes actual parameters are measured (open-circuit voltage (OCV) and the initial discharge voltage). Correct polarity and reliable contacts must be ensured; a violation of polarity correspondence may result in serious disorders and even in the breakdown of load circuitry, especially circuits involving transistors and electrolytic capacitors.

The electrical conditions of cell operation are determined by the load schedule. Typically, primary batteries are used with complicated, and often arbitrary, loading schedules. The current drain of a transistor radio battery, for example, is a function of the volume setting; the times of turning the set on and off are arbitrary. Batteries in electronic watches and pacemakers are loaded continuously but the discharge current is pulsed. The cases of continuous discharge to a constant load are rather infrequent.

As a rule, primary batteries are used until completely discharged. Sometimes, however, this should be avoided; leaving the discharged battery in the unit it has powered, especially, may be harmful. Leakage after complete discharge is observed in some, and in particular in the most widely used manganese–zinc batteries. The lost electrolyte may cause corrosion of the consuming unit. Explosions of sealed batteries, especially of the mercury–zinc batteries, may sometimes (although very rarely) be

Electrochemical Power Sources: Batteries, Fuel Cells, and Supercapacitors, First Edition.
Vladimir S. Bagotsky, Alexander M. Skundin, and Yurij M. Volfkovich.
© 2015 John Wiley & Sons, Inc. Published 2015 by John Wiley & Sons, Inc.

observed if spent batteries are stored. Batteries with a great number of cells connected in a series must be watched with special care for signs of cell reversal. In order to avoid reversal, a voltage check on each cell is desirable. This being impossible in most cases, the battery must be disconnected in advance.

Most of the primary batteries have high internal resistance, so that brief external short-circuiting is unlikely to produce extensive damage.

Certain types of primary batteries require more complicated handling. Thus, reserve batteries need to be activated before discharging may begin. In low-temperature conditions, a battery often has to be warmed up. Detailed manuals are supplied with cells and batteries for all these situations.

7.2 MAINTENANCE OF STORAGE BATTERIES

7.2.1 Operational Modes of Storage Batteries

Storage batteries are serviced in a much more complex way than primary ones, owing mostly to the charging procedure. While the discharge mode is determined, in the first place, by the specifics of the consumer circuits, the charging mode depends primarily on the specifics of the storage batteries and strongly affects cell lifetimes. Because of gassing at the charging phase, most storage batteries are not sealed, which entails additional complications in maintenance in comparison with primary batteries.

Storage batteries operate in three distinct modes: alternating charge–discharge mode, floating mode, and standby mode. In the first of them battery charge regularly alternates with discharge. This operation is typical for traction batteries and light portable batteries. In the floating mode, a battery is connected in parallel to another current source; when the load rises, the battery is partially discharged, to be recharged again when the load diminishes. The floating mode is characteristic for automobile starter batteries, batteries in aerospace systems, and some others. A storage battery in the standby mode is always maintained ready to take over but is connected to the load circuit only in emergency situations, when the main power source breaks down.

In addition to conventional charge–discharge cycles, some special cycles may be used: forming cycles, employed with freshly manufactured plates or after filling the battery with electrolyte, to achieve the optimum operating state of the plates (sometimes only a forming charge is used), and training cycles, employed when the battery is used on an irregular basis or is never discharged fully. Training cycles consist of a complete discharge followed by a complete charging at a rated current.

7.2.2 Charging of Storage Batteries

As a rule, charging devices have a decreasing current–voltage characteristic, that is, the charge voltage is the lower the higher the supplied current. This means that unless special control circuits are used, the charge current falls off as the battery voltage increases. Charging devices are in most cases equipped with systems that control and stabilize one of the electrical parameters, either voltage or current. Correspondingly, two basic methods exist: constant-current charge and constant-voltage charge.

Constant-current charging requires comparatively simple equipment, as current stabilization is normally less complicated than that of voltage. Another advantage is that the total electrical charge is easily found as the product of current and charging time. The method, unfortunately, has its shortcomings. Charging time is high if current is small. With high charge currents, chargeability suffers at the end of charging, because of nonuniform distribution of current over depth into the porous electrode: intensive gassing occurs in the outer, fully charged layers while the inner layers are still slowly accumulating charge.

In the case of constant-voltage charging, current is high at the beginning of the process and then slowly diminishes. The total charging time is quite high as the end-of-charge current is very small. The choice of the charge voltage becomes critical: lower voltage means suppressed side processes and gassing but longer charging time. The optimum charge duration is a function of temperature and the state of the battery so that this technique is rather difficult to use. Another shortcoming consists of battery overheating by high initial currents.

Various combined charging procedures are widely used to obviate the defects inherent to the simple constant-current and constant-voltage techniques. In all these procedures, charging time is shortened and gassing diminished by using large currents at the beginning of charging and small currents at the end of the process.

Stepwise-current charging is a comparatively simple procedure: a battery is first charged by rated current until the prescribed cutoff voltage is reached, after which the charge current is decreased by a factor of 2–3 and the process goes on until the prescribed voltage is reached again. Three-step and even four-step constant-current charging is possible.

Overheating of the battery in the case of constant-voltage charging is often prevented by restricting and stabilizing the initial charge current. After a certain interval, charging is switched over to the constant-voltage mode. In other cases, charging starts with constant voltage and is continued in the constant-current mode after current drops to a prescribed level. This shortens the total charge duration. Various other combinations of these modes are possible. Sometimes the transition from one method (or step) to another is done by hand, and sometimes it is realized automatically, when the preset current or voltage is reached. The dropping current–voltage curve of a charging unit can be used to diminish current in the process of charging, without resorting to control circuits. This requires, however, that the characteristics of the charging device and those of the battery be matched; this is a restriction on the universality of the method.

In some cases storage batteries (especially alkaline batteries) can be charged with asymmetric alternating current that is by superposition of DC and AC charging modes. This procedure modifies the structure of the active mass formed in the process, and therefore affects the operational characteristics of storage batteries (the discharge capacity may be slightly increased or discharge voltage stabilized). The observed effects greatly depend on the ratio of AC and DC current and on the frequency of the former. Asymmetric current charging intensifies heating and may increase gassing in the battery. This method results in sharply reduced service life of some types of storage batteries (in particular, some lead battery systems), and thus cannot be used for these. These batteries may be adversely affected even by insufficiently filtered current at the rectifier output.

We distinguish between complete charging of a battery, which was almost completely discharged previously, and partial recharging to compensate for capacity loss because of self-discharge or partial discharge. Such recharging may be periodic or continuous. In the latter case, recharging at very low current to compensate for self-discharge losses in the course of long-term storage of a standby battery is called compensation recharging.

In some types of storage batteries, so-called equalizing (leveling) charging is sometimes performed using small charge current ($j = 0.03-0.05$). Equalizing charges serve to regenerate active materials on all electrodes of all batteries of the battery, in order to level off the differences in the degree of charging.

In charging a battery, one faces the problem of determining the moment when the battery as a whole and its individual batteries are charged to completion. Long service life of batteries is realizable only if this moment is determined correctly. For instance, undercharge is harmful to lead batteries, and therefore they are charged until stable gas evolution. On the contrary, overcharging is unacceptable in the case of silver–zinc storage batteries.

Several techniques are used to determine the moment of charge completion. Often a battery is simply charged with a predetermined amount of electricity calculated on the basis of the known degree of discharge. It is sufficient to monitor the duration of charging if constant-current charging is employed. Coulometers (calibrated in Ah) are used to measure the cell capacity as a function of time in constant-voltage charging, with current diminishing continuously. In other cases, the completion of charging is derived from the voltage at the predetermined charge current. For example, charging of silver–zinc batteries is terminated immediately after the cutoff voltage is reached. In order to avoid overcharging, not only the net voltage of the battery but also that of each cell has to be monitored. Lead acid batteries require slight overcharging; consequently, an additional charge around 10–20% of the rated capacity is delivered to the battery after the voltage hike to 2.6–2.7 V. Some commercially available charging devices automatically cut off the charge current immediately after the end-of-charge state is diagnosed by one of the abovementioned methods.

If charging is conducted at low temperatures, one has to take into account that charge voltage is a function of temperature. Thus, the end-of-charge voltage of nickel–cadmium batteries increases, for rated-current charging, from 1.85 V at an electrolyte temperature of 20°C to 2.35 V at −40°C. The use of thermal insulation is recommended in winter charging of batteries, because of the decreasing chargeability at low temperatures. Conversely, summer charging has sometimes to be interrupted to avoid excessive heating of the battery.

Systematic use of booster charges works to reduce the capacity and cycle life of the batteries. One must therefore be very careful in applying such charging technique. It is advisable, for instance, to alternate booster charges with training cycles (see above).

Reduction of charge time and of battery sensitivity to booster charging constitutes one of the important problems in storage batteries development.

7.2.3 Storage Batteries in the Alternating Charge–Discharge Mode of Operation

This mode of operation is characterized by deep discharges. Monitoring of the batteries at the end of discharge must therefore be more careful than in the case of primary batteries. Reversals of storage batteries are intolerable as they result either in substantial shortening of service life or in complete failure of the unit (lead acid and silver–zinc storage batteries are especially sensitive to cell reversal). Excessively deep discharge, even if it does not cause cell reversal, may also be harmful to some battery types. For instance, regular deep discharges produce an extremely undesirable sulfation of lead acid battery plates. Consequently, the methods and devices for the determination of the degree of charging of a battery (calculation of the capacity delivered, the use of coulometers, determination of the electrolyte concentration in lead acid batteries, etc.) are very important for the operation and maintenance of storage batteries.

Alternating charge–discharge operation creates the most favorable conditions for monitored charging. Normally, batteries are charged in special rooms with normal temperature conditions, by optimal charging methods and with all the required monitoring.

Spare batteries have to be used if the battery-powered equipment cannot be switched off for sufficiently long intervals. Discharge of one battery is accompanied by the charging of another battery (or batteries).

7.2.4 Storage Batteries in the Floating Mode of Operation

In this mode (also referred to as "buffer") a battery is connected to the loading circuit parallel to another (main) source of electrical energy. The floating mode is used in two cases: (i) when the main energy source is working intermittently, such as a wind generator, or the car generator operating only when the engine is running and (ii) when the generator's power is not sufficient to meet periodically occurring extreme power demands (this is encountered in aerospace systems powered by solar panels). With small loads and the generator running, the generator voltage is somewhat higher than the battery OCV so that the battery is being charged; at high loads or with the generator off, the battery is discharging. Discharges in the floating mode are typically not deep. A system can operate in the floating mode for long periods only if the generator power corresponds to the mean integral consumed power (taking into account energy losses in the cycling of the storage batteries).

Typically, fluctuations of voltage are harmful for load circuits. In the case of the floating mode, voltage varies from the value characteristic for the battery discharge to that characteristic for its charging. Lead acid batteries, with their inherently low-slope charge and discharge characteristics, and with the charge and discharge voltages differing only by 10–15%, have definite advantages in this respect. The difference reaches up to 20–30% in nickel–cadmium batteries (this also reflects their lower rated voltage).

When a battery is operating in the floating mode, it must be protected from voltage hikes in the generator, which is from the possibility of overcharging (e.g., during idle runs of the generator, with all external circuits disconnected). On the other hand, the generator voltage at an average load must be sufficient to recharge the battery. Generators are therefore often equipped with additional voltage control devices. Furthermore, the battery must be protected (for instance, by diodes) from useless discharge to the generator winding when the generated voltage is lower than that of the battery, or when the generator is stopped.

7.2.5 Storage Battery in the Standby Mode of Operation

In this mode all consumer circuits are powered by the main power source—electric network or an autonomous generator. The storage battery starts delivering current only in an emergency, which is when the main source fails. Several circuit versions are possible in the standby mode. In the simplest case, the main source and the standby battery are connected in parallel (as in the floating mode). The main source provides continuous compensation recharging of the battery whose discharge starts when the main source fails. At this moment the voltage in the system drops jumpwise. This constitutes a significant shortcoming of this arrangement. A sharp drop of 10–20% is not crucial in some cases (such as emergency lighting), but is clearly unacceptable for more critical equipment. In the second version, the standby battery is not connected to the load circuits. The connection is realized by a relay only in the case of failure of the main source. In this case the discharge voltage of the battery is chosen equal to the rated voltage of the main source, so that switching causes only small changes in voltage. Between emergencies the standby battery is either not recharged or recharged through a converter that provides voltage higher than the rated voltage of the generator. This arrangement also has a shortcoming, namely, the finite time of switching. This time is equal to the pickup time of the switching device, when current is not delivered to the consumer circuits. Some types of equipment, such as computers, do not tolerate the interruption of power supply even for 1 ms.

Sometimes a combined scheme is used, with the battery consisting of two subsections. One subsection is constantly connected to the main electric source while the second subsection (its voltage is 10–20% of that of the first subsection) is connected to the first in series in the case of emergency to compensate for the drop in voltage.

7.3 GENERAL ASPECTS OF BATTERY MAINTENANCE

Batteries need certain servicing that includes preparation procedures, monitoring, periodic checks, and other operations. Following correct maintenance procedures is very important for storage batteries designed for long service life. Maintenance is aimed at keeping batteries in their "best shape" and at preventing premature breakdown.

The required maintenance operations may be different with different battery types, and sometimes the differences in regulation operations may be very substantial. In any case, these operations must never deviate from those specified for each battery type. In this section we discuss only some general aspects of maintenance.

The first requirement is that a battery must be kept clean. The outer surface must never have liquid electrolyte on it. Alkaline batteries must be visually checked with special care because of the tendency of alkaline solutions to "creepage." Electrolyte on a battery cover leads to shorting and wasteful discharge, as well as to corrosion. The parts that have to be kept especially clear are terminals, vents, valves, and filling orifices. Electrolyte and other contaminants getting into terminal contacts cause corrosion, which results (especially if tightening is insufficient) in the formation of transition resistances and in sparking. It is advisable to ensure corrosion-proof contacts with special acid-free greases.

Measuring of the OCV and on-load voltage is an important method of battery checking. In some cases even very crude measurement confirms that the battery is still operational. In other cases, however, such as in quantitative determination of the degree of charging of a lead acid battery, the OCV must be measured by means of high-resistance voltmeters (precision grade not worse than $1-5\%$). In order to minimize Ohmic-loss error when measuring the on-load voltage, the voltmeter must be connected to the battery terminals or very close to them. A very convenient device for checking a battery is the so-called load prong. This device comprises contact legs, a load resistor, and a voltmeter. The pointed contact legs provide reliable electrical connection, while the load resistance is such that the current is typical for the battery to be checked.

With the exception of sealed storage batteries, the main element of maintenance is the control of the amount and concentration of the electrolyte. When the electrolyte is poured into a battery, its purity must be checked. All the accessories required for maintenance operations, such as vessels, thermometers, funnels, hydrometers, and so on, have to be very clean because both acids and alkalis of battery grade are chemically very pure. Contacts of alkaline electrolytes with air must be minimized to avoid absorption of carbon dioxide (carbonization of the electrolyte).

In order to compensate for partial decomposition of water because of overcharge and corrosion, water must be regularly added. Normally addition of acid or alkali solutions is inadmissible as the resulting increase in electrolyte concentration impairs battery characteristics. Only deionized water (if unavoidable, pure rainwater or melted snow may be acceptable) is to be added to the electrolyte. This water must not be kept in metal containers as even very small amounts of iron ions (or ions of other metals) introduced into the electrolyte are very harmful to batteries. Chloride ions present in normal tap water are also very dangerous. It should be kept in mind that boiling of water does not remove either chloride or iron ions. When water is added, the electrolyte must come to the level indicated in the manual for each specific battery (usually it is $3-5$ mm above the upper edges of the plates). A certain time is required for the concentration to level off in the bulk of the electrolyte.

The electrolyte concentration should be periodically checked. Sometimes it has to be corrected by adding an acid or alkali, for instance, after spilling part of the liquid or after prolonged charges (in the last case gases carry away tiny droplets of the electrolyte fog). Sometimes the electrolyte concentration is changed in the transition from summer to winter conditions of operation (or vice versa).

Usually electrolyte concentration is determined by measuring its density with special hydrometer. Temperature dependence of density of solutions must not be overlooked. Battery servicing manuals always indicate the temperature recommended for density measurements. The monitoring and maintenance of the level and concentration of electrolyte in nonsealed storage batteries constitute fairly labor-consuming operations. New versions of storage batteries, the so-called low-maintenance and maintenance-free batteries have been developed in recent years; the rate of gas evolution and water decomposition in such batteries is very much reduced so that they can operate for a long period without regular additions of water (see Section 2.5).

The conditions in which batteries are stored must also satisfy certain requirements. Again, batteries must be clean, and especially their venting holes. Unfilled batteries must be stored with plugs tightly screwed. It is advisable to store unfilled lead acid batteries and all alkaline batteries in discharged state. Electrolyte-filled lead acid batteries should be stored after charging them to completion.

Certain safety regulations have to be followed when using storage batteries. Facilities with a large number of storage batteries, such as electric power stations, communication centers, transport depots, and so on, must have special space for servicing batteries. This space must have an appropriate layout and equipment (ventilation, water pipes, sewer system, electric power mains, fire-prevention equipment, etc.). As hydrogen is produced in battery charging, ventilation must be sufficiently intensive to prevent accumulation of detonating concentrations of gas. The electrolyte fog also enters the atmosphere. Therefore, the personnel of large battery maintenance stations must be supplied with individual protection kits. Conventional safety measures must be taken when electrolytes are prepared or anything else is done with acids or alkalis (rubber gloves, aprons, and protective goggles are compulsory; each working site is to be equipped with neutralizing agents and a first-aid kit; acids and alkalis are not to be poured into common sewers). It must be remembered that corrosive activity of these solutions is very high.

Alkaline and acid storage batteries must be serviced separately (this covers storage, charging, and repair). Alkali batteries of all types are damaged irrevocably by the addition of sulfuric acid and lead acid batteries—by the addition of alkaline solutions.

All rules covering operations with electrical units are valid in the case of storage batteries as they often have voltages above 36 V. Batteries must be reliably protected from shorts that could be caused not only by defects of the external circuit but also by tools, pieces of bare wire, or other metallic objects left around by careless personnel. Wiring of all electric circuits must be reliable, and contacts must be kept clean. Sparking in contacts may result in the explosion of the evolved hydrogen.

8

OUTLOOK FOR BATTERIES WITH AQUEOUS ELECTROLYTE

Metal–air batteries are unique among rechargeable batteries because one of the active materials, oxygen, is not stored in the battery. In this respect they resemble fuel cells. The mass advantage at the air electrode and the high electrode potentials of the oxygen electrode can result in very high-specific energy values when combined with reactive and lightweight metal anodes such as lithium, aluminum, magnesium, and iron. For example, the theoretical specific energy for lithium–air batteries is 11 kWh/kg. Even if a modest 10% of this value could be realized in practice it would be a five-fold increase over the state-of-art rechargeable energy storage systems. Large-scale energy storage for the integration of wind and solar generation require batteries to cost less than $100/kWh and last over 5000 cycles. These requirements have prompted renewed interest in improvements to rechargeable metal (zinc, iron)–air batteries with a traditional aqueous alkaline electrolyte.

The air electrode in rechargeable metal–air batteries must be "bifunctional," that is, it should support both oxygen evolution and oxygen reduction efficiently. This type of bifunctional performance requires optimal catalytic materials and electrode structure, in particular, the catalyst must resist degradation caused by oxygen evolution. Carbon-based support materials are unsuitable for achieving durability beyond 1000 cycles. Carbon dioxide absorption by the alkaline electrolyte must be avoided to prevent clogging of the pores in the cathode (see papers by Hamlen and Atwater, 2002; Abraham and Jiang, 1996).

One of the key challenges in achieving a rechargeable zinc electrode is retaining a stable morphology and preventing a "shape change" usually observed after a

Electrochemical Power Sources: Batteries, Fuel Cells, and Supercapacitors, First Edition.
Vladimir S. Bagotsky, Alexander M. Skundin, and Yurij M. Volfkovich
© 2015 John Wiley & Sons, Inc. Published 2015 by John Wiley & Sons, Inc.

few hundred cycles (this is commonly referred to as the "displacement" problem). Passivation and hydrogen evolution on zinc must also be suppressed. The iron electrode (in contrast to the zinc electrode) during continuous cycling does not undergo any morphological change because the iron hydroxide is only sparingly soluble in the electrolyte. As a result over 3000 cycles have been demonstrated with the iron electrode. However, the parasitic evolution of hydrogen during charge and poor electrode utilization at high discharge rates are its shortcomings, which contribute to low energy efficiency (see publications by Putt et al., 2004; Manohar et al., 2011).

REFERENCES

Putt R, Naimer N, Atwater T. Fourth-generation zinc-air batteries. Proceedings of the 41th Power Sources Conference; 2004 June; Philadelphia, PA.

Hamlen R, Atwater T. In: Linden D, Reddy TB, editors. *Handbook of Batteries*. McGraw-Hill; 2002.

Abraham KM, Jiang Z. J Electrochem Soc 1996;143:1.

Manohar A, Malkhandi S, Yang B, Prakash GKS, Narayanan SR. Electrochemical properties of carbonyl iron electrodes for iron-air batteries. Abstracts of Electrochemical Society Meeting; 2011. p 303.

PART II

BATTERIES WITH NONAQUEOUS ELECTROLYTES

9

DIFFERENT KINDS
OF ELECTROLYTES

Electrolytes are highly important components of all galvanic cells and electrochemical devices. In most of the electrochemical devices such as batteries, electrolyzers, and the like, aqueous solutions of acids and salts are used as electrolytes. Aqueous solutions are easy to prepare, convenient to handle, and as a rule made from readily available, relatively inexpensive materials. By changing the composition and concentration of the components, it is relatively easy to adjust the specific conductance and other physicochemical properties of these aqueous solutions.

Cases exist, however, where for fundamental reasons aqueous solutions cannot be used. One such case is devices in which electrochemical processes take place at elevated temperatures (above 180–200°C), for instance, the electrowinning of aluminum is performed at temperatures close to 1000°C. Another case is devices in which electrodes consisting of alkali metals are used, which are unstable in aqueous solutions. An example is the batteries with a lithium negative electrode. For this reason, other kinds of electrolytes are used in addition to aqueous solutions, *viz*, non-aqueous solutions of salts, salt melts, and a variety of solid electrolytes.

The problems of electrolyte selection have become particularly acute in connection with the miniaturization and sealing of a variety of electrochemical devices such as batteries, sensors, and the like. Apart from all their advantages, aqueous electrolyte solutions here exhibit certain defects, insofar as sealing of a device containing them is often difficult, and leaking of the liquid readily occurs (particularly when alkaline solutions are used). From devices that are not sealed, the electrolyte solvent

Electrochemical Power Sources: Batteries, Fuel Cells, and Supercapacitors, First Edition.
Vladimir S. Bagotsky, Alexander M. Skundin, and Yurij M. Volfkovich.
© 2015 John Wiley & Sons, Inc. Published 2015 by John Wiley & Sons, Inc.

can evaporate, which leads to changes in electrolyte concentration. In certain cases when gas evolution is strong, solutions may tend to become dispersed as an ultrafine spray.

9.1 ELECTROLYTES BASED ON APROTIC NONAQUEOUS SOLUTIONS

In nonaqueous electrolytes based on some organic solvents metallic lithium is stable and can be used as anodes in batteries. Lithium and other alkali metals have highly negative electrode potentials (see Table 1.1). Thus batteries with lithium anodes have much higher EMF and OCV values than batteries with aqueous electrolytes.

In Table 9.1 certain organic solvents are described, which can be used for preparing nonaqueous electrolytes with a high ionic conductivity. In order to avoid

TABLE 9.1. Physical Properties of Certain Solvents at 25°C

Solvent	Formula	ϵ	η (MPa.s)	d (kg/dm^3)	t_m(°C)	t_b(°C)
Protic						
Water	H_2O	78.5	0.89	0.997	0	100
Ammonia[a]	NH_3	20	0.26	0.67	−78	−33
Acetic acid	CH_3COOH	6.2	1.13	1.05	16.7	−118.1
Methanol	CH_3OH	32.6	0.547	0.792	−97.9	64.5
Aprotic						
Acetonitrile	CH_3CN	36	0.345	0.786	−45.7	81.6
Dimethylformamide	$(CH_3)_2NCHO$	37	0.796	0.944	−61	153
Propylene carbonate	$CH_3-CH-CH_2$ $O\ \ O$ $C{=}O$	66.1	2.53	1.198	−49.2	242
γ-Butyrolactone	$CH_2-CH-CH_2$ $O----C{=}O$	30.1	1.75	1.125	−43.5	204
Tetrahydrofuran	CH_2-CH_2 CH_2-CH_2 O	7.4	0.47	0.880	−65	64
Dimethoxyethane	CH_2-O-CH_3 CH_2-O-CH_3	5.5	0.46	0.869	−70	85
Dioxolane	CH_2-CH_2 $O\ \ O$ CH_2	7	0.57	1.038	−95	77

[a]In liquefied state, temperature −20°C, pressure 0.2 MPa.

an interaction with lithium, these solvents must be aprotic (i.e., in contrast to water, they must not dissociate with the formation of H^+ ions—solvated protons). In Table 9.1, the properties of water and of other possible protic solvents are included for comparison.

In this table ϵ denotes the relative permittivity, η—the viscosity coefficient, d—the density, t_m, and t_b respectively, the melting and boiling temperatures.

9.2 IONICALLY CONDUCTING MOLTEN SALTS

Many electrochemical devices and plants (batteries, electrolyzers, and others) contain electrolytes that are melts of various metal halides (particularly chlorides), nitrates, carbonates, and certain other salts with melting points between 150 and 1500°C. The salt melts can be single-component (neat) or multicomponent, that is, consist of mixtures of several salts (for their lower melting points in the eutectic region).

Melts are highly valuable as electrolytes, as processes can be realized in them at high temperatures, which would be too slow at ordinary temperatures or which yield products that are unstable in aqueous solutions (e.g., electrolytic production of the alkali metals).

The ionic conductivities of most solid crystalline salts and oxides are extremely low (exceptions are the so-called solid electrolytes, which are discussed in Section 9.3). The ions are rigidly held in the crystal lattices of these compounds and cannot move under the effect of applied electric fields. When melting, the ionic crystals break down forming free ions; the conductivities rise drastically and discontinuously, in some cases up to values of over 1 S/cm, that is, values much higher than those of the most highly conducting electrolyte solutions.

A typical special feature of the melts of ionic crystals (ionic liquids) are their high concentrations of free ions, of about 25 mole/l. Because of the short interionic distances, considerable electrostatic forces act between the ions, so that melts exhibit pronounced tendencies for the formation of different ionic aggregates: ion pairs, triplets, complex ions, and so on.

Another special feature of ionic liquids is the lack of a foreign ("inert") molecular medium, and particularly a solvent, between the ions. Hence, they lack ion–molecule and the many kinds of nonelectrostatic interactions.

The results of X-ray studies show that on melting, the ionic crystals retain some short-range order; an anion is more likely to be found in the immediate vicinity of a cation, and vice versa. The interionic distances do not increase on melting; they rather decrease somewhat. Yet the volume of a salt markedly increases on melting, usually by 10–20 %. This indicates that melts contain a rather large number of voids (holes). These holes constantly form and perish owing to fluctuation phenomena attending the kinetic-molecular motion of the ions. Their mean size is between one and two interionic distances; they are uniformly distributed throughout the entire liquid volume.

9.3 IONICALLY CONDUCTING SOLID ELECTROLYTES

9.3.1 Ionic Semiconductors

The conductivity of solid salts and oxides was first investigated by M. Faraday in 1833. It was not yet known at that time that the nature of conduction in solid salts is different from that in metals. A number of fundamental studies were performed between 1914 and 1927 by Carl Tubandt in Germany and from 1923 onward by Abram F. Ioffe and coworkers in Russia. These studies demonstrated that a mechanism of ionic migration in the lattice over macroscopic distances is involved. It was shown that during current flow in such a solid electrolyte, electrochemical changes obeying Faraday's laws occur at the metal/electrolyte interface.

In some cases (particularly at elevated temperatures) mixed electronic and ionic conduction is observed in solid salts. Typical materials with purely ionic conduction are the halides and sulfides of a number of metals, viz., $AgBr$, Ag_2S, $PbCl_2$, $CuCl_2$, and many others.

The conductivity σ of such materials is usually low at room temperature. The values of σ strongly increase with temperature (Fig. 9.1). The conductivities of ionic crystals strongly depend on their purity. Impurities in the crystals markedly raise the values of σ, particularly at lower temperatures when the intrinsic conductivity of the pure material is still low.

All these features: low values of σ, a strong temperature dependence, and the effect of impurities, are reminiscent of the behavior of p-type and n-type semiconductors. By analogy, we can consider these compounds as ionic semiconductors with intrinsic or impurity-type conduction.

As a rule (though not always), ionic semiconductors have unipolar conduction because of ions of one sign. Thus, in compounds $AgBr$, $PbCl_2$, and others the cation transport number t_+ is close to unity. In the mixed oxide $ZrO_2 \cdot nY_2O_3$ pure O^{2-} anion conduction $t_- = 1$) is observed.

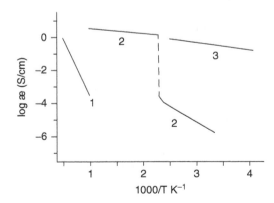

Figure 9.1. Conductivity of various solid electrolytes as a function of temperature. (1) $ZrO_2 \cdot CaO$; (2) AgI; (3) $RbAg_4I_5$.

In an ideal ionic crystal all ions are rigidly held in the lattice sites where they perform only thermal vibratory motion. Transfer of an ion between sites under the effect of electrostatic fields (migration) or concentration gradients (diffusion) is not possible in such a crystal. Initially, therefore, the phenomenon of ionic conduction in solid ionic crystals was not understood. Yakov I. Frenkel showed in 1926 that ideal crystals could not exist at temperatures above the absolute zero. Part of the ions leave their sites under the effect of thermal vibrations and are accommodated in the interstitial space leaving vacancies at the sites.

9.3.2 Ionic Conductors

It had been discovered long ago that the character of conduction in AgI changes drastically at temperatures above 147°C, when β- and λ-AgI change into α-AgI. At the phase transition temperature, the conductivity σ increases discontinuously by almost four orders of magnitude (from 10^{-4} to 1 S/cm). At temperatures above 147°C, the conductivity increases little with temperature, in contrast to its behavior at lower temperatures (see Fig. 9.1).

Starting in the 1960s many compounds with such properties were discovered, that is, with high conductivities and low temperature coefficients of conductivity. Some of them are double salts with silver iodide (nAgI · mMX) or other silver halides where MX has either the cation or the anion in common with the silver halide. The best-known example is $RbAg_4I_5(= 4AgI · RbI)$, where this sort of conduction arises already at −155°C and is preserved up to temperatures above 200°C. At 25°C this compound has a conductivity of 0.26 S/cm, that is, the same value as found for 7% KOH solution. Another example is Ag_3SI, which above 235°C forms a α-phase with a conductivity of 1 S/cm.

The same conduction type is found for another class of compounds, the so-called sodium polyaluminates or ß-aluminas $Na_2O · nAl_2O_3$, where n has values between 3 and 8. Polycrystalline samples of these materials have a room temperature conductivity of about 0.005 S/cm, but at 300°C the conductivity is about 0.1 S/cm.

Because of the high values of conductivity, which, in individual cases, are found already at room temperature, such compounds are often called superionic conductors or ionic superconductors; but these designations are unfounded, and a more correct designation is "solid ionic conductors."

Strictly unipolar conduction is typical for all solid ionic conductors; in the silver double salts, conduction is because of silver ion migration, while in the sodium polyaluminates, conduction is because of sodium ion migration.

The discovery of the various ionic conductors has elicited strong interest, as they can be used in batteries and other devices. It could be shown after numerous studies that the high ionic mobility in these compounds are the result of particular lattice structures. In such lattices, the immobile ions of one kind (most often the anions) are fixed at their lattice sites and form a rather rigid, nondeformable sublattice. The sublattice of the other ions (most often the cations), to the contrary, is disordered: the cations are not bound to particular sites but can occupy any of a large number of equally probable sites. As at any particular time an ion physically occupies just one site; the other available sites function as the vacancies for the ion's motion. The differences between sites and interstitials are obliterated here, and a peculiar, highly mobile cation fluid is formed. The conductivity in the crystal will be anisotropic, that is, depend on the direction in space, when the vacancies have a particular spatial configuration relative to the rigid anionic sublattice. As in the case of ionic semiconductors, this state is characteristic for certain structures, and can exist only within the temperature range where the particular crystal structure is stable.

REFERENCES

Frenkel YI. Z Phys 1926;35:652.
Ioffe AF. Ann Phys Ser 4 1923;72:461.
Tubandt C, Lorenz E. Z Phys Chem 1914;87:513.

10

INSERTION COMPOUNDS

As is shown in Chapter 12, the operation principle of lithium ion batteries consists of the extraction of lithium ions from the active material of positive electrode and their intercalation into the negative electrode material under charging. Under discharge, these processes are reversed. Thus, the electrode materials must possess the ability to reversibly acquire and release lithium ions with the corresponding charge transport. Technically, such materials can be designated as insertion materials. It is often said that these materials represent a host matrix, into which guest species, here, lithium ions, are inserted. In general, the electrode process can be described by the following equation:

$$H + xLi^+ + xe \leftrightarrow Li_xH \tag{10.1}$$

where H is the host material.

Generalized Equation (10.1) is applicable to all electrodes, but the finer mechanism of lithium insertion and extraction is different in different cases.

In the case of lithium insertion into the oxide material of the positive electrode (e.g., $Li_{(1-x)}CoO_2$) the lithium ion is embedded into the crystal lattice. Herewith, the valency of the main metal (here, cobalt) decreases by one. In fact, under the operation of such an electrode, that is, in the case of the following process:

$$Li_{(1-x)}CoO_2 + xLi^+ + xe \leftrightarrow LiCoO_2 \tag{10.2}$$

Electrochemical Power Sources: Batteries, Fuel Cells, and Supercapacitors, First Edition.
Vladimir S. Bagotsky, Alexander M. Skundin, and Yurij M. Volfkovich
© 2015 John Wiley & Sons, Inc. Published 2015 by John Wiley & Sons, Inc.

the redox couple of Co^{+3}/Co^{+4} is realized, which determines the high positive working potential, and its dependence on the lithiation degree (i.e., on index x) follows the Nernst equation (in terms of activity rather than concentration).

When lithium is inserted into such materials as lithium iron phosphate, a separate phase of lithium iron phosphate is formed that is insoluble in the initial lithium iron phosphate:

$$FePO_4 + Li^+ + e \leftrightarrow LiFePO_4 \qquad (10.3)$$

Thus, no variable composition compounds are formed at an increase in the lithiation degree; it is the ratio of the amounts of the $FePO_4$ and $LiFePO_4$ phases that changes and the potential remains unchanged in the course of charging or discharge.

When lithium is intercalated into graphite, lithium ions are intercalated into the crystal lattice and occupy the sites in the gaps between graphene sheets. An intercalation compound similar to intercalation compounds with anions is formed. Herewith, ions lose practically the whole positive charge and therefore the Li/Li^+ redox couple is implemented on an electrode of lithiated graphite in a solution of lithium salt. In the case of the maximum lithiation degree, the activity of intercalated lithium in this material comes to the activity of metallic lithium, that is, the potential of such an electrode corresponds to the potential of metallic lithium in the same lithium salt solution. At lower lithiation degrees, the potential of such an electrode is more positive. Technically, the intercalation compound of lithium and carbon can be considered as a certain counterpart of a lithium solution in carbon; herewith, the activity coefficient of "dissolved" lithium significantly differs from unity.

The maximum amount of lithium that can be intercalated into carbon is a single lithium atom per six carbon atoms, that is, the maximum lithiated carbon has the formula of LiC_6. All intermediate compounds can be represented as $Li_xC_6 (0 < x < 1)$. The mechanism of lithium intercalation into graphite consists of the successive filling of the space between graphene layers by lithium. This stepwise process can be described by a stage index corresponding to the amount of graphene layers between the two neighboring layers of lithium. In the case of the maximum filling, there would be only a single graphene layer between the lithium layers. Each stage is characterized by its reversible potential and corresponds to a certain lithium concentration in the graphite matrix. The charging–discharge curve is stepped (Fig. 10.1).

$$LiC_{72} + Li \leftrightarrow 2\, LiC_{36}$$

(stage 8)   (stage 4)

$$3\, LiC_{36} + Li \leftrightarrow 4\, LiC_{27}$$

(stage 4)   (stage 3)

$$2\, LiC_{27} + Li \leftrightarrow 3\, LiC_{18}$$

(stage 3)   (stage 2)

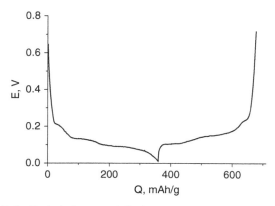

Figure 10.1. Typical charge and discharge curves of a graphite electrode.

$$2\,LiC_{18} + Li \leftrightarrow 3\,LiC_{12}$$

(stage 2)   (stage 2)

$$2\,LiC_{12} + Li \leftrightarrow LiC_6$$

(stage 2)   (stage 1)

The mechanism of lithium intercalation into nongraphitized carbon materials is as yet unclear. But at least three types of interaction between lithium and the carbon material have been suggested: interaction with graphene layers, with the surface of polynuclear aromatic planes, and lithium intercalation into microcavities on the front surface of the carbon material.

In the case of lithium intercalation into nongraphitized materials, the filling by lithium occurs simultaneously throughout the carbon material, so the charging–discharge curve is smoothed and contains no pronounced steps (Fig. 10.2).

When such elements as tin or silicon are used as the active material of the negative electrode, their lithiation (i.e., electrode charging) is reduced to the formation of alloys, more specifically, a sequence of intermetallic compounds ($LiSn_2$–$LiSn$–Li_2Sn–Li_5Sn_2–Li_7Sn_2–$Li_{22}Sn_5$; Li_2Si–$Li_{21}Si_8$–$Li_{15}Si_4$–$Li_{22}Si_5$). Activity of lithium in such alloys is sufficiently high, so that the potential of alloys with the maximum lithium content is close to the potential of metallic lithium.

Irrespective of the particular mechanism of formation of such "insertion compounds," they must satisfy a number of requirements to be used as electrode materials in lithium ion batteries.

1. The material must be capable of intercalating as high an amount of lithium per unit mass or volume as possible. This means that the structure of such a material must provide enough sites for the arrangement of lithium ions; the valency of the main metal must change as much as possible under lithiation and the material itself must have the minimum molecular mass.

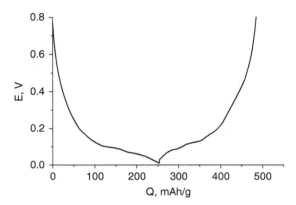

Figure 10.2. Typical charge and discharge curves for an electrode made from carbon paper.

2. Lithium intercalation and extraction must result in minimum structural changes to provide sufficient cycling life. Reactions of lithiation and delithiation must be reversible.

3. Lithium insertion and extraction into/from the positive electrode material must occur at a higher potential and the same reactions on the negative electrode material must occur at the maximum negative potential. The difference between the lithium insertion and extraction potentials for the same material must be minimum.

4. The material must possess sufficient chemical and thermal stability.

5. The material must not interact with electrolyte.

6. The material must possess sufficient electron conductivity and a high lithium diffusion coefficient.

7. The material must not be expensive; it must be environmentally friendly.

MONOGRAPHS AND REVIEWS

Beguin F, Frackowiak E, editors. *Carbons for Electrochemical Energy Storage and Conversion Systems.* CRC Press; 2010. p 532.

Huggins RA. Advanced Batteries. Materials Science Aspects. Springer; 2009. p 472.

Pyun S-I, Shin H-C, Lee J-W, Go J-Y. *Electrochemistry of Insertion Materials for Hydrogen and Lithium.* Heidelberg, New York, Dordrecht, London: Springer; 2012.

11

PRIMARY LITHIUM BATTERIES

11.1 GENERAL INFORMATION: BRIEF HISTORY

Primary cells with a lithium anode and nonaqueous (aprotic) electrolyte were developed in the 1970s. In particular, a Japanese company, Matsushita, launched on the market in 1973 cells with the active material of the positive electrode made of perfluorinated carbon, which was unusual for the cell practice. The electrolyte in these cells was a lithium perchlorate solution in propylene carbonate. Later, in 1975, another Japanese company, Sanyo, achieved success in developing a cell with a lithium anode and a cathode of customary manganese dioxide. Using metallic lithium characterized by a record high specific capacity (3.83 Ah/g) and an extremely negative standard potential (-3.045 V) resulted in a significant increase in the energy density as compared to all known variants of chemical power sources.

In all primary lithium cells, the negative electrode is made of metallic lithium. Thus, different types of lithium cells differ in the positive electrode material and in the type of electrolyte. A variety of oxidant materials was offered as the active material of the positive electrode. These included different oxides, sulfides, selenides, oxysulfides, oxychlorides, and some other substances: perfluorinated carbon and sulfur. However, only a small number of electrochemical systems in the cells actually reached the industrial production stage. The electrochemical systems of the cells produced industrially are given in Table 11.1. This Table also presents the values of open circuit voltage (OCV) of these cells and the theoretical values of their energy density.

Electrochemical Power Sources: Batteries, Fuel Cells, and Supercapacitors, First Edition.
Vladimir S. Bagotsky, Alexander M. Skundin, and Yurij M. Volfkovich.
© 2015 John Wiley & Sons, Inc. Published 2015 by John Wiley & Sons, Inc.

TABLE 11.1. Varieties of Lithium Cells with Nonaqeous Electrolyte

System	OCV, V	Theoretical Energy Density, Wh/kg
(1) Lithium – thionyl chloride ($SOCl_2$)	3.66	1475
(2) Lithium – sulfuryl chloride (SO_2Cl_2)	3.91	1405
(3) Lithium – sulfur dioxide (SO_2)	2.91	1098
(4) Lithium – manganese dioxide (MnO_2)	3.50	998
(5) Lithium – fluorocarbon (CF_x)	2.82	1989
(6) Lithium – copper oxide (CuO)	2.24	1285
(7) Lithium – copper sulfide (CuS)	2.12	1033
(8) Lithium – iron disulfide (FeS_2)	1.75	1268
(9) Lithium – iodine (I_2)	2.77	554
(10) Lithium – titanium disulfide (TiS_2)	2.45	552
(11) Lithium – vanadium pentoxide (V_2O_5)	3.50	2512
(12) Lithium – molybdenum trioxide (MoO_3)	3.30	586

Apart from the main classification of lithium power sources based on the nature of the electrochemical system used, there are other types of classification.

As regards the character of electrolyte, the following power sources can be distinguished:

1. with liquid electrolyte based on aprotic organic solvents;
2. with liquid aprotic oxidant electrolyte;
3. with solid (polymer) electrolyte;
4. with electrolyte in the form of molten salts;
5. with aqueous electrolyte (lithium hydroxide solution).

As regards the design features, the following cells are distinguished:

1. prismatic ("canned");
2. cylindrical reels (rammed);
3. cylindrical wound coil;
4. disks ("coin");
5. thin-film (foil or "paper").

Finally, as regards the operating features, the following are distinguished:

1. high-current cells (hot standby);
2. low-current cells (hot standby);
3. reserve cells.

In high-current cells, liquid oxidants are used that simultaneously serve as electrolytes. In Table 11.1, these are systems (1), (2), and (3).

In low-current cells, cathodes of solid active substances and electrolytes based on aprotic solutions are used. These are the other systems mentioned in Table 11.1, apart from the cells of the "lithium–iodine" system, in which solid inorganic electrolyte is used. Low-current cells differ in the operating voltage and nominally fall into two groups: 3 V cells and 1.5 V cells. Of course, the actual discharge voltage values differ from 3.0 to 1.5 V, but still, such division into two nominal groups is rather convenient. One of the advantages of 1.5 V cells consists of the possibility of their application in equipment initially designed for supply by cells with conventional electrochemical systems: manganese–zinc, mercury–zinc, and so on.

11.2 CURRENT-PRODUCING AND OTHER PROCESSES IN PRIMARY POWER SOURCES

11.2.1   Processes on the Negative Electrode

The current-producing anodic process in all cells is reduced to lithium dissolution with the formation of its ions:

$$Li \rightarrow Li^+ + e \qquad (11.1)$$

The rate of this process in aprotic electrolytes is rather high; the exchange current density is fractions to several mA/cm^2. As pointed out already, the first contact of metallic lithium with electrolyte results in practically the instantaneous formation of a passive film on its surface conventionally denoted as solid electrolyte interphase (SEI). The SEI concept was formulated yet in 1979 and this film still forms the subject of intensive research. The SEI composition and structure depend on the composition of electrolyte, prehistory of the lithium electrode (presence of a passive film formed on it even before contact with electrode), time of contact between lithium and electrolyte. On the whole, SEI consists of the products of reduction of the components of electrolyte. In lithium thionyl chloride cells, the major part of SEI consists of lithium chloride. In cells with organic electrolyte, SEI represents a heterogeneous (mosaic) composition of polymer and salt components: lithium carbonates and alkyl carbonates. It is essential that SEI features conductivity by lithium ions, that is, it is solid electrolyte. The SEI thickness is several to tens of nanometers and its composition is often nonuniform: a relatively thin compact primary film consisting of mineral material is directly adjacent to the lithium surface and a thicker loose secondary film containing organic components is turned to electrolyte. It is the ohmic resistance of SEI that often determines polarization of the lithium electrode.

11.2.2   Processes on the Positive Electrode

The cathodic process is determined by the nature of the active material. As a rule, processes on positive electrodes are much more complicated than processes on lithium.

Desolvated lithium ions are intercalated into the crystalline lattice under discharge of solid cathodes, including oxides and chalcogenides. The charge of lithium ions remains herewith unchanged and the atoms of the active substance metal are reduced.

Meanwhile, an intercalation compound of the active substance and lithium is formed at first. Examples of such reactions are shown further.

$$MnO_2 + xLi^+ + xe \rightarrow Li_xMnO_2 \qquad (11.2)$$

$$MoO_3 + xLi^+ + xe \rightarrow Li_xMoO_3 \qquad (11.3)$$

$$V_2O_5 + xLi^+ + xe \rightarrow Li_xV_2O_5 \qquad (11.4)$$

$$TiS_2 + xLi^+ + xe \rightarrow Li_xTiS_2 \qquad (11.5)$$

If the formed intercalation compound is crystallographically stable, that is, if no significant change in the elementary cell volume occurs under lithium intercalation, then such a compound is the actual discharge product. An example of such a process is reaction (11.5). Active substances forming stable lithium intercalation compounds can be the basis of not only primary cells, but also rechargeable cells.

In other cases, crystal lattice decomposition occurs under cathodic insertion of lithium, and the cathodic process becomes irreversible. In this case, the product of the cathodic reaction is a mixture of a lithium compound (oxide, chalcogenide) with a reduced form, particularly, directly with a metal. It is such processes that occur when iron sulfide or copper oxide are used:

$$FeS_2 + 4Li^+ + 4e \rightarrow 2Li_2S + Fe \qquad (11.6)$$

$$CuO + 2Li^+ + 2e \rightarrow Li_2O + Cu \qquad (11.7)$$

Reduction of fluorocarbon results in the irreversible formation of lithium fluoride and elementary carbon:

$$(CF_x)_n + xnLi^+ + xne \rightarrow xnLiF + nC \qquad (11.8)$$

Thionyl chloride is also irreversibly reduced with the formation of a solid lithium chloride deposit and a certain amount of elementary sulfur along with the formation of sulfur dioxide dissolved in electrolyte:

$$2SOCl_2 + 4Li^+ + 4e \rightarrow 4LiCl + SO_2 + S \qquad (11.9)$$

Sulfuryl chloride is reduced according to a similar scheme:

$$SO_2Cl_2 + 2Li^+ + 2e \rightarrow 2LiCl + SO_2 \qquad (11.10)$$

The main product of sulfur dioxide reduction is lithium dithionite:

$$2SO_2 + 2Li^+ + 2e \rightarrow Li_2S_2O_4 \qquad (11.11)$$

The actual positive electrode in cells with a liquid oxidant is made of an inert porous carbon material, on the surface of which cathodic processes (11.9)–(11.11) occur.

The cells of the lithium–iodine system stand somewhat apart. In such cells, the oxidant is not elementary iodine as such, but it is complex with poly-2-vinylpyridine $[CH_2CH(C_5H_5N)]_n$ (P2VP). Its composition can be expressed as P2VP mI_2. The content of iodine in the complex decreases under discharge, and solid lithium iodide is formed:

$$P2VP\ mI_2 + xLi^+ + xe \rightarrow P2VP\ (m - x/2)\,I_2 + xLiI \qquad (11.12)$$

11.2.3 Electrolytes of Primary Lithium Cells

In all lithium cells with solid cathodic materials, to the exclusion of cells of the lithium–iodine systems, solutions of some lithium salts in aprotic solvents are used as electrolytes. The solvents used are generally mixtures of substances with relatively high dielectric permeability ϵ (which ensures solubility of salts, salvation of their ions, and reaching of high specific conductivity) and one or several substances with relatively low viscosity (which also contributes to an increase in conductivity according to the well-known Walden rule). An example of solvents with high dielectric permeability can be propylene carbonate ($\epsilon = 65.1$) and γ-butyrolacton ($\epsilon = 39$). An example of a solvent with low viscosity (diluter) is dimethoxyethane. Physical properties of some organic solvents are presented in Table 9.1. Lithium salts used most often are perchlorate ($LiClO_4$), hexafluoroarsenate ($LiAsF_6$), hexafluorophosphate ($LiPF_6$), tetrafluoroborate ($LiBF_4$), and also salts of perfluoroalkyl sulfoacids: trifluoromethyl sulfonate (triflate) ($LiCF_3SO_3$), bis-trifluoromethyl sulfonyl imide (imide) ($Li[N(CF_3SO_2)_2]$), and tris-trifluoromethyl sulfonyl methide (methide) ($Li[C(CF_3SO_2)_3]$).

Solid lithium iodide is used in the cells of the lithium–iodine system. Its specific conductivity at the room temperature is several $\mu S/cm$. Such cells are characterized by very long-term discharge (years) with very low currents (tens of μA).

The oxidant is a part of liquid electrolyte in the cells of the lithium–thionyl chloride and lithium–sulfuryl chloride systems. In the first case, such oxidant electrolyte is the lithium tetrachloroaluminate ($LiAlCl_4$) solution in pure thionyl chloride or in thionyl chloride with a sulfur dioxide additive; in the second case, it is the solution of the same salt in sulfuryl chloride.

11.3 DESIGN OF PRIMARY LITHIUM CELLS

Individual cells with a lithium anode are designed with the capacity of about 1/20 to 20 Ah. Such cells are released both in conventional nominal sizes (AAA, AA, C, D) and in original alternates, which prevents their assembly in equipment designed for manganese–zinc cells or nickel–cadmium batteries. Nonstandard tabs in the form of flat blades, axial acicular pins, and so on are usually provided in lithium cells to eliminate the possibility of the application of a 3 V lithium cell instead of a 1.5 V manganese–zinc cell. Cylindrical (or prismatic) lithium cells, same as the cells of other electrochemical systems, are designed to obtain the maximum energy density (the so-called rammed design), and to obtain the maximum power (wound coil design with thin electrodes).

The design of all lithium cells without distinction provides their absolute leakproofness. It is clear that failure of airtightness of the lithium cell results in ingression of air and water vapors, which puts the cell completely out of action and can bring about an explosion or inflammation. Besides, leakage of liquid electrolyte results in an adverse effect on the powered equipment and on men (which is especially important when thionyl chloride cells are used). It is for this reason that the technology of lithium cell manufacturing includes complex procedures regarding welding and other types of coupling of dissimilar materials.

11.4   FUNDAMENTALS OF THE TECHNOLOGY OF MANUFACTURING OF LITHIUM PRIMARY CELLS

The regularities of the manufacturing technology of lithium cells are determined by the requirement of eliminating the contact between metallic lithium and water (and water vapors), oxygen and nitrogen, and hygroscopic properties of the electrolyte components and requirements of minimization of electrolyte humidification. All process stages in the production of lithium anodes and cell assembly are carried out in airproof glove boxes with a dry argon atmosphere. Sometimes, not pure argon, but its mixture with carbon dioxide is used in such glove boxes: the presence of carbon dioxide in the atmosphere promotes formation of SEI of higher quality with enhanced content of lithium carbonate. Process steps in the manufacturing of positive electrodes and electrolytes can be carried out in "dry rooms," as these materials must not contact moisture, but may contact oxygen.

In the manufacturing of lithium anodes, thin metallic lithium sheets are pressed or rolled on gauze or foil current collectors of copper, nickel, or stainless steel. Positive electrodes containing a solid oxidant are manufactured by pressing (pasting, spreading) of the active mass on gauze or foil current collectors of nickel or another stable material. The active material of positive electrodes represents a mixture of the active substance (oxidant) powder, conducting additive, and binder. The conducting additive used is generally carbon black and the binder represents fluorinated polymers. The content of each additive varies from 3 to 15%. Positive electrodes in cells with a liquid oxidant are made of carbon materials.

Opposite electrodes in cells with liquid electrolyte are separated by a porous separator. Separators of fiberglass are used in thionyl chloride–lithium cells and separators of polypropylene are used in cells with organic electrolytes.

11.5   ELECTRIC CHARACTERISTICS OF LITHIUM CELLS

In most cases, discharge of primary lithium cells does not result in the formation of compounds with variable composition, so the discharge occurs at a practically constant voltage. In other words, primary lithium cells are characterized by a slanting discharge curve. The discharge voltage is practically always lower than the open circuit voltage. Herewith, in the case of discharge at a constant current or switching on to constant resistance, the change in the voltage from OCV to the established value is not instantaneous, but takes several minutes. The difference between OCV and discharge voltage is especially high for cells of the lithium−copper oxide system.

A characteristic "voltage dip" appears at the very beginning of discharge (several seconds) under operation of thionyl chloride−lithium cells or cells of the lithium−sulfur dioxide system. This voltage dip is especially pronounced after prolonged storage at reduced temperatures. Under such storage conditions, a relatively thick and compact passive film with high ohmic resistance is formed on the lithium anode surface. In the initial period of discharge, lithium is etched from under this film; the film deteriorates and flakes.

A special case is the cells of the lithium−iodine system. The electrolyte in these cells is solid lithium iodide. Its conductivity under usual operation conditions is several orders of magnitude lower than the conductivity of usual liquid electrolytes. It is for this reason that such cells are discharged by very low currents and the main voltage drop under discharge corresponds to ohmic losses in electrolyte. Lithium iodide is formed under discharge of lithium−iodine cells as a result of the current-producing reaction, so that the thickness of electrolyte grows under discharge in proportion to the charge consumed. The resistance of electrolyte and ohmic voltage drop increase simultaneously. Therefore, the discharge curve of lithium−iodine cells is an inclined line.

As a rule, a decrease in the discharge temperature results in a decrease in the discharge voltage (following an increase in the polarization and ohmic losses) and a decrease in capacity.

Specific electric characteristics of lithium cells considerably exceed specific characteristics of cells with aqueous electrolytes. The maximum actual energy is typical for lithium thionyl chloride cells: above 600 Wh/kg and above 1100 Wh/dm^3. Specific energy of cells with organic electrolyte is roughly 200–350 Wh/kg and 500–600 Wh/dm^3. It is characteristic that specific energy, same as specific capacity of a single electrochemical system, is practically independent of absolute cell capacity (Fig. 11.1).

The maximum power density is also developed in cells with liquid oxidants: 50–100 W/kg and 70–200 W/dm^3. The minimum power density is characteristic for lithium iodine cells designed for continuous operation for 10 years and more.

11.6 OPERATIONAL CHARACTERISTICS OF LITHIUM CELLS

Primary lithium cells compare favorably with cells with aqueous electrolytes because of their very good shelf life; in other words, very low self-discharge. The best shelf life is characteristic for lithium iodine cells, in which the capacity loss under storage for 10 years at the temperature of 40°C does not exceed 10%! The guaranteed shelf life (under due conditions) of lithium cells with other electrochemical systems is

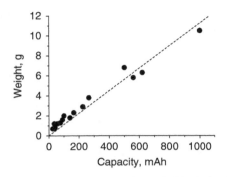

Figure 11.1. Weight versus capacity for cylindrical Li–MnO$_2$ cells.

5–15 years, which corresponds to the annual capacity loss because of self-discharge of 0.5–3%.

High activity of metallic lithium determines the importance of the problem of safety of lithium cells. The design of such cells provides for a reliable protection against internal faults and in general, against current increase that would result in the excessive heating of the cell and melting of lithium. Operation of lithium cells requires strict adherence to elementary safety requirements, including the guarantee against polarity nonobservance, cell heating, attempts of their disassembly, and attempts of charging primary cells similar to rechargeable batteries. There are still limitations in the transportation of lithium cells (including those in equipment) by air.

All lithium cells are absolutely airproof, which provides their operation in any space attitude. Most of lithium cells possess sufficient mechanic strength and can be used in military, aerospace, and other critical equipment.

Lithium cells with liquid oxidants can operate in a rather wide temperature range: −60 to +70°C. The operating temperature range of cells with solid cathodes is approximately −20 to +50°C.

11.7 FEATURES OF PRIMARY LITHIUM CELLS OF DIFFERENT ELECTROCHEMICAL SYSTEMS

11.7.1 Cells with the "Lithium Thionyl Chloride" System

The current-producing reaction of thionyl chloride–lithium cells is described to a certain approximation by Equation (11.9). The electrolyte (and simultaneously, the active substance of the positive electrode) in such cells is the solution of lithium chloroaluminate LiAlCl$_4$ in thionyl chloride. Stoichiometrically, lithium chloroaluminate can be considered as a complex compound of LiCl AlCl$_3$ or, in a more general form, as xLiCl AlCl$_3$. Solutions with index $x < 1$ are usually applied in the production of cells. In the course of discharge, the generated product, LiCl, forms a complex. Herewith, index x grows and can exceed unity. Under further discharge, the solubility limit of LiCl is reached and it precipitates in the form of an insoluble deposit.

Cathodes of thionyl chloride–lithium cells are made of carbon black (or of some other carbon material). The theoretical specific capacity of thionyl chloride–lithium cells as per mass of the initial active substances according to Equation (11.9) without taking into account electrolyte is 403 Ah/kg. At OCV of 3.66 V, this corresponds to theoretical energy density of 1475 Wh/kg. The actual specific energy that can be obtained in thionyl chloride–lithium cell is always lower than the theoretical value not only by taking into account the mass of structural parts (cathode, case, lid, current collectors), electrolyte, and so on, but also by considering the fact that the precipitating lithium chloride deposit blocks the cathode surface. Not only the overall porosity of the cathode, but also the porosity corresponding to large and medium pores, in which solid discharge products are deposited, is of great importance for obtaining high specific capacity. In the case of an unfavorable cathode design, the actual cell capacity is determined not by the mass of the initial reagents, but by the volume of the cathode pores accessible for solid products. It is herewith important to provide it that lithium chloride generated in the current-producing reaction would not be deposited in a thin compact layer on the cathode pore walls, but would be sufficiently loose and would occupy the whole pore volume. It is important to provide high power thionyl chloride cells that the cathode would lead to high microporosity; in other words, it would possess developed internal surface.

The density of sulfur is close to that of lithium chloride. The volume of solid sulfur formed in reaction (11.9) is less than 16% of the overall volume of solid products. Sulfur dioxide formed in reaction (11.9) is rather soluble in electrolyte, so that the intrinsic pressure in the cells is enhanced negligibly. Despite this, the cases of thionyl chloride–lithium cells are made to be rather strong and the extreme corrosion activity of thionyl chloride enforces application of high-alloy steels or nickel. Sintered glass–metal pressure seals are used in thionyl chloride–lithium cells, same as in cells with some other systems.

As already pointed out, because of the fact that the cathodic process in thionyl chloride–lithium cells occurs in the liquid phase and no variable composition compounds are formed, the discharge voltage remains constant during the whole discharge and drops drastically at the end of discharge (Fig. 11.2). Thus, the discharge voltage cannot be an indicator of the discharge degree and this poses a certain problem.

Though self-discharge of thionyl chloride–lithium cells is low, still, it manifests itself under discharge by very low currents (i.e., long-term discharge) and specific capacity of such cells diminishes at a decrease in the discharge current (in the range of very low currents). At a discharge by not very low currents, specific capacity of such cells decreases at an increase in current because of nonuniformity of distribution of the process in a porous cathode (see Chapter 3). Thus, the dependence of actual capacity of thionyl chloride–lithium cells on the discharge current corresponds to a curve with a maximum (Fig. 11.3).

Such a type of dependence of capacity on the discharge current is also characteristic for cells with other electrochemical systems. Because of toxicity and corrosive activity of thionyl chloride (which can become apparent in various emergency situations), civilian use of thionyl chloride–lithium cells is limited.

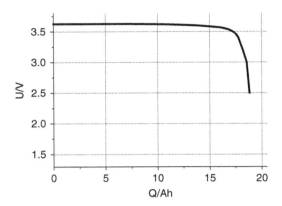

Figure 11.2. Typical discharge curve for thionyl chloride–lithium cell. Room temperature. C/5000.

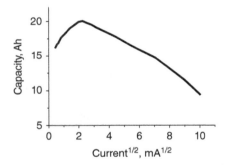

Figure 11.3. Dependence of actual capacity of thionyl chloride–lithium cells on discharge current.

They are mainly used in military and special equipment, but the sphere of their application has lately extended to alarms and security wireless devices, utility metering, identification, and tracking systems. Individual cells with the capacity of 0.3–250 Ah are produced.

11.7.2   Cells of "Lithium–Sulfur Dioxide" System

The cells of "lithium–sulfur dioxide" system are largely similar to thionyl chloride–lithium cells. Similar to thionyl chloride, liquefied SO_2 can simultaneously play the role of an oxidant and a solvent. However, the vapor pressure over liquefied SO_2 is much higher than the vapor pressure over thionyl chloride, and dielectric permeability of SO_2 is, on the contrary, lower than that of thionyl chloride. It is for this reason that the solvent used is not pure SO_2, but its mixture with acetonitrile or acetonitrile and propylene carbonate. The lithium salt used is commonly bromide and, in some variants, chloroaluminate, perchlorate, tetrafluoroborate, or hexafluoroarsenate.

The main current-producing reaction is described by Equation (11.11). The product of this reaction is lithium dithionite that is insoluble in the electrolyte. Therefore, a reasonable porous cathode structure is of great importance for provision of high capacity and power characteristics of sulfur dioxide–lithium cells. The cathodes in such cells are similar to cathodes of thionyl chloride–lithium cells. The cells of the "lithium–sulfur dioxide" system are also produced using the rammed and wound coil designs. Porous polypropylene is applied as a separator.

Freshly made cells of the "lithium–sulfur dioxide" system are characterized by enhanced intrinsic pressure: up to 0.4 MPa. Only by the end of discharge or rather by the time liquid sulfur dioxide is exhausted, this pressure somewhat decreases. Therefore, such cells are produced in rather robust steel nickel-plated cases of a not very large size. As a rule, the maximum capacity of individual cells does not exceed 30 Ah. The cases of sulfur dioxide–lithium cells are usually equipped with relief valves preventing fracture of cases at an increase in pressure (e.g., because of the cell overheating).

The discharge characteristics of the cells of the "lithium–sulfur dioxide" system are close to the characteristics of the cells of the "lithium thionyl chloride" system: the discharge voltage remains constant up to the end of the discharge. The discharge characteristics also feature a voltage drop after storage, especially under discharge at low temperatures. In general, the operating temperature range of sulfur dioxide–lithium cells is very wide: from -60 to $+70°C$. They are also characterized by a good shelf life (at least 10 years).

The theoretical specific capacity of sulfur dioxide–lithium cells is 377 Ah/kg, the theoretical specific energy is 1170 Wh/kg. In practice, the values of specific energy of up to 350 Wh/kg and power density of up to 100 W/kg are obtained.

11.7.3 Cells of "Lithium–Manganese Dioxide" System

The cells of "lithium–manganese dioxide" system are most abundant and most popular lithium cells. They are produced by many companies all over the world. Cells of cylindrical, disk, and prismatic form are produced. The main advantages of manganese dioxide–lithium cells amount to their relatively low cost (which is related to the application of not very expensive materials and simplicity of manufacturing) at a sufficiently high energy density.

The current-producing reaction in manganese dioxide–lithium cells is described by Equation (11.2); here, indicator x is generally close to unity. In this case, the theoretical specific capacity of such a cell is 285 Ah/kg, which corresponds to the theoretical energy density of 998 Wh/kg at OCV of 3.5 V. The actual energy density for disk and cylindrical cells of a not too low capacity (above 0.1 Ah) is 200–350 Wh/kg and strongly depends on the discharge mode.

As the product of the current-producing reaction in manganese dioxide–lithium cells is a variable composition compound, the discharge voltage changes in the course of discharge and this change depends on the discharge current (Fig. 11.4).

As well known, there are several crystallographic modifications of manganese dioxide. The γ-modification of MnO_2 is most often used as the active material of

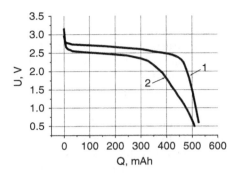

Figure 11.4. Typical discharge curves of lithium–manganese dioxide cell. C-rate: 1 – C/5, 2 – C/2.

the positive electrode in manganese dioxide–lithium cells. The active mass of the positive electrode has the usual composition (active substance, conductive additive, binding agent). The electrolyte is usually a $LiClO_4$ solution in propylene carbonate or its mixtures with dimethoxyethane.

As a rule, manganese dioxide–lithium cells are discharged to the final voltage of 2 V; the rated voltage is 3 V. The cells function in the temperature range of −20 to +50°C and can be stored for 10 years without any significant self-discharge.

11.7.4 Cells of "Lithium–Fluorocarbon" System

Cells of the "lithium–fluorocarbon" system are very close by their characteristics to the cells of the "lithium–manganese dioxide" system. Though the working voltage of fluorocarbon–lithium cells is lower than the voltage of manganese dioxide–lithium cells, such cells can in many cases be interchangeable.

The active substance of the positive electrode is the polymer of fluorinated carbon with the overall formula of $(CF_x)_n$. As a rule, subscript x in this formula is close to unity, polymerization degree n exceeds 1000. The polymer of fluorinated carbon is a layered compound obtained by fluorination of carbon (graphitized or nongraphitized) in the form of a powder, fibers, or even fabrics by elementary fluorine at the temperatures of 350 − 600°C. As polyfluorocarbon is characterized by negligible electron conductivity, a certain amount of a conductive additive (carbon black) is introduced into the active mass of cathodes. Elementary carbon is formed in the course of discharge and the overall conductivity of the cathode increases.

The overall current-producing reaction (11.8) describes the process of cell discharge only approximately. The intermediate product of cathodic polyfluorocarbon reduction is assumed to be a certain solvated intercalation compound C-F–Li decomposed in the course of discharge.

The electrolyte generally used in fluorocarbon–lithium cells is solutions of $LiBF_4$ in γ-butyrolacton. $LiClO_4$ solutions in a mixture of propylene carbonate with dimethoxyethane are used less frequently.

11.7.5   Cells of "Lithium–Copper Oxide" System

These are most widespread 1.5 V cells. The active mass of such cells is pressed of a mixture of copper oxide powder, graphite (or carbon black), and Fluoroplast binder. The electrolyte used in copper oxide–lithium cells is lithium perchlorate solutions in dioxolane or in a mixture of propylene carbonate with tetrahydrofuran. Discharge of such cells is described by Equation (11.7). Therefore, no variable composition compounds are formed under discharge. Thus, the discharge occurs under constant voltage. A significant fault of copper oxide–lithium cells is that their open circuit voltage (2.2–2.3 V) considerably exceeds the working discharge voltage (about 1.5 V). At the same time, no voltage drops at the beginning of discharge are observed in such cells even after prolonged storage. One should point out very long shelf life of copper oxide–lithium cells.

11.7.6   Cells of "Lithium–Iron Sulfide" and "Lithium–Iron Disulfide" Systems

These cells also belong to the 1.5 V category. Also, no variable composition compounds are formed under discharge of such cells, and the discharge curve of the "lithium–iron sulfide" cells contains a single horizontal plateau at the voltage of about 1.4 V and the discharge curve of the "lithium–iron disulfide" cells consists of two horizontal plateaus with the voltage of 1.6 and 1.4 V. The electrolyte in the cells of both systems is lithium perchlorate solution in a mixture of propylene carbonate with dimethoxyethane. Contrary to the cells of the "lithium–copper oxide" system, iron sulfide–lithium and iron disulfide–lithium cells have no increased open circuit voltage.

MONOGRAPHS

Lithium Batteries. J.-P. Gabano. L: Academic Press; 1983; p 43

Dahlin GR, Strom KE. *Lithium Batteries: Research, Technology and Applications.* Nova Science Pub Inc; 2010. p 226. ISBN:607417227.

Linden D, Reddy TB, editors. *Handbook of Batteries.* 3rd ed. McGrawHill; 2002. p 1453.

Lithium Batteries, New Materials, Development and Perspectives. In: Pistoia G. Elsevier; 1994; p 137.

12

LITHIUM ION BATTERIES

12.1 GENERAL INFORMATION: BRIEF HISTORY

While the development of primary cells with a lithium anode has been crowned by
relatively fast success and such cells have filled their secure rank as power sources
for portable devices for public and special purposes, the history of development
of lithium rechargeable batteries was full of drama. Generally, the chemistry of
secondary batteries in aprotic electrolytes is very close to the chemistry of primary
ones. The same processes occur under discharge in both types of batteries: anodic
dissolution of lithium on the negative electrode and cathodic lithium insertion into the
crystalline lattice of the positive electrode material. Electrode processes must occur in
the reverse direction under charge of the secondary battery with a negative electrode
of metallic lithium. Already at the end of the 1970s, positive electrode materials were
found, on which cathodic insertion and anodic extraction of lithium occur practically
reversibly. Examples of such compounds are titanium and molybdenum disulfides.

The main problem arose with respect to the negative electrode. Complications
typical for galvanic practice appear under its charge, that is, under cathodic deposition
of lithium. As pointed out in Chapter 11, the surface of lithium in aprotic electrolytes
is covered by a passive film (SEI) because of the chemical interaction with the components of electrolyte: the organic solvent and anions. This film has the properties
of solid electrolyte with conductivity by lithium ions. The film is sufficiently thin
(its thickness does not exceed several nanometers) and it protects lithium safely from

Electrochemical Power Sources: Batteries, Fuel Cells, and Supercapacitors, First Edition.
Vladimir S. Bagotsky, Alexander M. Skundin, and Yurij M. Volfkovich.
© 2015 John Wiley & Sons, Inc. Published 2015 by John Wiley & Sons, Inc.

self-discharge, that is, from interaction with electrolyte. A very active fresh surface is formed under cathodic lithium deposition. A passive film grows on it and as lithium is deposited in the form of dendrites, the film envelopes separate lithium microparticles in many cases in charge–discharge cycles and prevents their electronic contact with the support. Such a phenomenon received the name of "encapsulation." The result of encapsulation is that a part of lithium drops out of further operation under each charging. Therefore, secondary batteries with a metallic lithium electrode have to be provided with an excess amount of lithium as compared to the stoichiometric amount. This excess is 4- to 10-fold. Thus, effective specific capacity of lithium decreases from the theoretical value of 3828 mA-h/g to the values of 380–800 mA-h/g. Besides, dendrite formation leads to the danger of short-circuit failures, that is, in fire and explosion hazard of such devices. A lot of effort has been put in with the aim of searching for various methods of surface treatment or introduction of the corresponding additives into electrolyte that would hinder dendrite formation under cathodic deposition of lithium. However, these attempts are as yet unsuccessful.

Attempts were made to circumvent the problems related to the application of metallic lithium by using a suitable lithium alloy as the negative electrode. The most popular alloy was that of lithium and aluminum. Discharge of such an electrode results in the etching of lithium from the alloy, that is, a decrease in its concentration, and the concentration of lithium in the alloy increases under charge. Activity of lithium in the alloy is somewhat lower than it is in pure metallic lithium, so that the alloy electrode potential is somewhat more positive (approximately by 0.2–0.4 V). This fact results, on the one hand, in a decrease in the working voltage and, on the other hand, in a decrease in the interaction between the alloy and electrolyte, that is, in a decrease in self-discharge. The key fundamental problem of using lithium–aluminum alloys consists of a great change in the specific volume of the alloy under variation of the alloy composition (under cycling). Under deep discharge, electrode embrittlement and shattering occur, so this direction never evolves. From the viewpoint of specific volume, lithium alloys with heavy metals (Wood type alloys) are preferable, but specific characteristics of electrodes based on heavy metal alloys are very low, so this line is considered to be of little promise.

The communication on batteries developed in Japan with a negative electrode of carbon materials causing a revolution in the development of secondary lithium power sources. Carbon proved to be a very convenient matrix for the intercalation of lithium. The specific volume of many carbon graphitized materials changes under intercalation of a rather large amount of lithium by not more than 10%. The potential of carbon electrodes containing a not too high amount of intercalated lithium can be more positive than the potential of a lithium electrode by 0.5–0.8 V. To reach a sufficiently high battery voltage, Japanese researchers applied cobalt oxides as the active material of the positive electrode. Lithiated cobalt oxide has the potential of about 4 V versus the lithium electrode, so that the characteristic working battery voltage is more than 3 V.

Under discharge of the battery, deintercalation of lithium from the carbon material occurs on the negative electrode and lithium is inserted into the oxide on the positive electrode. Under charging, the processes are reversed. Thus, there is no lithium metal in the whole system and the discharge and charging processes are reduced to the transfer of lithium ions from one electrode to another. For this reason, such batteries received the name of "lithium ion" or "rocking chair" batteries.

Lithium ion batteries proved to be very advantageous. Their development was very rapid and their specific characteristics more than doubled in 10–15 years, while their cost decreased fivefold. Their production scale grew at a tremendous rate. Current annual production is several billions of pieces. It was the large-scale production of lithium ion batteries that determined the modern production rate of such devices as cell phones, digital photo and video cameras, notebooks, and wireless power tools.

12.2 CURRENT-PRODUCING AND OTHER PROCESSES IN LITHIUM ION BATTERIES

At present, the negative electrode in all industrially produced lithium ion batteries is manufactured of carbon materials. The potential in the range of 2.5–3.5 V versus the lithium electrode is established in this material at the first contact between the carbon material and the aprotic electrolyte. This steady-state potential is determined by chemical groups (primarily, the oxygen-containing ones) on the carbon surface. The potential shifts under cathodic polarization in the negative direction and lithium intercalation starts. Simultaneously, the electrolyte is reduced (both the solvent and various impurities). Reduction of electrolyte on the carbon surface occurs much more easily (at less negative potentials) than on the surface of metallic lithium. The products of this reduction are partly insoluble and form a film on the carbon material surface with its composition and properties similar to those of a passive film on the lithium surface (SEI). Thus, charge is consumed in the first cathodic run both in the intercalation of lithium and the formation of a passive film. After this film is fully formed, it prevents the direct contact between the electrolyte and carbon and virtually no further reduction of electrolyte occurs.

In the case of lithium intercalation into well-resolved graphite structures, the thermodynamically stable LiC_6 compound can be obtained. Activity of lithium in this compound is unity, as the potential of such a compound is usually equal to the potential of a lithium electrode. Therefore, the equation of intercalation–deintercalation is usually as follows:

$$xLi^+ + xe + 6C \leftrightarrow Li_xC_6 \tag{12.1}$$

Several almost horizontal steps corresponding to compounds LiC_6, LiC_{12}, LiC_{18}, and so on are observed in the curve of the dependence of the potential of the Li_xC_6 compound on intercalation degree x, that is, in the charging curve of a graphite electrode (Chapter 10). Lithium is intercalated into graphite structures under sufficiently negative potentials. The larger part of lithium is intercalated under the potentials to the negative of 0.5 V, that is, the activity of lithium in such intercalates is rather high. This is good as regards the battery voltage, but bad in that reduction of electrolyte on such intercalates occurs more intensively than on nongraphitized samples. Besides, intercalation of lithium into graphite structures is often accompanied by solvent cointercalation, graphite swelling, and exfoliation.

Lithium is intercalated into amorphous structures at more positive potentials; generally, in the range of potentials from 0.9 to 0.0 V. The dependence of potentials on the value of x in these cases is a smooth curve corresponding to a uniform system. Nongraphitized structures are not laminated under intercalation of lithium.

Attempts were made to discover the correlation between the crystalline structure of carbonaceous materials and their capability to reversibly intercalate lithium. This correlation has not been definitely established, but still, one can assume as a certain general principle that the optimum materials would contain an amorphous matrix with inclusions of a mesophase: nuclei of graphite crystallites. Such materials are various cokes, pyrographite, and products of pyrolysis (carbonization) of various polymers. For practical purposes, the industry mastered some special materials providing high characteristics of negative electrodes in lithium ion batteries. The most popular material is manufactured by the Japanese company of Osaka Gas Co. under the name of mesocarbon microbeads, MCMB; it represents the carbonization product of pitchy resins under a certain temperature regime.

While a diversity of active materials are applied in the positive electrodes of primary lithium cells, the choice of the positive electrode in lithium ion batteries is limited. The positive electrodes of lithium ion batteries are made almost exclusively of lithiated cobalt or nickel oxides and of lithium–manganese spinels.

The operation of the positive electrode amounts to lithium extraction under battery charging (under positive electrode oxidation) with a simultaneous increase in the metal valency and to insertion of lithium under discharge:

$$LiCoO_2 \leftrightarrow xLi^+ + xe + Li_{1-x}CoO_2 \qquad (12.2)$$

Though materials based on cobalt oxides are noticeably more expensive than other materials, their industrial synthesis is characterized by relative simplicity and reproducibility. Electrodes based on lithiated cobalt oxides are noted for the lowest polarization and highest specific capacity. Lithiated nickel oxide works at a somewhat less positive potential than cobalt oxide, which abates the requirements toward the choice of electrolyte that is stable toward oxidation under charging. At the same time, the discharge curve of nickel oxide electrodes is steeper, that is, the change in the voltage in the course of the charging is higher than in cells with cobalt oxide electrodes. Materials based on manganese oxides (lithium–manganese spinels) are considered to be the best from the commercial and environmental viewpoints. Various mixed lithiated oxides, for example, $LiNi_xCo_{1-x}O_2$, $LiNi_{1-x-y}Co_xAl_yO_2$, and so on enjoy the greatest popularity.

Electrochemical characteristics of positive electrodes depend largely on the technology of their manufacturing. The technology of synthesis of lithiated oxides is based on various high-temperature (sintering) and low-temperature (sol–gel, ion exchange, precipitation from solutions) processes. Lithiated cobalt oxide is mainly obtained using the low-temperature technology. Lithiated nickel oxide is most often obtained by sintering.

The principal fault of lithiated cobalt oxide as the positive electrode material consists of the impossibility of full delithiation; as a rule, the material with the formula of Li_xCoO_2 is cycled only in the range of $0.5 < x < 1.0$. Deeper delithiation results in irreversible structural changes. Besides, significant faults of lithiated cobalt oxide are its high cost and toxicity.

As regards lithiated manganese oxides, spinels with the composition close to $LiMn_2O_4$ are most often considered. Compounds with the composition of $Li_{1+x}Mn_2O_4$ are formed under insertion of lithium into such a material (i.e., under discharge of a lithium ion battery) and such electrodes feature the rated potential value of about 3 V. Under lithium extraction, that is, when the $Li_{1-x}Mn_2O_4$-type compounds are obtained, the potential is close to 4 V. The main emphasis has lately been placed on precisely such spinels. Many variants of the technology for spinel synthesis are described. They differ both in the stoichiometry ($LiMn_2O_4$, $Li_2Mn_4O_9$, $Li_4Mn_5O_{12}$, etc.) and synthesis method. The best results are observed when low-temperature "wet" technologies are used that allow obtaining sufficient highly dispersed materials.

The main fault of lithium–manganese spinels (alongside with a somewhat lower specific capacity as compared to cobaltites and nickelates) is the relatively high degradation of capacity under cycling, especially at elevated temperatures.

Practically all conventional positive electrode materials do not allow increasing the potential above a certain critical value. This may, on the one hand, result in irreversible structural changes and, on the other hand, may cause oxidation of electrolyte with the formation of badly conducting products on the surface of active materials. It is for this cause that lithium ion batteries are generally charged using the stagewise scheme: at first, in the constant charging current mode (usually, from C/5 to 1C), until the battery voltage reaches the given limit (usually from 4.1 to 4.3 V) and then in this constant voltage mode, until the charging current decreases to negligible values. In most cases, at least 80% of total capacity are imparted to the battery in the first, galvanostatic, charging stage (Fig. 12.1).

As not all conventional positive electrode materials are designed for the charging at relatively high positive potentials, these materials pose certain problems regarding operational safety of lithium ion batteries. That is why much attention has been lately

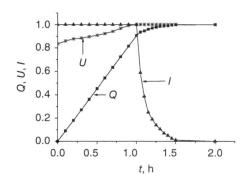

Figure 12.1. Typical charge pattern of lithium ion battery. Normalized values of voltage (U), current (I), and charge (Q) are shown.

paid to search for materials that would operate at somewhat less positive potentials. The most probable variant of replacement of conventional materials is considered to be lithiated iron phosphate that has already reached the commercialization stage. Anodic oxidation of this materials leads to the formation of iron phosphate that is practically isostructural to the initial $LiFePO_4$.

$$LiFePO_4 \rightarrow FePO_4 + Li^+ + e \qquad (12.3)$$

Thus, the cycling of lithium iron phosphate electrodes brings about no significant structural changes, which assures a very large cycle life of such electrodes (several thousand full cycles). As no compounds with variable composition are formed in reaction (12.3), the discharge and charging occur at a practically unchanged potential. The main fault of $LiFePO_4$ is its very low electron conductivity (less than 10^{-9} S/cm). It is for this reason that the electrode material is manufactured in the form of particles with the size of 20–30 nm with their surface practically completely covered by a carbon layer with the thickness of several nm. The equilibrium potential of the $LiFePO_4/FePO_4$ system is 3.5 V versus the lithium electrode, that is, it is considerably lower than in the case of conventional positive electrode materials. This simplifies substantially the problem of oxidative degradation of electrolyte and enhances the safety of batteries. Correctly manufactured electrodes based on the $LiFePO_4/C$ composites feature the capacity close to the theoretical value even under appreciable loads.

As to electrolyte, in the most cases it is $1–1.3$ M $LiPF_6$ in mixed solvent containing ethylene carbonate, dimethyl carbonate, and diethyl carbonate.

12.3   DESIGN AND TECHNOLOGY OF LITHIUM ION BATTERIES

The design and technology of the manufacturing of lithium ion batteries are very close to those in the manufacturing of primary lithium cells. The vast majority of lithium ion batteries are manufactured in the prismatic design, as the main purpose of such batteries is to support cell phones and notebooks. The design of such batteries is generally not standardized and most of the companies manufacturing cell phones and other devices rule out using in them foreign batteries. At the same time, the production volume of disk and cylindrical lithium ion batteries is also rather large.

Lithium ion batteries are produced both as wound coil and as bobbins as dependent on the required power. Negative electrodes are manufactured by the pasting or pressing of the active material consisting of a carbon material (e.g., MCMB) and a binder, most frequently, polyvinylidene fluoride dissolved in N-methylpyrrolidone. In some cases, carbon black additives are introduced within the active material of the negative electrode. The aim consists not in increasing conductivity, but in improving the plastic properties of the active mass and enhancing the buffer capacity of the electrode by the electrolyte.

Positive electrodes of lithium ion batteries, same as the positive electrodes of primary cells, are manufactured by pressing (pasting, spreading) or some other method of active material application on the current collectors (woven or expanded mesh, foil or porous plates, etc.). The active mass of positive electrodes represents a mixture of the active material, conducting additive, and binder. The conducting additives used are almost exclusively carbon materials: carbon black, graphite, carbon fibers, and so on. The binders are generally fluorinated polymers. The content of each additive in the active material varies from 3% to 15%.

Opposite electrodes in lithium ion batteries are separated by a porous polypropylene separator, particularly, of the Celgard material.

The design of lithium ion batteries, same as the design of all primary cells with a lithium anode, is characterized by complete air tightness. The requirement of complete tightness is stipulated both by the inadmissibility of the leaking of liquid electrolyte (producing a harmful effect on the powered equipment) and the inadmissibility of ingression of oxygen and water vapor from the air into the battery; oxygen and water vapor interact with the electrode and electrolyte materials and put the battery completely out of action. Naturally, lithium ion batteries can operate in any space orientation.

As a rule, the design of lithium ion batteries provides their vibration resistance. In this respect, lithium ion batteries are appropriate for the power supply of the equipment in which they are used.

Technological procedures in the manufacturing of electrodes and other components and the assembly of batteries are carried out in special dry rooms or in sealed glove boxes in the atmosphere of pure argon. Assembly of batteries involves complex state-of-the-art welding technologies, complex designs of sealed tabs, and so on.

Lithium ion batteries are generally manufactured (assembled) in the discharged state, that is, the negative electrode is a pure carbon material and the positive electrode is made of lithiated cobalt or nickel oxides or of lithium–manganese spinel. This is because of the fact that lithiated cobalt and nickel oxides, same as lithium–manganese spinels, are much more stable toward exposure to the atmosphere than lithiated carbon materials that interact with oxygen and water from the air vigorously.

Lithium–polymer batteries, that is, batteries with polymer electrolyte are generally manufactured in the form of thin flat elastic products with a simple plastic casing ("coffee bag") instead of a metallic case.

In development and manufacturing of lithium ion batteries, much attention is paid to the issues of operational safety, storage, and transportation. The cause for ignition and even explosion of a lithium ion battery can be sudden local heating because of internal short circuits or because of occurrence of exothermal reactions of reduction of electrolyte on the negative electrode or oxidation of electrolyte on the positive electrode. The probability of the latter case is considerably enhanced when the charge voltage is exceeded. A multilayer separator is used for protection from internal short circuits in many batteries. One of the layers of such a separator is made of a more low melting material than polypropylene, for example, polyethylene. In the case of local overheating (e.g., because of growth of lithium dendrites toward the positive electrode when the charging mode is disrupted), polyethylene melts and fills the pores of the separator, thus interrupting the current. Many lithium ion batteries are equipped with a microcircuit preventing overcharge and overdischarge of the battery and switching the battery off under overheating. Such a microcircuit is often installed directly in the power-consuming equipment.

12.4   ELECTRIC CHARACTERISTICS, PERFORMANCE, AND OTHER CHARACTERISTICS OF LITHIUM ION BATTERIES

Technical characteristics of lithium ion batteries of various manufacturers are rather close and are described by the following overall indicators:

Discharge voltage 3.5–3.7 V
Cycle life 500–1000 cycles
Energy density 150–200 Wh/kg and 250–400 Wh/l
Self-discharge 6–10% per year
Operating temperature range −20 to +60°C
Maximum discharge current 2C.

The abovementioned characteristics should be considered as certain rated reference values. All indicators largely depend on the conditions. For every particular item, the discharge voltage depends on the discharge current, depth of discharge, temperature; the cycle life depends on the discharge and charging modes (currents), temperature, depth of cycling; the working temperature range depends on the wear degree, allowable operating voltages, and so on. Thus, an increase in the discharge current results in a negligible decrease in the overall capacity, but the discharge voltage decreases noticeably (Fig. 12.2).

At the temperatures above 0°C, the discharge curve is practically independent of the temperature. A decrease in the temperature at negative temperatures, however, strongly affects the discharge characteristics. The faults of lithium ion batteries also include sensitivity toward overcharging and overdischarging, so that they must have charging and discharge limiters.

It is characteristic that specifications of products of different companies largely coincide, which is manifested, for example, by the data of Table 12.1, where the key

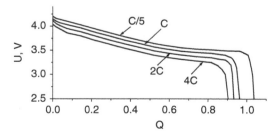

Figure 12.2. Typical discharge curves (voltage vs nominal capacity) of a lithium ion battery with different rates.

TABLE 12.1. Key Characteristics of Cylindrical Batteries 18650 Produced by Different Companies

Manufacturing Company	Capacity (mAh)	Weight (g)	Energy Density (Wh/kg)
EEMB	2200	49	166
Quallion	2500	47	197
Hitachi (Maxell)	2250	46	181
LG	2400	47	189
Panasonic	2250	44	189
Panasonic	2250	43	194

characteristics of cylindrical batteries of standard size 18650 produced by different manufacturers are shown.

Characteristics in Table 12.1 correspond to certain "universal" batteries. In some cases, special batteries with increased capacity (but such batteries do not allow forced charging and discharge) and with increased power (their energy density is relatively low, but they allow operation under forced modes) are manufactured.

The functioning of lithium ion batteries is generally allowed in the temperature range of -20 to $+40°C$. Storage of lithium ion batteries is allowed in a somewhat wider temperature range.

In discharge by currents corresponding to the 1C mode and below, the capacity of lithium ion cells depends little on the current. Under faster discharges, the capacity drops at an increase in the discharge current.

It is recommended to store lithium ion batteries in an almost discharged state (at the depth of discharge of 50–70%).

12.5 PROSPECTS OF DEVELOPMENT OF LITHIUM ION BATTERIES

Despite the fact that lithium ion batteries reached a certain degree of perfection, development of state-of-the-art technologies requires further improvement of these power sources. Whatever the obtained energy density, it always has to be increased. Many consumers demand expanding the operating temperature range of the batteries. There is an urgent problem regarding transition to more forced charging modes, especially, in the development of batteries for general electric transport. A very significant issue is the problem of operation safety of lithium ion batteries. This is particularly important, although the probability of inflammation or explosion of lithium ion batteries is generally very low. About two dozen incidents a year are registered at the annual world production of several billions of individual lithium ion batteries, which corresponds to the probability of 10^{-7} to 10^{-6} %. Finally, a vital issue consists of the extension of the capacity range of individual batteries: development of microminiaturized batteries, including thin film ones, with the capacity of below 10 mA-h and large batteries with the capacity of 1–10 kA-h. All the abovementioned problems cannot be solved without changing the electrochemical system of lithium ion batteries. So far, energy density of such batteries was increased solely by design improvements (particularly, enhancement of the amount of active materials per electrode surface area) and application of lighter structural materials. At present, these reserves are exhausted.

The attention of developers in search for new electrochemical systems was in the first place turned to materials with a higher reversible capacity of lithium intercalation than in the case of conventional carbon materials (negative electrode) and traditional oxide materials (positive electrode).

It is known that such elements as aluminum, tin, and especially silicon feature very high lithium intercalation capacity. (It is for this reason that aluminum was suggested as the negative electrode matrix even before lithium ion batteries were invented.) Tin and silicon can form compounds with lithium with the composition of $Li_{4.4}Sn$ and $Li_{4.4}Si$, which corresponds to specific capacity of lithium insertion of 991 and 4200 mA-h/g, accordingly. At the same time, it is well known that formation of such compounds (i.e., alloying with lithium) is accompanied by a great (more than threefold) increase in specific volume, which results in the appearance of enormous internal stresses and electrode destruction (pulverization). It was found a while ago that nanostructured objects could withstand repeated cycles of lithium insertion–extraction without significant destruction. A fine example of such materials is thin amorphous silicon films. Such films of submicron thickness withstand several hundreds of cycles at the capacity close to the theoretical value. At an increase in the thickness of amorphous silicon films, the problem of their destruction arises because of their flaking from the support. Another example of nanostructured silicon electrodes is electrodes with nanofibers (whiskers) with the diameter of 20–100 nm and the length of several micrometers. Finally, a lot of hopes are placed on nanocomposites of silicon and carbon. Such composites can be prepared by the codeposition of both components (e.g., magnetron sputtering or gas-phase deposition) or thermal decomposition of organosilicon compounds or mixtures of polymers with silicon compounds. In silicon–carbon nanocomposites, the carbon matrix plays the role of a certain buffer smoothing the volume changes of the silicon component.

Electrodes based on tin oxides work in a similar way. In the first cathodic polarization of such materials, tin nanoparticles incorporated into a matrix of lithium oxide are formed. When lithium is further intercalated into these nanoparticles, the oxide matrix levels the volume variation and ensures prolonged cycling.

In general, very intensive search for alternative materials for negative electrodes has been performed in the past two decades, but all the suggested materials are much inferior to silicon as regards specific capacity.

Replacement of carbon materials by silicon has yet another advantage. It was found that in certain cases, reduction of electrolyte on carbon electrodes occurs with significant heat evolution, which is one of the fire risk factors. In this respect, silicon electrodes are quite safe.

As regards improvement of positive electrodes, the possibilities of enhancement of energy density are much lower. The valency of the base metal in the active material of the positive electrode changes under operation by not more than unity for all the existing materials. It is this that limits the specific capacity of such materials. Great hopes are put into materials operating at higher positive potentials than conventional $LiCoO_2$ (such materials belong to the 5 V class).

One of the examples of "5 V" materials is the $LiNi_{0.5}Mn_{1.5}O_4$-substituted spinel. Its discharge occurs at the potentials above 4.5 V. The specific capacity of such a material does not differ from the specific capacity of conventional materials, but an increase in the discharge voltage results in an increase in energy density by almost 25%. Such materials have two significant faults: firstly, as all manganese spinels, these materials can suffer corrosion (with the formation of soluble Mn^{+2} ions) at the temperatures above 45°C; secondly, the charging of such electrodes occurs at the potentials of about 5 V, which results in a very complicated problem of searching for a stable electrolyte.

Of the materials with a fundamentally higher specific capacity, great hopes are placed on materials based on vanadium oxide. Vanadium pentoxide V_2O_5 can be reduced to bivalent vanadium oxide, which corresponds to insertion of six lithium ions per formula unit or the theoretical capacity of 883 mA-h/g. When vanadium pentoxide is used in primary cells, it can intercalate about four lithium ions per formula unit. However, intercalation of lithium into crystalline V_2O_5 is accompanied by structural changes. In the intercalation of a single lithium ion, such changes are quite reversible and allow stable cycling of such electrodes with the capacity of about 140 mA-h/g. In the case of intercalation of more than two lithium ions, the structural changes become irreversible. However, it was found lately that certain nanostructured forms of vanadium pentoxide can feature stable cycling with the capacity of more than 400 mA-h/g. Vanadium oxide is reduced in the range of potentials by about 1 V to the negative than that of lithiated cobalt oxide, so that the actual increase in energy density would not be so high in the transition to lithiated vanadium oxide.

MONOGRAPHS

Beguin F, Frackowiak E, editors. *Carbons for Electrochemical Energy Storage and Conversion Systems*. CRC Press; 2010. 532 p.

Belharouak I. Lithium ion batteries – new developments; InTech; 2012.

Ozawa K, editor. *Lithium Ion Rechargeable Batteries*. Wiley-VCH Verlag GmbH & Co: Weinheim; 2009.

Nazri G-A, Pistola G. *Lithium Batteries: Science and Technology*. Springer; 2009. 716 p.

Park CR. Next generation lithium ion batteries for electrical vehicles. In-Tech; 2010. 140 p.

Balbuena PB, Wang Y. *Lithium-Ion Batteries. Solid-Electrolyte Interphase*. Imperial College Press; 2004.

Yoshio M, Brodd RJ, Kozawa A, editors. *Lithium-Ion Batteries: Science and Technologies*. Springer; 2009. 452 p. ISBN: 0387344446.

Wakihara M, Yamamoto O, editors. *Lithium Ion Batteries: Fundamentals and Performance*. Wiley-VCH Verlag GmbH; 1998. 247 p.

13

LITHIUM ION BATTERIES: WHAT NEXT?

Development of lithium ion batteries proved to be a power factor of technical advance. While at present such batteries form the base for portable electronics, in the near future, one could look forward to wide application of larger devices based on lithium ion batteries, including their application in electric transport and smart grids. However, many researchers at present have already started attempting to predict the further development of batteries that fundamentally differ from lithium ion batteries. One can identify three electrochemical systems against various possible new battery variants: (i) lithium–air batteries, (ii) lithium–sulfur batteries, and (iii) sodium ion batteries.

13.1 LITHIUM–AIR BATTERIES

On the face of it, the concept of a lithium–air battery seems if not crazy, at least strange. Of course, using air as the active substance is very attractive. Indeed, if the air weight is not taken into account, the theoretical power density of a lithium–oxygen system is more than 11 kWh/kg! However, problems with the development of a reversible air (oxygen) electrode even in the systems with aqueous electrolyte (Chapter 4) must call into question the feasibility of development of an air electrode in nonaqueous electrolytes. Retardation of oxygen ionization and evolution reactions, in other words, high overpotential of these reactions result in a very high difference

Electrochemical Power Sources: Batteries, Fuel Cells, and Supercapacitors, First Edition.
Vladimir S. Bagotsky, Alexander M. Skundin, and Yurij M. Volfkovich
© 2015 John Wiley & Sons, Inc. Published 2015 by John Wiley & Sons, Inc.

in the potentials of the positive electrode charging and discharge and therefore in low energy efficiency (W-h efficiency) under cycling. Besides, as pointed out in Chapter 12, it is the abandoning of the application of metallic lithium as the negative electrode and its replacement by an intercalation compound that brought about the success of lithium ion batteries. And still, despite these obvious obstacles, the prospects of widespread development of lithium ion batteries appear optimistic to many experts.

A reversible lithium–air system was first implemented on a laboratory scale in 1996. In this cell, the gel–polymer electrolyte was pressed between lithium foil on the one side and an air electrode on the other. (Later, usual liquid electrolyte in a porous, for example, glass fabric, separator was often used in lithium–air batteries). The whole cell was sealed into a plastic container ("coffee bag") and small holes were made in the container wall adjacent to the air electrode to supply air under discharge and remove oxygen under charging. The air electrode was made of a mixture of particles of polymer electrolyte and carbon black with the catalyst supported on its surface (cobalt phthalocyanine).

Two reactions may basically occur in the case of discharge on the positive electrode:

$$O_2 + 2e + 2\,Li^+ \rightarrow Li_2O_2 \tag{13.1}$$

and

$$O_2 + 4e + 4\,Li^+ \rightarrow 2\,Li_2O \tag{13.2}$$

Reaction (13.2) is essentially irreversible, so that only process (13.1) is implemented in practice in the cycling of the abovementioned device. The discharge product is lithium peroxide insoluble in electrolyte that fills the pores of the positive electrode. In this respect, the positive electrode of a lithium–air battery is similar to the positive electrode of a thionyl chloride–lithium cell (Chapter 11) and it is exactly the amount of hydrogen peroxide that can be located in the electrode pore space that determines the actual capacity of such a battery. It is quite a rare occasion that this amount should correspond to the total electrode porosity (total pore volume). In many cases, actual specific capacity of the positive electrode proves to be much lower than the value calculated on the basis of the pore volume. This is explained by the fact that the pore outlets are often clogged when lithium peroxide deposition starts and the current-forming process in the pore stops long before it is completely filled by the lithium peroxide deposit. It was observed that the so-called biporous electrodes, that is, electrodes containing both micro- and macropores are capable of embedding much higher amounts of deposit than homoporous ones. In fact, the ratio of rates of reactions (13.1) and (13.2) determines the stability of cycling: an increase in the fraction of reaction (13.2) causes a decrease in the amount of cycles up to complete failure of the battery because of irreversible accumulation of lithium oxide (Li_2O).

In order to enhance the air electrode capacity, it was suggested to introduce certain additives into electrolyte that would enhance solubility of Li_2O_2 because of complexation. An example of such an additive is tris-(pentafluorophenyl)-borane (TPFPB). However, such additives reduce oxygen solubility in electrolyte and enhance the overpotential of the anodic process under battery charging.

The air electrode operation is largely determined by the catalyst. In the above first version of the lithium–air battery, the catalyst was pyrolyzed cobalt phthalocyanine. Later, manganese dioxide applied on a carbon support became the most popular catalyst. As a rule, carbon black is treated by a solution containing potassium permanganate and a bivalent manganese salt to obtain the catalytically active material of the positive electrode. Because of the disproportionation reaction,

$$2KMnO_4 + 3MnSO_4 + 2H_2O \rightarrow 5MnO_2 + K_2SO_4 + 2H_2SO_4 \qquad (13.3)$$

the catalyst that is active both in the cathodic and anodic processes (bifunctional catalyst) is deposited on the carbon black pore surface. The positive electrode catalyst affects not only the rate of the oxygen reduction reactions (process (13.1)) under discharge and oxygen evolution under charging

$$Li_2O_2 \rightarrow O_2 + 2e + 2Li^+ \qquad (13.4)$$

but also the character of crystallization of lithium peroxide deposited in the cathodic process. In the presence of the manganese dioxide catalyst, lithium peroxide is deposited not in the form of a compact pore-free layer, but in the form of a loose porous deposit posing no hindrance to oxygen transport and causing no passivation of pore walls. Consequently, the catalyst promotes enhancement of power density, specific capacity, and cyclability of the air electrode. An increase in the rates of the cathodic and anodic reactions is equivalent to a decrease in the cathodic and anodic overpotential and therefore an increase in energy efficiency. In the technical respect, the best bifunctional catalyst is a platinum–gold system supported on a carbon carrier, but its application is naturally restricted because of economic considerations.

An essential feature of air electrodes in contact with nonaqueous electrolyte is that the pore walls in such an electrode are generally well wetted by electrolyte and therefore all electrode pores are filled by liquid (lyophobic gas pores are absent). In this situation, the discharge current will be the higher: the greater the solubility of oxygen in electrolyte and the higher its diffusion coefficient. Oxygen solubility depends on the composition of electrolyte (both on the nature of the solvent and the nature and concentration of the lithium salt). Besides, it turned out that the current efficiency of irreversible process (13.2) is high in electrolytes with relatively low oxygen solubility.

All problems of reversible lithium electrodes in contact with liquid electrolyte mentioned in Chapter 12 are preserved (dendrite formation and encapsulation) in the case of the operation of a metallic lithium electrode in lithium–air batteries. Besides, additional problems arise in the operation of lithium–air batteries: as the positive electrode contacts the atmosphere, the processes of absorption of water vapor, oxygen, and CO_2 and their transport through electrolyte to the lithium electrode surface; also, organic solvent evaporation through the pores of the positive electrode occurs inevitably. To eliminate the evaporation of liquid electrolyte and prevent water vapor transport through electrolyte, it was suggested to use electrolytes based on hydrophobic, that is, immiscible with water, ionic liquids with a practically zero vapor pressure. An example of such an ionic liquid may be 1-ethyl-3-methylimidazolium bis(trifluoromethylsulfonyl)imide (EMITFSI). Still, the catalytic activity of usual catalysts in ionic liquids is noticeably lower than in usual liquid electrolytes.

A fundamental solution for the problems of lithium–air batteries is assumed to be the development of a device with two electrolytes: aqueous, contacting with the air electrode, and nonaqueous, contacting with metallic lithium. In this case, it is essential to provide reliable separation of the aqueous and nonaqueous electrolytes. For such separation, the metallic lithium electrode is covered by a bilayer membrane. It is suggested to manufacture the inner layer adjacent to the lithium of the LiPON material (lithium phosphoroxynitride) and the outer layer of an amorphous or vitriform material with the overall formula of $Li_{1+x}M_xM'_{2-x}(PO_4)_3$, where M is Al or Fe, and M' is Ti, Hf, or Ge. The outer layer material is a counterpart of the NaSICON superionic conductor and is sometimes denoted as LiSICON. The materials of the both layers possess sufficient conductivity by lithium ions. The thickness of each layer may range from fractions to several micrometers. It is important that adhesion of the inner layer both to metallic lithium and the outer layer is very high, so that both layers follow the lithium surface without losing contact with it under cycling when the volume of lithium changes.

As well known, air electrodes in systems with aqueous solutions feature the best characteristics in alkaline solutions. Unfortunately, the LiSICON-type materials are destroyed and lose ionic conductivity in alkaline solutions. Therefore, using neutral solutions, particularly, buffer solutions containing acetic acid and lithium acetate in lithium–air batteries is suggested. Certainly, the processes on the positive electrode in this case are not described by Equations (13.1) and (13.4), but by equations

$$4CH_3COOH + O_2 + 4e \rightarrow 4CH_3COO^- + 2H_2O \qquad (13.5)$$

$$4CH_3COO^- + 2H_2O \rightarrow 4CH_3COOH + O_2 + 4e \qquad (13.6)$$

13.2   LITHIUM–SULFUR BATTERIES

The theoretical energy density of a lithium–sulfur electrochemical system is 2500 Wh/kg or 2800 Wh/l, which makes it immensely attractive for the development of a chemical power source. This attractiveness is also enhanced by the ready availability and cheapness of sulfur and the absence of environmentally harmful components. And, indeed, attempts of developing a battery using this electrochemical system were made yet in the end of the 1960s of the previous century, at the rise of the studies of electrochemical lithium systems. It was suggested in the beginning to use the negative electrode made of metallic lithium and the positive one of elementary sulfur supported directly on the current collector. The characteristics of these first layouts were clearly unsatisfactory, partly, because sulfur is an insulator. Later, the positive electrode came to be made of a mixture of sulfur and a carbon material (carbon black).

The current-producing process in a lithium–sulfur battery is very simple as regards stoichiometry: lithium and sulfur form lithium sulfide Li_2S. However, the mechanism of this process, same as the mechanism of the reverse charging process is very complicated.

Elementary sulfur exists in the form of linear and cyclic molecules of different composition and different degree of stability. S_8 molecules are most stable. Under discharge, sulfur is reduced on the positive electrode to polysulfide S_n^{2-} ions. Lithium polysulfides Li_2S_n with $n = 8, 6, 4,$ and 3 are soluble in organic solvents; polysulfide Li_2S_2, same as sulfide Li_2S is not dissolved in the electrolyte. The potential of a fresh-made sulfur electrode is close to 3 V versus the lithium electrode. In the beginning of the discharge of the sulfur electrode, the potential is very soon shifted by about 0.5 V in the negative direction and then it starts shifting at a relatively slow rate. Soluble polysulfides are formed at this first stage:

$$S_n + 2e \rightarrow S_n^{2-} \tag{13.7}$$

Herewith, the n value decreases in the course of discharge approximately from 8 to 3, which formally corresponds to a change in the average valency of sulfur in polysulfide, which explains the negative shift in the potential. In the further reduction, insoluble Li_2S_2 and Li_2S are formed and the cathodic process occurs under a practically constant potential of about 2 V (the second stage of discharge). Ultimately, when soluble polysulfides are quite depleted, the potential drops drastically in the negative direction (Fig. 13.1). The final discharge potential is commonly 1.5 V.

Under charging (after discharge), the sulfur electrode potential jumps very fast to the value of about 2.2 V and then two, not very pronounced though, steps can also be distinguished in the discharge curve: the first one is almost horizontal at the potential in the range of 2.2–2.3 V and the second one corresponds to the potentials of about 2.4 V.

The processes on the metallic lithium electrode are simple: anodic dissolution with the formation of Li^+ ions under discharge and cathodic deposition under charging. In a freshly assembled battery, the lithium surface is covered by SEI (see Chapter 11) that includes the products of interaction between lithium and components of electrolyte. Under cycling, the SEI composition changes: in the course of the first several

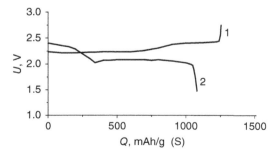

Figure 13.1. Typical charge (1) and discharge (2) curves of lithium–sulfur battery.

cycles, it is enriched by lithium sulfide formed under interaction between lithium and dissolved polysulfides. As pointed out in Chapter 12, dendrites are usually formed under cathodic deposition. This is not so when a lithium–sulfur battery operates: polysulfides hinder dendrite formation.

Reduction of dissolved polysulfides on the negative (lithium) electrode is equivalent to internal short circuit and results in a decrease in the Coulombic efficiency under cycling and self-discharge under storage.

There is as yet no consolidated opinion as to the optimum electrolyte for lithium–sulfur batteries. Experiments with solid polymer electrolyte are described, but aprotic electrolyte in a Celgard-type separator commonly used in lithium ion batteries is applied more frequently. A large number of electrolytes has been studied that differ both in solvents and the lithium salt. The greatest acceptance was gained by lithium imide solutions in dioxolane (or in a mixture of dioxolane and dimethoxyethane) and also lithium perchlorate solutions in sulfone. Dissolution of polysulfides in electrolyte is accompanied by a noticeable increase in viscosity and specific resistance of electrolyte. It is the great complexity of the composition of the electrochemical system and that of the processes occurring therein that prevent as yet commercialization of lithium–sulfur electrolytes.

One of the ways for the improvement of the positive electrode is to use a composite obtained by the impregnation of multiwall carbon nanotubes (MWNTs) by liquid sulfur. MWNTs with a high true surface area and high electron conductivity, on the one hand, promote retention of polysulfides in the positive electrode bulk (thus decreasing the shuttle transfer and self-discharge) and, on the other hand, enhance the electrode performance across its depth.

Of extraordinary importance is the structure of the positive electrode accumulating in its pores the discharge product, sulfide. In the case of untoward crystallization, it passivates the carbon matrix surface.

At present, lithium–sulfur batteries are at the pilot production stage. Flat batteries in flexible cases ("coffee bag") are produced with the capacity of several ampere-hours and energy density of 350 – 400 Wh/kg.

13.3 SODIUM ION BATTERIES

The concept of sodium ion batteries as an alternative to lithium ion batteries was voiced for the first time in 1993, but interest toward it has drastically increased only recently. The main cause for such interest is the relative scarcity of lithium resources. The estimated content of sodium in the lithosphere is 2.5%, which is almost three orders of magnitude higher than the content of lithium. (The content of sodium in the World Ocean exceeds the content of lithium by five orders of magnitude.) The world prices of the main raw material, lithium carbonate, are 20–30 times the prices of sodium carbonate. Besides, it is assumed that sodium ion batteries would operate at lower voltages as compared to lithium ion ones, which would provide an increase in their stability and safety.

Despite the fundamental similarity of sodium ion batteries and lithium ion batteries, they differ considerably in electrode materials. The materials well suitable for reversible intercalation of lithium do not allow reversible intercalation of sodium. This difference is primarily because of a difference in the size of sodium and lithium ions and is applicable to the materials of both negative and positive electrodes.

It was established already in the first period of works on sodium ion batteries that contrary to the lithium ion, the sodium ion is not intercalated into the interlayer space of graphite. Sodium ions penetrate nongraphitized carbon materials, but the nature of this penetration is not intercalation. In the case of oil coke, the capacity values of $90 - 95$ mAh/g were obtained, which approximately corresponds to the composition of NaC_{24}. In the case of carbon black electrodes, capacity of about 200 mAh/g was obtained. Quite suitable materials for the negative electrode could be different varieties of nanoporous hard carbon (obtained, e.g., by pyrolysis of glucose). In this case, intercalation of sodium ions is provided not only by their intercalation into the interlayer space, but also by their adsorption on the inner nanopore surface. The capacity of electrodes of nanoporous hard carbon reaches 300 mAh/g. Most recently, negative electrodes of carbon nanotubes with nearly similar sodium intercalation capacity were described.

Such noncarbon materials as various oxide materials (spinel oxide $NiCo_2O_4$, thin films of Sb_2O_4, TiO_2 nanotubes), $NaTi_2(PO_4)_3$, and different titanates, for example, $Na_2Ti_6O_7$, were suggested as negative electrodes of sodium ion batteries. However, the capacity of all the studied noncarbon materials did not exceed 200 mAh/g, so that they had no pronounced advantages as compared to hard carbon. It appears, however, more promising to use tin as a matrix for the sodium electrode. There is information on the implementation of thin–film tin electrodes and pasted electrodes of tin powder with a capacity of more than 600 mAh/g. It is pointed out that cycling stability of such electrodes depends on the composition of electrolyte. Additives, such as fluoroethylene carbonate, appear to be useful.

In the operation of all negative electrodes of sodium ion batteries, SEI is also formed on their surface as a result of the reduction of the components of electrolyte.

The set of compounds considered as active materials of the positive electrode of sodium ion batteries is much more extensive than in the case of negative electrodes. Here, one should primarily mention such counterparts of active materials of lithium ion batteries, as Na_xCoO_2, $NaMPO_4$, and $Na[Ni_{1/3}Fe_{1/3}Mn_{1/3}]O_2$, $Na_x[Ni_{0.5}Mn_{0.5}]O_2$, $Na_4Mn_9O_{18}$, $Na_{1.5}VOPO_4F_{0.5}$, $NaFeF_3$, Na_2FePO_4F, $NaM_2(PO_4)_3$ (M = Ti, V, or Fe), and especially $NaCrO_2$.

The electrode behavior of $NaCoO_2$ resembles the behavior of $LiCoO_2$: anodic sodium ion extraction (i.e., charging) results in structural changes that become irreversible when a large amount of sodium is extracted. Full extraction of the sodium ion from $NaCoO_2$ corresponds to a capacity of 235 mAh/g; the actual electrodes with $NaCoO_2$ have a capacity of not more than 120 mAh/g. The properties of $NaCoO_2$ strongly depend on the synthesis method; the sol–gel method allows obtaining better materials than the solid-phase synthesis.

The similar material of $NaCrO_2$ with a layered structure (layered rock salt structure like a-$NaFeO_2$) is considered more promising. In this case, full extraction of sodium would correspond to the capacity of 250 mAh/g, but the actual values do not exceed 120 mAh/g, that is, the composition of Na_xCrO_2 is reversibly cycled in the range of $0.5 < x < 1$. Electron conductivity of $NaCrO_2$ does not allow performing charging and discharge under forced modes and results in a too fast degradation, so it is recommended to cover the $NaCrO_2$ particles by a layer of carbon. Discharge of such electrodes occurs at the potential of about 3 V. The final charge voltage cannot exceed 3.5 V, which eliminates oxidation of electrolyte (leading to ignition and explosion) under charging of a sodium ion battery.

A NASICON-like material, $Na_3V_2(PO_4)_3$, is capable of reversibly releasing and intercalating the sodium ion under the potentials of 3.3−3.4 V at the specific capacity of about 100 mAh/g and with a flat charging and discharge curve. The capacity of electrodes of such a material decreases considerably at an increase in the current because of high ohmic resistance. Application of carbon coating on $Na_3V_2(PO_4)_3$ particles allows performing discharge and charging in the modes up to 20°C. Close characteristics are also typical for fluorophosphates of the Na_2FePO_4F type.

Similarly to lithium ion batteries, it is intended to use in sodium ion batteries conventional aprotic electrolytes impregnating a porous separator (Celgard), solid polymer electrolytes, and much attention has been paid lately to ionic liquids. As the electrochemical window of sodium ion electrochemical systems is somewhat narrower than in the case of lithium ion systems, the probability of successful application of ionic liquids in sodium ion batteries is rather high. Of liquid electrolytes, $NaPF_6$ solutions in pure propylene carbonate (PC) and in mixtures of ethylene carbonate (EC) with diethyl carbonate (DEC), solutions of $NaClO_4$ in PC, mixtures of EC−DEC, EC—dimethyl carbonate, and so on have been described.

The prospects of development of sodium ion batteries are very uncertain. The developers of such batteries remember the numerous efforts directed at the commercialization of batteries with a sodium negative electrode and ceramic electrolyte of β-alumina. Intensive development of batteries with the system of sodium−sulfur has been carried out since 1966 (for almost half a century!) and development of batteries with the system of sodium−nickel chloride (ZEBRA batteries) has been performed since 1978. It was assumed that these high-temperature batteries would form a basis for electric transport, but these systems are still referred to in the future tense.

REVIEWS

Kraytsberg A, Ein-Eli Y. J Power Sources 2011;196:886−893.

Mikhaylik Y, Kovalev I, Schock R, Kumaresan K, Xu J, Affinito J. ECS Trans 2010;25:23−34.

14

SOLID-STATE BATTERIES

14.1 LOW-TEMPERATURE MINIATURE BATTERIES WITH SOLID ELECTROLYTES

A number of batteries with solid electrolytes based on simple silver halides were developed in the 1950s. The feasibility of developing miniature batteries of high mechanical strength, with good shelf life (even at high temperatures) and without the danger of electrolyte leakage was shown. However, a significant drawback of these batteries is the high electrolyte resistance: even with very thin electrolyte films the discharge current densities of the batteries are not higher than several microamperes per square centimeter. The trend to miniaturization and the need to decrease the thickness of the electrolyte films led to the use of sputtering and other technological processes developed for the manufacture of solid-state devices.

When high-conductivity electrolytes, in particular $RbAg_4I_5$, were discovered they opened the way for the development of batteries for higher discharge current. A battery with such an electrolyte can have a silver anode and an iodine cathode. In the course of the operation of batteries, silver is dissolved and the silver ions migrate through the solid electrolyte and react with the iodine ions produced in the cathodic reaction; the final reaction product is silver iodide, AgI.

These simple batteries have a number of significant drawbacks. The iodine vapor has a relatively high pressure necessitating careful sealing of the batteries. Iodine can be dissolved in electrolyte and migrate toward the anode causing self-discharge and

Electrochemical Power Sources: Batteries, Fuel Cells, and Supercapacitors, First Edition.
Vladimir S. Bagotsky, Alexander M. Skundin, and Yurij M. Volfkovich.
© 2015 John Wiley & Sons, Inc. Published 2015 by John Wiley & Sons, Inc.

formation of a poorly conducting film of AgI on the anode. Iodine also facilitates the decomposition of $RbAg_4I_5$, giving rise to separate phases of AgI and RbI_3. However, the primary drawback of these batteries is the accumulation of the reaction product AgI near the cathode and the associated increase in the internal resistance of the battery.

To eliminate these drawbacks other cathode materials are employed: polyiodides of the type of RbI_3, $N(CH_3)_4I_3$, for instance, and others, free of such undesirable properties. Reactions with these materials produce high-conductivity compounds, for instance:

$$2 \ Ag + RbI_3 \rightarrow 2AgI + RbI \rightarrow (RbAg_4I_5; \ Rb_2AgI_3; \ \text{etc.}) \qquad (14.1)$$

and the resistance of batteries does not increase. Since no free iodine is present in the batteries, the abovementioned side reactions do not occur.

Batteries of this type have been developed with film, tablet, and cylindrical designs. In the first type, the electrolyte film is applied onto the metal anode or cathodic current collector by sputtering or vaporization. The tablet and cylindrical batteries designed for comparatively high drain rates employ porous electrodes manufactured by pressing a mixture of the powders of the active materials (silver or polyiodide), electrolyte, and conductive additive (carbon black, etc.).

The open circuit voltage (OCV) of the $Ag|RbAg_4I_5|RbI_3$ batteries is 0.67 V and their discharge current density can be as high as 50 mA/cm^2. At low-current drains, the slope of their discharge curve is small and the coefficient of utilization of the active materials can be as high as 90% or more. Polarization is mainly ohmic in nature and even at high current densities it is due almost entirely to electrolyte resistance. At current densities of about 0.1 mA/cm^2, the batteries are operable at temperatures down to $-55°C$. Owing to low OCV and the large equivalent mass of the reactants the theoretical specific energy (i.e., not taking into account the masses of electrolyte and structural components) is low, about 53 Wh/kg. In practice, the specific energy of the miniature batteries is approximately a factor of 10 lower. The range of application of low-temperature batteries with solid electrolyte is thereby limited to some special fields requiring longtime storage, operability in a wide range of temperatures, and high mechanical strength.

14.2 SULFUR–SODIUM STORAGE BATTERIES

Development of a high-temperature sulfur–sodium storage batteries employing solid sodium polyaluminate as electrolyte was first reported in 1966 by the Ford Company, United States. This battery type immediately attracted attention and considerable research and development efforts in many countries were started. The main advantages of these batteries are the high specific power and energy, good reversibility, the absence of side reactions, hermetic sealing and (the most important) cheapness, and free availability of the main reactants: sulfur and metallic sodium. A drawback of the batteries is their high working temperature in the range of 300–350°C. In contrast to other battery types, these batteries employ a solid electrolyte and liquid reactants (the melting points of sulfur and sodium are 119°C and 97.5°C, respectively). The solid electrolyte also serves as the separator between the reactants of both electrodes.

14.2.1 Current-Producing Reactions

In the course of discharge, sodium is anodically oxidized to sodium ions Na^+, which penetrate the solid electrolyte and act as current carriers in it. Sulfur is reduced on the positive electrode and reacts with sodium ions from the electrolyte giving rise to various sodium polysulfides, Na_2S_m. The overall current-producing reaction can be divided into two stages:

$$5S + 2Na \underset{ch}{\overset{disch}{\rightleftharpoons}} Na_2S_5 \tag{14.2}$$

$$3Na_2S_5 + 4Na \underset{ch}{\overset{disch}{\rightleftharpoons}} 5Na_2S_3 \tag{14.3}$$

Stage (14.2) gives rise to molten Na_2S_5, which does not mix with molten sulfur so that a two-phase liquid system is formed. At stage (14.3) when free sulfur has previously been consumed, the system consists of one phase. The melting points of polysulfides with m between 3 and 5 lie in the range of 235–285°C.

In principle, Na_2S_3 can be reduced further giving rise ultimately to the simple sulfide Na_2S. However, sodium sulfide and sodium disulfide, Na_2S_2, have higher melting points and, hence, form solid phases in the batteries considerably reducing its reversibility. Therefore, the batteries capacity corresponding to complete conversion of pure sulfur to pure sodium trisulfide is regarded as the limiting (theoretical) batteries capacity.

The high working temperature of the sulfur–sodium batteries is necessitated not only by the desire to increase the conductivity of electrolyte but also by the need to operate with molten reactants and intermediate compounds.

14.2.2 Electrolyte

There are two crystallographic modifications of sodium polyaluminates, $Na_2O \cdot nAl_2O_3$, with different sodium contents. One is β-alumina with $n = 9$–11 and the other is β″-alumina with $n = 3$–5. At 300°C the conductivity of the β-phase is 0.03–0.1 S/cm and the conductivity of the β″-phase (which has a higher sodium content but is less stable) is approximately twice as great. The conductivity depends on the manufacture conditions of the electrolyte. It can be raised by a factor of 2–3 by adding magnesium or lithium oxide. However, the additions impair the stability of battery operation and give rise to hygroscopicity complicating the manufacture of the batteries. The existing electrolytes with optimum composition have conductivities of 0.1–0.3 S/cm at 300°C.

The stability of the electrolyte during operation presents considerable difficulties. In the early battery prototypes, the electrolyte developed microcracks; metallic sodium leaked through them resulting in a battery failure. The cracks were caused not only by mechanical stresses but also by nonuniformities of the electrolyte structure and the presence of defects. Electrolyte stability is greatly improved owing to careful regulation of the manufacturing process and control of the microstructure of the electrolyte. The important factor in this respect is the uniformity in size of the primary electrolyte particles; their optimum size is $2-4 \, \mu m$.

The electrolyte is a poreless ceramic material shaped either as a thin disk (for flat batteries) or a tube with one sealed end (similar to test tube) employed in cylindrical batteries. The process of electrolyte manufacture consists of the following stages: (i) preliminary calcination and grinding of the starting materials (alumina, α-Al_2O_3, and sodium carbonate); (ii) careful mixing of the components with a binder; (iii) isostatic compacting of the powder at a pressure of about 400 MPa; (iv) sintering at a temperature of about 1600°C. Sintering is the most critical stage; since sodium oxide, Na_2O, is volatile the electrolyte is sintered in closed platinum crucibles. On the whole, the technological process is rather complicated and its efficiency so far is not higher than $50-60\%$.

14.2.3 Battery Design

The batteries are mostly of cylindrical construction, with the electrolyte shaped as a tube of length 20–50 cm, diameter 1.5–3.5 cm, and wall thickness about 1 mm. One reactant is inside the tube and the other is on the outside. Molten sulfur and polysulfides are typically inside the tube. Molten sodium is in the gap between electrolyte and the batteries' container wall. A stock of sodium is stored in a container in the upper and lower part of the batteries.

Since both sulfur and polysulfides lack electron conductivity, a very loose felt-like graphite material is employed as the current-collector of the positive electrode; its mass is only $3-10\%$ of the reactant masses. The graphite felt also has another function; in the course of charging at stage (14.2) sulfur is formed on the electrolyte surfaces. Since sulfur lacks ionic conductivity (in contrast to polysulfides), it can block the polysulfide melt, thus hampering further charging. The graphite felt has good wettability by molten sulfur but not by the polysulfide melt. When the large-pore felt layer is close to the electrolyte and the fine pore layer is farther from it, capillary forces tend to draw molten sulfur away from the electrolyte surface thus reducing the blocking effect of sulfur. This arrangement contributes to a significant increase in the chargeability of the sulfur electrode at stage (14.2) so that the coefficient of utilization of sulfur is raised to $80-90\%$.

14.2.4 State-of-the-Art

Extensive R&D activities in the field of sodium–sulfur batteries started in 1966 (almost half a century ago). At that time such batteries were thought to be of basic importance for electric vehicles. Nowadays they are considered only for yet unknown possible future applications.

MONOGRAPHS AND REVIEWS

Churmann AT. Electr Vehicles 1975;61:12.

Fischer W, Haar W, Hartmann B, Meinhold H, Weddigen G. J Power Sources 1978;3:299.

van Gool, W. (ed.), In: *Solid State Batteries and Devices*. North-Holland Publishing Company, Amsterdam (1973).

Mahan GD, Rofh WL. *Superionic Conductors*. New York and London: Plenum Press; 1976.

May GJ. J Power Sources 1978;3:1.

15

BATTERIES WITH MOLTEN SALT ELECTROLYTES

15.1 STORAGE BATTERIES

Considerable attention has been focused on the development of long-lived storage batteries with molten electrolyte designed for high current densities and high specific powers with sufficiently high specific energies. A large number of studies on this topic has been reported. However, a major problem with these batteries is that of materials as structural components in such batteries are attacked by highly aggressive agents (strong oxidants and high-temperature melts). Considerable difficulties are associated also with thermal cycling—variation of the battery temperature from the ambient temperature to the working temperature and back.

15.1.1 Negative Electrode

Most proposed battery types employed lithium-negative electrodes. In such batteries the electrolyte must contain a lithium salt and the electrode processes on the lithium electrode consist of simple transfer of the lithium ions from the crystal lattice of the metal to the melt and back.

At the working temperature of the battery (400–600°C) pure lithium is liquid. Two main construction types of lithium electrodes have been reported: liquid lithium in a porous matrix and a solid alloy of lithium with another metal. The matrix of the

---

Electrochemical Power Sources: Batteries, Fuel Cells, and Supercapacitors, First Edition.
Vladimir S. Bagotsky, Alexander M. Skundin, and Yurij M. Volfkovich.
© 2015 John Wiley & Sons, Inc. Published 2015 by John Wiley & Sons, Inc.

lithium electrode can be manufactured from the porous plates of stainless steel, from felt-like stainless steel, nickel or a similar material possessing high porosity (up to 90%), small pore diameter, and the required elasticity.

Polarization of the lithium electrode is negligible both during charging and discharge; record-breaking current densities, up to 40 A/cm^2, have been obtained with lithium electrodes.

One of the main drawbacks of liquid lithium is its noticeable solubility in the salt melt and its capability to expel potassium from the melt:

$$Li + KCl \rightarrow K + LiCl \qquad (15.1)$$

The vapor pressure of potassium is much higher than that of lithium, and potassium can form an undesirable gas phase. Lithium dissolved in the electrolyte migrates toward the positive electrode where it is consumed in an unproductive chemical reaction. For an interelectrode distance of 1 mm, the associated self-discharge is equivalent to a leakage current density of 1–10 mA/cm^2. Moreover, dissolved lithium causes disintegration of the ceramic separators. The solubility of lithium greatly increases with increasing temperature.

Lithium alloys have been suggested to replace lithium in order to reduce the activity and to decrease the solubility in electrolyte. Electrodes with lithium alloys that are liquid at the working temperature (alloys with zinc, tin, etc.) have the same construction as pure lithium electrodes. The solid lithium alloys (with silicon or aluminum) are used in the form of porous plates pressed onto current-collecting meshes. Replacement of pure lithium with alloys results in a decrease in the open current voltage (OCV) and working voltage by 0.2–0.3 V but it is justified by a considerable reduction of self-discharge and a longer service life.

15.1.2   Positive Electrode

The first prototypes of batteries with molten electrolyte employed gas diffusion chlorine electrodes similar to the electrodes of fuel cells. Even the early battery prototypes had discharge current densities of up to 4 A/cm^2 without significant polarization. The chlorine electrode is manufactured from porous graphite, carbides of boron and silicon, or similar materials. The difficulty encountered in the development of such electrodes is choosing the technique for storing molecular chlorine. At first, the use of activated carbon for adsorbing chlorine was suggested. A carbon with higher developed true surface can provide for an adsorption capacity for chlorine of up to 0.3–0.5 Ah/cm^3. Another suggestion was to store chlorine in vessels under elevated pressure. However, neither of the suggestions was put into practice.

Later there was a suggestion to employ sulfur and then chalcogenides, primarily sulfides, as active materials for the positive electrode. Sulfur is liquid at the working temperature of the battery. The sulfur electrode was manufactured from a mixture of sulfur and carbon or in the form of a niobium box with niobium filler packed with sulfur. The high volatility of sulfur (at 507°C the pressure of sulfur vapor is

2 atm) and its solubility in the molten electrolyte lead lo self-discharge. The greatest success was obtained with iron sulfides, FeS and FeS_2. Both these sulfides have a high ampere-hour capacity and are cheap and nontoxic. The process of electrode manufacture is simple (Ivins RO et al., 1976).

The discharge and charging processes in sulfide electrodes can be described as follows:

$$FeS + 2e^- \underset{ch}{\overset{disch}{\rightleftarrows}} Fe + S^{2-} \tag{15.2}$$

and

$$FeS_2 + 4e^- \underset{ch}{\overset{disch}{\rightleftarrows}} Fe + 2S^{2-} \tag{15.3}$$

The reaction with FeS_2 involves intermediate formation of FeS which is why the charge−discharge curve of the batteries with FeS_2 electrodes consists of two gently sloping parts of approximately the same length corresponding to discharge voltages of 2.05 and 1.65 V. The theoretical specific consumption of FeS_2 is somewhat lower than that of FeS (11.2 g/Ah and 1.64 g/Ah) but FeS_2 has a higher corrosive activity, which can result in a shorter service life.

The sulfide electrodes are manufactured from a mixture of FeS or FeS_2 with some additives (sulfides of lithium, copper, and cobalt) placed into a porous frame of molybdenum, tungsten, graphite, and so on, and electrolyte is added to the positive electrode. For example, the positive electrode can have the following composition: 60% FeS, +2.2% Li_2S + 29.3% LiCl + KC1 eutectic +7% carbon +1.5% iron powder (percentages by mass). Iron is added to prevent the formation of elementary sulfur in case of overcharge; it reacts with Li_2S giving rise to FeS_2.

Rather high current densities (up to 0.4 A/cm^2) are obtained with the sulfide electrodes.

15.1.3   Electrolyte

The maximum specific capacity of the lithium−chlorine and lithium−sulfide batteries could be obtained with electrolytes consisting of pure LiCl or Li_2S, which are the products of the current-producing reactions. However, their melting temperatures are too high (613°C and 950°C, respectively), so that mixtures LiCl + KC1 and LiCl + KC1 + Li_2S with lower melting points are typically used as electrolytes to obtain working temperatures of not more than 400°C (the melting point of the LiCl + KCl eutectic is 352°C). The presence of the inert additives in the electrolyte results in a certain decrease of the specific battery capacity. Moreover, the composition of electrolyte varies in the course of operation as the content of LiCl or Li_2S increases with discharge and decreases with charge, and the melting point of the electrolyte changes accordingly.

Most battery types employed are immobilized or matrix electrolytes. Fine powders of boron nitride, lithium aluminate, and so on, are used as immobilizing agents and matrixes are manufactured from ceramic fabrics, such as boron nitride, stabilized zirconia, and so on.

15.1.4 Prototypes

Large-scale research and development projects on the Li-FeS and Li-FeS$_2$ storage batteries are under way in the Argonne National Laboratory and some other US organizations. Two types of batteries were developed, one for electric vehicles and military applications and the other for load-leveling at power plants. The batteries of the first type employed positive electrodes of FeS$_2$, providing for higher performance (the rated specific energy was 150 Wh/kg) but with shorter service and cycle life (the planned cycle life was 1000 cycles and service life was 3 years) and necessitating the use of more expensive materials. The batteries of the second type employ FeS electrodes. The specific energy is about half that of the FeS$_2$ batteries but their cost is lower by almost 50% and the planned cycle life and service life are 3000 cycles and 10 years. The batteries of the second type are designed for lower rates of charge and discharge and have higher efficiency.

When the batteries are not working they can be cooled to room temperature, that is, the electrolyte can be frozen. The batteries can then be heated up to the working temperature without any loss of capacity or deterioration of parameters. This greatly simplifies the long-term storage of the batteries.

One of the battery prototypes for electric vehicles had a volume of 320 l and mass of 820 kg. The positive electrode is manufactured from FeS with the addition of CoS$_2$. A few layers of the active material alternating with graphitized fabric are placed into a basket of molybdenum mesh welded to the central molybdenum current collector. The positive electrode is wrapped into a two-layer separator. The inner layer consists of ZrO$_2$ fabric and the outer layer of BN fabric. The negative electrode consists of a lithium–silicon alloy in the porous nickel matrix. The container and the cover are manufactured from stainless steel and electrically connected to the negative electrode. The prototype was drained with current up to 50 A, and the specific power was as high as 53 W/kg (Martino FJ et al, 1978).

A larger battery designed for a submarine had a sealed container with six positive and six negative electrodes. The negative electrodes were manufactured from a lithium–aluminum alloy. Separators made from BN fabric were inserted between the electrodes. The battery had the following dimensions: diameter, 30.5 cm; height, 21.1 cm; mass 43 kg; rated discharge voltage 1.45 V; and the specific capacity about 150 Wh/kg. The battery was designed for normalized current drain $j_d = 0.08$.

15.2 RESERVE-TYPE THERMAL BATTERIES

In Section 4.5.3, reserve-type thermal batteries were mentioned that are activated after an instantly melting solid-state salt electrolyte. In this section more details about these batteries are described.

15.2.1 Electrolyte

Eutectic salt mixtures are used as electrolyte in these batteries (most often LiCl + KC1 or LiBr + KBr). The molten electrolytes have conductivity of $1 - 2$ S/cm at a temperature of 450°C. The high voltage of electrolyte decomposition (about 3 V) makes it possible to employ strong reducers and oxidants. Sometimes, various components are added to the electrolyte (up to 15% of chromates, bromates, and perchlorates) which stabilize the discharge voltage of the battery. To increase the mechanical strength of the battery, and to reduce shrinkage with melting, immobilizing agents—such as silica or asbestos—are added to the electrolyte, or fiber glass fabric is impregnated with electrolyte and serves as a matrix.

15.2.2 Electrodes

Anodes are manufactured from calcium or magnesium. The anodic process consists of the transfer of the metal ions from the crystal lattice to the melt.

Oxides of heavy metals, chromates, and other oxidants are used as materials for the cathodes. Typically, the cathode mix contains 85% of calcium chromate and 15% electrolyte forming a mixture whose composition is close to eutectic. In such batteries, the cathode's current collector is just the battery cover.

Design and Performance. The main type of thermal battery's single cells have a thickness varying from 1.5 to 3 mm and a diameter between 30 and 70 mm. Thermal batteries are assembled from alternating single cells and heaters of approximately equal thickness. The assembled battery is placed into a sealed container with reliable thermal insulation made from asbestos, glass wool, silica, alumina, and so on. The heaters of thermal batteries consist of pyrotechnic compounds with high caloric values (thermite powders, mixtures from barium peroxide with aluminum powder, etc.). Electric fuses are used to ignite these mixtures. Different varieties of thermal cells (depending on application) have Wh capacities from a fraction of a watt-hour to tens of watt-hours and masses from 40 g to 3 kg; the voltage of batteries varies from 12 to 500 V.

The rates of discharge of thermal batteries can be as high as 400 mA/cm^2 and more with the discharge voltages of a single cell up to 2.5–3.0 V. For short-time discharges (less than 1 min) the specific power of thermal batteries can be as high as 600 W/kg. Commercially available batteries have capacities from a fraction of a watt-hour to a few tens of watt-hours and masses from 40 g to 3 kg. The amount of active material in thermal batteries is sufficient to operate for about half an hour or longer. In practice, the batteries do not operate for more than 15 min owing to cooling and solidification of electrolyte. Therefore, the specific energy of thermal batteries is low, usually not more than 10 W/kg. This parameter can be improved by employing better heating compounds and more efficient heat insulation.

REFERENCES

Ivins RO, Gray EC, Walsh WJ, Chilenskas AA. Design and performance of Li-Al/FeS batteries, *Proceedings of 27th Power Sources Symposium*; 1976. p 8.

Martino FJ, Kaun TD, Shimotake H, Gay EC. Advances in the development of lithium-aluminum-metalsulfide batteries for electric vehicle batteries. *Proceedings of 13th Intersociety Energy Conversion Engineering Conference*; 1978. p 709.

PART III

FUEL CELLS

PART IV

16

GENERAL ASPECTS

16.1 THERMODYNAMIC ASPECTS

16.1.1 Limitations of the Carnot Cycle

Up to the middle of the twentieth century, all energy needs of mankind have been satisfied by natural kinds of fuel: coal, oil, natural gas, wood, and a few others. The thermal energy, Q_{react}, set free on combustion (a chemical reaction of oxidation by oxygen) of natural fuels is called the *reaction enthalpy*, or also the lower heat value (LHV), because usually the heat of condensation of water vapor as one of the reaction products is disregarded. Large parts of this thermal energy serve to produce mechanical energy in heat engines (steam turbines and different kinds of internal combustion engines).

According to one of the most important laws of nature, the second law of thermodynamics, the conversion of thermal to mechanical energy W_m is always accompanied by the loss of a considerable part of thermal energy. For a heat engine working along a Carnot cycle within the temperature interval defined by an upper limit T_2 and a lower limit T_1, the highest possible efficiency, $\eta_{theor} \equiv W_m/Q_{react}$, is given by

$$\eta_{theor} = \frac{(T_2 - T_1)}{T_2} \tag{16.1}$$

Electrochemical Power Sources: Batteries, Fuel Cells, and Supercapacitors, First Edition.
Vladimir S. Bagotsky, Alexander M. Skundin, and Yurij M. Volfkovich.
© 2015 John Wiley & Sons, Inc. Published 2015 by John Wiley & Sons, Inc.

T_2 and T_1 being the temperatures (in Kelvin) of the working fluid entering into and leaving the heat engine, respectively. The Carnot heat Q_{Carnot} (or "irretrievable heat")—for thermodynamic reasons known as the Carnot-cycle limitations is given by $Q_{Carnot} = (T_1/T_2) \cdot Q_{react}$. There is no way to reduce this loss. For a steam engine operated with superheated steam of 350°C ($T_2 = 623$ K) and release of the exhausted steam into a medium having the ambient temperature of 25°C ($T_1 = 298$ K), the maximum efficiency according to Equation (16.1) is about 50%, so half of the thermal energy is irretrievably lost. As a matter of fact, the efficiency that can be realized in practice is even lower on account of various other kinds of thermal losses Q_{loss} (because of heat transfer out of the engine, friction of moving parts, etc.), that is, the total losses ($Q_{exh} = Q_{Carnot} + Q_{loss}$) are even higher. The efficiency η_{theor} can be raised by working with a higher value of T_2 (see Fig. 16.1), but losses because of nonideal heat transfer will also increase.

In part, the mechanical energy produced in heat engines is used in turn to produce electrical energy in generators of stationary and mobile power plants. This additional step of converting mechanical into electrical energy involves additional energy losses, but these could be as low as 1–2% in a large modern generator. Thus, for a modern thermal power generating plant, a total efficiency η_{total} of about 40% is regarded as a good performance figure.

16.1.2 Electrochemical Energy Conversion

In Section 16.2, it was shown that in batteries an electric current is generated as the result of a chemical reaction between two reactants, an oxidant and a reducing agent. This reaction proceeds not by direct interaction of both reactants, which would lead only to evolution of the reaction energy Q_{heat} as heat into the ambient medium (chemical mechanism), but by an electrochemical mechanism in the form of two separate

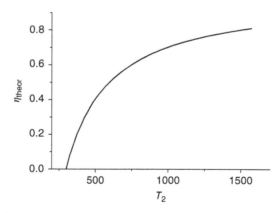

Figure 16.1. Limitations of the Carnot cycle. Theoretical efficiency η_{theor} versus the upper operating temperature T_2 of the heat engine at a lower temperature T_1 of 298 K (25°C) according to Equation 16.1.

partial reactions on different electrodes: cathodic reduction of the oxidant on the cathode and anodic oxidation of the reducing agent on the anode. By the electrochemical mechanism, a substantial part of the reaction energy, the Gibbs energy, can be transformed directly into electrical energy (Eq. 16.6).

16.1.3   First Definition of the Term "Fuel Cell"

In 1894, the German physical chemist Wilhelm Ostwald formulated the idea that an electrochemical mechanism can be used instead of combustion (chemical oxidation) of natural kinds of fuel, such as those used in thermal power plants. Because in this case the reaction will bypass the intermediate stage of heat generation, this would be "cold combustion," the direct conversion of chemical energy of a fuel to electrical energy not being subject to Carnot-cycle limitations. A device to perform this direct energy conversion was named a *fuel cell*.

The electrochemical mechanism of cold combustion in fuel cells has analogies in living beings. In fact, the conversion of the chemical energy of food by men and other biological species into mechanical energy (blood circulation, muscle activity, etc.) also bypasses the intermediate stage of thermal energy. The physiological mechanism of this energy conversion includes stages of an electrochemical nature. The average daily output of mechanical energy by human body is equivalent to the electrical energy of a few tens of watt-hours.

The work and teachings of Ostwald were the start of a huge research effort in the field of fuel cells.

Ostwald had considered only the thermodynamic aspects of fuel cells, but had entirely disregarded kinetic aspect, that is, the question of whether electrochemical reactions involving natural types of fuels are feasible, and how efficient they would be. Even the first experimental studies performed after the publication of Ostwald's paper showed that it is very difficult to build devices for direct electrochemical oxidation of natural kinds of fuel. Soon, it became obvious that the electrochemical mechanism can be successfully applied only for the oxidation of hydrogen and some natural fuels' gasification products.

16.1.4   A New Meaning of the Term "Fuel Cell"

About the mid-1980s, apart from work toward building powerful plants for efficient grid power generation (according to Ostwald's ideas) a second application of fuel-cell work was discerned, that of building autonomous power sources of intermediate or small capacity. Such power sources are intended for applications where grid energy is inaccessible, such as in means of transport, in portable devices, and in remote areas. The term *fuel cell* began to lose its original meaning as a power source with a current-producing reaction involving natural kinds of fuels and began to be used for an electrochemical power source that, in contrast to ordinary batteries, works continuously as long as the reactants, a reducing agent (the fuel) and an oxidizer, are supplied and the reaction products are removed. This definition implies that both reactants are either gases and/or liquids.

The most widely used example of a fuel cell with gaseous reactants are hydrogen–oxygen (air) fuel cells. An example of a fuel cell with at least one liquid reactant is the methanol–oxygen (air) fuel cell in which methanol is supplied as an aqueous solution.

16.2   SCHEMATIC LAYOUT OF FUEL-CELL UNITS

16.2.1   An Individual Fuel Cell

Fuel cells, like batteries, are a variety of galvanic cells, that is, devices in which two or more electrodes (electronic conductor) are in contact with an electrolyte (ionic conductors). Another variety of galvanic cells are electrolyzers in which electric current is used to generate chemicals in a process that is the opposite of those occurring in fuel cells and involving the conversion of electrical to chemical energy.

In the simplest case, a fuel cell consists of two metallic electrodes (say, platinum electrodes) dipping into an electrolyte solution (Fig. 16.2). In an operating fuel cell, the negative electrode or anode produces electrons by oxidizing ("burning") the fuel. The positive electrode or cathode absorbs electrons to reduce an oxidizing agent. The fuel and the oxidizing agents are supplied, each to "its" electrode. It is important at this point to create conditions that would exclude direct mixing of the reactants or their supply to the "wrong" electrode. In these two undesirable cases, direct chemical interaction of the reactants would start, and yield thermal energy, lowering or completely stopping the production of electrical energy.

So as to exclude accidental contact between anode and cathode (which would produce an internal short of the cell), an electronically insulating porous separator holding an electrolyte solution that supports current transport by ions is often placed into the gap between these electrodes. The solid ion-conducting electrolyte may serve at once as a separator. In any case, the cell circuit continues to be closed.

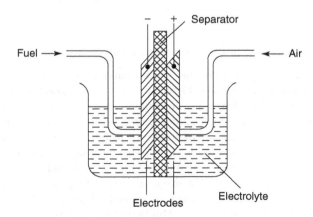

Figure 16.2. Schematic of an individual fuel cell.

For continued work of the fuel cell, provisions must be made to ensure continuous reactant supply to each of the electrodes and continuous withdrawal of reaction products from the electrodes, as well as removal and/or utilization of the heat being evolved.

16.2.2   Fuel-Cell Stacks

As a rule, any individual fuel cell has a low working voltage of less than 1 V. Most users need a much higher voltage of, for instance, 6, 12, 24 V or more. In a real fuel-cell plant, therefore, the appropriate number of individual cells is connected in series, forming stacks (batteries).

A common design is the so-called filter-press design of stacks built up of bipolar electrodes, one side of such electrodes working as the anode of one cell and the other side working as cathode of the neighboring cell (Fig. 16.3).

The active (catalytic) layers of each of these electrodes face the separator, the pores of which are filled with an electrolyte solution. A bipolar fuel-cell electrode is usually built up from two separate electrodes, their backs resting on the opposite sides of a separating plate, known as the *bipolar plate*. These plates are

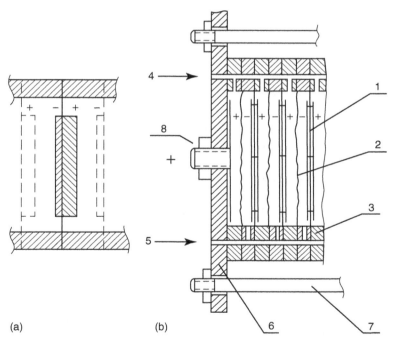

Figure 16.3. Fuel-cell components. (a) Bipolar electrode and (b) filter-press battery: (1) bipolar electrode; (2) separator; (3) gaskets; (4) air supply; (5) fuel supply; (6) end plate; (7) tie bolts; (8) positive current collector.

electron-conducting and function as cell walls and as intercell connectors (i.e., the current between neighboring cells merely crosses this plate, which forms a thin wall having negligible resistance). This implies considerable savings in the size and mass of the stack. The bipolar plates alternate with electrolyte compartments and both must be carefully sealed along the periphery to prevent electrolyte overflow and provide a reliable separation of the electrolyte in neighboring compartments. The stacks formed from the bipolar plates (with their electrodes) and the electrolyte compartments (with their separators) are compressed and tightened with the aid of relatively thick end plates and tie bolts. Sealing is realized with the aid of gaskets compressed when tightening the assembly. After sealing, the compartments are filled with electrolyte via manifolds and special narrow channels in the gaskets or electrode edges. Gaseous reactants are supplied to the electrodes via manifolds and grooves in the bipolar plates.

16.2.3 Power Plants Based on Fuel Cells

The heart of any fuel-cell power plant (in Russia, it is called *electrochemical generator*) or direct energy converter is one or a number of stacks built up from individual fuel cells. Such plants include a number of auxiliary devices needed to secure stable, uninterrupted functioning of the stacks. The number or type of these devices depends on the fuel cell type in the stacks and the intended use of the plant. In the following, we list the basic components and devices.

(a) *Reactant Storage Containers.* Gas cylinders, recipients, vessels with the reactants (e.g., petroleum products), cryogenic vessels for refrigerated gases, gas-absorbing materials, and so on.

(b) *Fuel Conversion Devices.*
   (i) For reforming of hydrocarbons yielding technical hydrogen,
   (ii) For gasification of coal yielding water gas (syngas), or
   (iii) For chemical extraction of the reactants from other substances. This includes devices for reactant purification from harmful contaminants or separate particular reactants from mixtures.

(c) *Devices for Thermal Management.* In most cases, the working temperature is distinctly above ambient temperature. In these cases the working temperature is maintained by exhaust heat (Q_{exh}) from the heat evolved during fuel-cell operation. A cooling system must be provided when excess heat is evolved in fuel-cell stacks. Difficulties arise when starting up the plant while its temperature is below the working temperature (such as after interruptions). In these cases, external heating of the fuel-cell stack must be made possible. In certain cases, sufficient heat may be generated in the stack by shorting with a low-resistance load, where heating is started with a low current, and leads to a larger current producing more heat, and so on, until the working temperature is attained (self-heating).

(d) *Regulating and Monitoring Devices.*
   (i) For securing uninterrupted reactant supply at the required rate,

(ii) For securing product removal (where applicable), with a view to its further utilization,

(iii) For securing the removal of excess heat and for maintaining the correct thermal mode,

(iv) For maintaining other operating parameters of the fuel cells needed for continuous operation.

(e) *Power Conditioning Devices.* Voltage converters, DC–AC converters, electricity meters, and so on.

(f) *Internal Electrical Energy Needs.* Many of the devices listed include components working with electric power (such as the pumps for gas supply or heat-transfer fluid circulation, electronic regulating and monitoring devices, etc.). As a rule, the power needed for these devices is derived from the fuel-cell plant itself. This leads to a certain decrease in the power level available to consumers. In most cases, these needs are not very significant. In certain cases, such as when starting up a cold plant, heating with external power supply may be required.

16.3 TYPES OF FUEL CELLS

Different attributes can be used to distinguish fuel cells:

(a) *Reactant Type.* As the fuel (reducing agent), fuel cells can use hydrogen, methanol, methane, carbon monoxide, and some inorganic substances (hydrogen sulfide H_2S, hydrazine N_2H_4). As the oxidizing agent fuel cells can use pure oxygen, air oxygen, hydrogen peroxide H_2O_2, and chlorine. Versions with other exotic reactants have also been proposed.

(b) *Electrolyte Type.* Apart from the common liquid electrolytes (aqueous solutions of acids, alkalis, and salts; molten salts), fuel cells often use solid electrolytes (ion-conducting organic polymers, inorganic oxide compounds). The solid electrolytes reduce the danger of leakage of liquids from the cell (which may lead to corrosive interactions with the construction materials and also to prevent electric shorts owing to contact between electrolyte portions in different cells of a battery). Solid electrolytes also serve as separators keeping reactants from reaching the "wrong" electrode space.

(c) *Working Temperature.* One distinguishes low-temperature fuel cells (having a working temperature of no more than 120–150°C, intermediate-temperature fuel cells (105–250°C), and high-temperature fuel cells (over 600°C). The low-temperature fuel cells include membrane-type fuel cells as well as most alkaline fuel cells. Intermediate-temperature fuel cells are those with phosphoric acid electrolyte as well as the alkaline cells of the Bacon type. High-temperature fuel cells include the fuel cells with molten carbonate (working temperature 600–700°C), and the solid-oxide fuel cells (working temperature above 900°C). In recent years, interim temperature fuel cells with a working temperature in the range between 200 and 650°C have been introduced. These include certain varieties of solid-oxide fuel cells developed recently. The temperature ranges are stated conditionally.

16.4   LAYOUT OF A REAL FUEL CELL: THE HYDROGEN–OXYGEN FUEL CELL WITH LIQUID ELECTROLYTE

At present, most fuel cells use either pure oxygen or air oxygen as the oxidizing agent. The most common reducing agents are either pure hydrogen or technical hydrogen, produced by steam reforming or with the water–gas shift reaction from coal, natural gas, petroleum products, or other organic compounds.

As an example of a real fuel cell, therefore, we consider the hydrogen–oxygen fuel cell with an aqueous acid electrolyte and its special features. The special features of other types of fuel cells are described in the following chapters.

16.4.1   Gas Electrodes

In a hydrogen–oxygen fuel cell with liquid electrolyte, the reactants are gases. Under these conditions, porous gas-diffusion electrodes are used in the cells. These electrodes (Fig. 16.4) are in contact with a gas compartment on their backside and with the electrolyte on their front, facing the other electrode.

A porous electrode offers a far higher true working surface area and thus a much lower true current density (current per unit surface area of the electrode). Such an electrode consists of a metal or carbon-based screen or plate serving as the body or frame, current collector, and support for active layers, containing a highly dispersed catalyst for the electrode reaction. The pores of this layer are filled in part with the liquid electrolyte, and in part with the reactant gas. The reaction itself occurs at the walls of these pores along the three-phase boundaries between the solid catalyst, the gaseous reactant, and the liquid electrolyte.

For efficient operation of the electrode, it is important to secure a uniform distribution of said reaction sites throughout the porous electrode. With pores having hydrophilic walls (walls well wetted by the aqueous electrolyte solution), there exists the risk of flooding of the electrode or complete displacement of gas from the pore space. There are two possibilities to prevent this flooding of the electrode.

(a) The electrode can be made partly hydrophobic, by adding water-repelling material. It is important to maintain the optimum degree of hydrophobicity. When there is an excess of hydrophobic material, the aqueous solution will be displaced from the pore space.

(b) The porous electrode can be left hydrophilic, but from the side of the gas compartment, the gas is supplied with a certain excess pressure, so that the liquid electrolyte is displaced in part from the pore space. To prevent gas bubbles from breaking through the porous electrode (and reaching the counter electrode) the front side of the electrode is in contact with the electrolyte and is covered

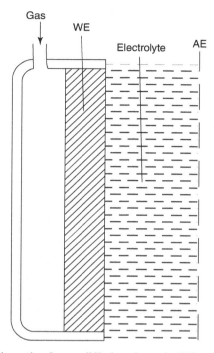

Figure 16.4. Schematic of a gas-diffusion electrode. WE, working electrode.

with a hydrophilic blocking layer having fine pores with a capillary pressure too high to be overcome by the gas so that the electrolyte cannot be displaced from this layer. Here it is important to select an excess gas pressure that is sufficient to partly fill the active layer with gas, but insufficient to overcome ("break through") the blocking layer.

16.4.2 Electrochemical Reactions

At the negative electrode (anode) hydrogen is oxidized:

$$2\,H_2 \rightarrow 4\,H^+ + 4\,e^- \tag{16.2}$$

while at the positive electrode (cathode) oxygen is reduced to water:

$$O_2 + 4\,H^+ + 4\,e^- \rightarrow 2\,H_2O \tag{16.3}$$

The hydrogen ions being formed in the electrolyte layer next to the anode in Reaction (16.2) are then transferred through the electrolyte toward the cathode, where they undergo Reaction (16.3). In this way a closed electrical circuit is obtained. In the electrolyte (positive) electrical current flows from the anode to the cathode; in the external circuit it flows in the opposite direction from the cathode terminal to the anode terminal. The overall chemical reaction producing the current is

$$2\,H_2 + O_2 \rightarrow 2\,H_2O \tag{16.4}$$

that means that by reaction of 2 mol of hydrogen and 1 mol of oxygen (at atmospheric pressure and a temperature of 25°C, 1 mol of gas takes up a volume of 22.4 l), 2 mol of water (36 g) are formed as the final reaction product.

The thermal energy Q_{react} (or reaction enthalpy $(-\Delta H)$) set free in Reaction (16.4), when this occurs as a direct chemical reaction, amounts to 285.8 kJ/mol. The Gibbs free energy $(-\Delta G)$ of the reaction amounts to 237.1 kJ/mol. This value corresponds to the maximum electrical energy W_e^{max} that theoretically could be gained from the reaction when following the electrochemical mechanism. This means that the maximum attainable thermodynamic efficiency η_{therm} of energy conversion in this reaction is 83%.

For practical purposes, it is convenient to state these energy values in electron volt (1 eV = $n \cdot$ 96.43 kJ/mol, where n is the number of electrons taking part in the reaction per mole of reactant, in this case per mole of hydrogen). In these units, the enthalpy of this reaction (with $n = 2$ per mol) is 1.482 eV, the Gibbs free energy is 1.229 eV. In the following, the heat of reaction expressed in electron volts is denoted as q_{react}.

The different fuel-cell systems differ in the nature of the components selected, and thus in the nature of the current-producing chemical reaction. Each reaction is associated with a particular value of enthalpy and Gibbs free energy $(-\Delta G)$ of the reaction and thus also with a particular value of the heat of reaction Q_{reac} and of the thermodynamic electromotive force (EMF) ε.

16.5   BASIC PARAMETERS OF FUEL CELLS

Most electrical parameters of fuel-cell plants are analogous to those of conventional batteries described in Chapter 3.

16.5.1   Electrode Potentials

At each electrode j in contact with an electrolyte, a defined value of electrode potential E_j is set up. It can be measured only relative to the potential of another electrode. By convention, in electrochemistry the potential of any given electrode is referred to as the potential of the so-called standard hydrogen electrode (SHE), which in turn by convention is taken as zero. A practical realization of the SHE is an electrode made of platinized platinum dipped into an acid solution having a mean ionic activity of hydrogen ions of unity and is washed by gaseous hydrogen at a pressure of 1 bar.

In our example, the potential $E_{h.e.}$ of the hydrogen electrode to which, according to Equation 16.2, electrons are transferred from the hydrogen molecule is more negative than the potential $E_{o.e.}$ of the oxygen electrode that, according to Equation 16.3, gives off electrons to an oxygen molecule.

The potentials of electrodes can be equilibrium (reversible ones) and nonequilibrium or irreversible. An electrode's equilibrium potential (which will be denoted as E_j^0 in the following) reflects the thermodynamic properties of the electrode reaction occurring at it (thermodynamic potential). The hydrogen electrode is an example of an electrode at which the equilibrium potential is established. When supplying hydrogen to the gas-diffusion electrode mentioned above, a value of electrode potential $E_{h.e.}^0$ is established at it (when it is in contact with the appropriate electrolyte) that corresponds to the thermodynamic parameters of Reaction (16.2). On the SHE scale, this value is close to zero (depending on the pH value of the solution, it differs insignificantly from the potential of the SHE itself).

An example of an electrode having a nonequilibrium value of potential is the oxygen electrode. The thermodynamic value of potential $E_{o.e.}^0$ of an oxygen electrode at which Reaction (16.3) takes place is 1.229 V (relative to the SHE). When supplying oxygen to a gas-diffusion electrode, the potential actually established at is 0.8–1.0 V, (i.e., 0.3–0.4 V less positive than the thermodynamic value) (steady-state potential).

The degree to which electrode potentials are nonequilibrium values depends on the relative rates of the underlying electrode reactions. Under comparable conditions, the rate of Reaction (16.3)—cathodic oxygen reduction—is 10 orders of magnitude lower than that of Reaction (16.2)—anodic hydrogen oxidation.

In electrochemistry, reaction rates usually are characterized by values of the exchange current density i^0 in units of milliampere per square centimeter, representing the (equal values of) current density of the forward and reverse reaction at the equilibrium potential when the net reaction rate or current is zero.

The reaction rates themselves strongly depend on the conditions under which the reactions are conducted. In particular, cathodic oxygen reduction that, at temperatures below 150°C, is far from equilibrium comes closer to the equilibrium state as the temperature is raised. The reasons why the real value of the oxygen electrode's potential at low temperatures is far from the thermodynamic value and why cathodic oxygen reduction is so slow are not clear so far, despite a large number of studies that have been conducted to examine it.

16.5.2   Voltage of an Individual Fuel Cell

As was said before, the electrode potential of the oxygen electrode is more positive than that of the hydrogen electrode. The practical voltage existing between them, the open-circuit voltage (OCV) U of a fuel cell, is

$$U = E_{o.e.} - E_{h.e.} \qquad (16.5)$$

When the two electrodes are linked by an external electrical circuit, electrons flow from the hydrogen to the oxygen electrode through the circuit, which is equivalent to (positive) electrical current flowing in the opposite direction. The fuel cell operates in a discharge mode, in the sense of Reactions (16.2) and (16.3) taking place continuously so long as reactants are supplied.

The thermodynamic value of voltage (i.e., the difference between the thermodynamic values of the electrode potentials) has been termed the *cell's electromotive force* (EMF) ε.

For a hydrogen–oxygen fuel cell the OCV is lower than the EMF owing to the lack of equilibrium of the oxygen electrode. Depending also on technical factors, it is 0.85–1.05 V.

The working voltage of an operating fuel cell U_d is even lower, the reasons being: (i) the internal ohmic resistance of the cell and (ii) the shift of potential of the electrodes occurring when current flows (also called *electrode polarization*) caused by slowness or lack of reversibility of the electrode reactions. The effects of polarization can be reduced by the use of suitable catalysts applied to the electrode surface that accelerate the electrode reactions.

The voltage of a working cell will be the lower, the higher the current I (i.e., the higher the current density $i = I/S$ at the electrode's working surface area S). The current–voltage and the current–power relations are basic fuel-cell characteristics and have the same shape as those of conventional batteries (see Fig. 3.2).

As for batteries the current–voltage relation can be expressed by the simplified linear equation:

$$U_d = U_0 - I \cdot R_{app} \tag{16.6}$$

where the apparent internal resistance R_{app} is conditionally regarded as constant. This is a rather rough approximation, as R_{app} includes not only the fuel cell's internal ohmic resistance but also components associated with polarization of the electrodes. These components are a complex function of current density and other factors. Often the U_d versus I relation is S-shaped. Sometimes it is more convenient to describe the relation in the coordinates of U_d versus log I.

At moderately high values of the current, the voltage of an individual hydrogen–oxygen fuel cell, U_d, is about 0.7 V.

The discharge current of a fuel cell at any given voltage U_d across an external load with the resistance R_{ext} is determined by Ohm's law:

$$I_d = \frac{U_d}{R_{ext}} \tag{16.7}$$

With increasing current (decreasing R_{ext}), the voltage decreases, hence the power–current relation goes through a maximum.

Neither the discharge current nor the power output are the sole characteristics of a fuel cell, as both are determined by the external resistance (load) selected by the user. However, the maximum admissible discharge current I_{adm} and the associated maximum power $P_{max,adm}$ constitute an important characteristic of all fuel cell types. These performance characteristics place a critical lower bound U_{crit} on cell voltage; certain considerations (such as overheating) make it undesirable to operate at discharge currents above I_{adm} or cell voltages below U_{crit}. To a certain extent the choice of values for I_{adm} and U_{crit} is arbitrary. Thus, in short-duration (pulse) discharge higher currents can be sustained than in long-term discharge.

For sustainable thermal conditions in an operating fuel cell, it will often be necessary for the discharge current not to fall below a certain lower admissible limit $I_{\text{min,adm}}$. The range of admissible values of the discharge current and the ability of a cell to work with different loads are important characteristics for each fuel cell type.

16.5.3 Operating Efficiency of a Fuel Cell

Operating efficiency of a fuel cell is its efficiency in transforming a fuel's chemical energy Q_{react} to electrical energy, W_e, or the ratio between the electrical energy produced and the chemical energy of oxidation of a fuel supplied. For the above-mentioned reasons the thermodynamic efficiency $\eta_{\text{therm}} = W_e / Q_{\text{react}}$ in some cases is somewhat lower than unity (100%), in other cases can be even higher than unity.

The maximum value of energy that can be produced when an amount of (electrical) charge λ_e (in coulombs) flows in the external circuit is $\varepsilon^0 \lambda_e$. For a hydrogen–oxygen fuel cell, the value of ε at the temperature of 25°C is 1.229 V.

A number of factors influence the overall conversion efficiency:

(a) *Voltage Efficiency.* As the real discharge voltage U_d is lower than the OCV (and in turn the OCV is mostly lower than the EMF) the overall voltage efficiency is given by

$$\eta_V = \frac{U_d}{\varepsilon} \tag{16.8}$$

(b) *Efficiency of Reactant Utilization: The Coulombic Efficiency η_{Coul} (Also Often Called Faradaic Efficiency).* Usually, not all of the mass or volume of the reactants supplied to a fuel-cell stack is used for the current-producing reaction or production of electric charges. External reasons for incomplete utilization include trivial leakage from different points in the stack. Intrinsic reasons include (i) diffusion of reactant through the electrolyte (possibly, a membrane) from its "own" to the opposite electrode where it undergoes direct chemical reaction with the other reactant; (ii) use of reactant for certain auxiliary purposes, such as circulation of (excess) oxygen serving to remove water vapor from parts of a membrane fuel cell, and its subsequent venting to the ambient air; (iii) incomplete oxidation of individual organic fuel types, for instance, an oxidation of part of methanol fuel to formic acid, rather than to carbon dioxide.

(c) *Design Efficiency η_{design}.* Often part of the electrical energy generated in a fuel cell is consumed for the (internal) needs of auxiliary equipment such as pumps supplying reactants and removing products, and devices for monitoring and controlling. The leakage of reactants mentioned above as a possibility also depends on the design quality. If the fuel cells constituting an electric power plant work with a secondary fuel derived on site from a primary fuel (such as with hydrogen made by steam reforming), then the efficiency of such processing must also be taken into account.

(d) *Overall Efficiency*. The overall efficiency of the power plant will depend on all of the factors listed:

$$\eta_{\text{total}} = \eta_{\text{therm}} \cdot \eta_{\text{Vol}} \cdot \eta_{\text{Coul}} \cdot \eta_{\text{design}} \qquad (16.9)$$

The overall efficiency is a very important parameter for fuel-cell-based power plants, both the centralized plants of high-capacity and the medium- or small-capacity plants set up in large numbers in a distributed fashion. The basic goal of these setups is that of reducing the specific consumption of primary fuels for power generation.

16.5.4 Heat Generation

The amount of thermal energy liberated during the operation of a fuel cell bears a direct relation to the value of the discharge or operating voltage. According to what had been said in the previous section, when passing a charge of λ_e coulombs, the total heat of reaction is given by $\lambda_e \cdot q_{\text{react}}$ joules (where the heat of reaction q_{react} is expressed in electron volts). The electrical energy produced is given by $\lambda_e \cdot U_d$ joules. The thermal energy produced will then be (in joules):

$$Q_{\text{exh}} = (q_{\text{react}} - U_d) \cdot \lambda_e \qquad (16.10)$$

This includes both the latent heat Q_{lat} and all kinds of energy loss Q_{loss} incurred because of the efficiencies mentioned above being less than unity.

For hydrogen–oxygen fuel cells, $q_{\text{react}} = 1.48$ eV. With a discharge voltage of $U_d = 0.75$ V, heat generation amounts at 0.73 λ_e joules, which is close to the value of electrical energy produced. Also, η_{Volt} can be expected to be about 0.5 at this discharge voltage.

16.5.5 Ways of Comparing Fuel Cell Parameters

Often a need arises to compare electrical and other characteristics of fuel cells that differ in their nature or size or to compare fuel-cell-based power generators with others. This is most readily achieved when using reduced or normalized parameters.

A convenient measure for the relative rates of current-producing reactions of fuel cells of a given type but differing in size is by using the current density, that is, the current per unit surface area S of the electrodes): $i = I/S$, with the units in milliampere per square centimeter. The power density $p_S = P/S$ (the units in milliwatt per square centimeter) is a convenient measure of relative efficiency of different varieties of fuel cells.

For users of fuel cells, important performance figures are the values of power density referred to unit mass M: $p_m = P/M$ (the units in watt per kilogram) or unit volume V: $p_v = P/V$ (the units in watt per liter), and also the energy densities per unit mass (in watt-hour per kilogram) or unit volume (in watt-hour per liter), both including the reactant supply. The power density usually is reported, merely by referring to the mass or volume of the fuel-cell battery itself but not to those of the power plant as a whole, as the mass and volume of reactants including their storage containers depend on the projected operating time of the plant. The energy density is usually reported for the power plant as a whole.

For stationary fuel-cell-based power plants, the most important parameter is the energy conversion efficiency, inasmuch as this will define the fuel consumption per unit of electric power generated. For portable and other mobile power plants, the most important parameters are the power density and the energy density, inasmuch as they reflect the mass and volume of the mobile plant.

16.5.6 Lifetime

Theoretically, a fuel cell should work indefinitely, that is, so long as reactants are supplied and the reaction products and heat generated are duly removed. In practice, however, the operating efficiency of a fuel cell decreases somewhat in the long run. This is seen from a gradual decrease of the discharge or operating voltage occurring in time at any given value of the discharge or operating current. The rate of decrease depends on many factors: the type of current load (constant, variable, pulsed), observation of all operating rules, conditions of storage between assembly and use, and so on. It is usually stated in microwatt per hour. If for a cell operated under constant load the lifetime may be stated in hours, a better criterion for the lifetime of cells operated under a variable load is the total of energy generated, in watt-hours, while the rate of decrease of the voltage would then be given in microvolt per watt-hour.

The major reason for this efficiency drop is a drop in the activity of the catalysts used to accelerate the electrode reactions. This activity drop may be because of

(a) spontaneous recrystallization of the highly disperse catalyst, its gradual dissolution in the electrolyte, or deposition of contaminants (inhibitors or catalytic poisons) on its surface;

(b) drop in ionic conductivity of the electrolyte, for instance, of the polymer membrane in proton exchange membrane fuel cell (PEMFC) and direct methanol fuel cell (DMFC) that is caused by its gradual oxidative destruction;

(c) corrosion of different structural parts of the fuel cells leading to its partial destruction and/or the formation of corrosion products lowering the activity of the electrodes—particularly so in high-temperature fuel cells;

(d) loss of sealing of the cells, for instance, on account of aging of packing, so that it becomes possible for reactants to reach the "wrong" electrode.

The rate of drop of fuel-cell efficiency depends largely on the mode and conditions of use: periodic interruptions, periodic temperature changes of idle cells from their operating temperature to ambient temperature and back when reconnected may have ill effects, and sometimes the documentation mentions an admissible number of load or temperature cycles.

On relatively rare occasions, a fuel cell may suddenly fail, its voltage falling to almost zero. This kind of failure is usually caused by an internal short that could occur when electrolyte leaks out through a defective packing, or when metal dendrites form and grow between electrodes.

It should be pointed out that as fuel-cell problems are relatively new, few statistical data are available from which to judge the expected lifetime of different types of fuel cells under different operating conditions. The largest research effort goes into finding the reasons for the gradual efficiency drop of fuel cells and finding possibilities to make it less significant.

REFERENCE

Ostwald W. Z Elektrochem 1894;1:122.

MONOGRAPHS

Bagotsky VS. *Fuel Cells: Problems and Solutions*. 2nd ed. Hoboken, NJ: John Wiley & Sons, Inc.; 2012.

Baker BS, editor. *Hydrocarbon Fuel Cell Technology*. New York: Academic Press; 1965.

Breiter MW. *Electrochemical Processes in Fuel Cells*. New York: Springer; 1969.

Bockris JO'M, Srinivasan S. *Fuel Cells: Their Electrochemistry*. New York: McGraw Hill; 1969.

Kordesch K, editor. *Brennstoffbatterien*. Berlin: Springer; 1984.

Justi EW, Winsel A. *Fuel Cells. Kalte Verbrennung*. Wiesbaden: Franz Steiner Verlag GmbH; 1962.

Liebhafsky HA, Cairns EJ. *Fuel Cells and Fuel Cell Batteries*. New York: John Wiley & Sons, Inc.; 1968.

Vielstich W. *Brennstoffelemente*. Verlag Chemie: Weinheim; 1965.

17

THE DEVELOPMENT OF FUEL CELLS

This chapter reflects the major points in the development of fuel cells but it does not describe in greater detail the contributions of the many research workers who have worked in this field.

17.1 THE PERIOD PRIOR TO 1894

The Volta pile first described in 1800 was of extraordinary significance for the developments both in the science of electricity and in the science of electrochemistry, as a new phenomenon, a continuous electric current, hitherto not known could now be realized. Soon, various properties and effects of the electric current were discovered, including many electrochemical processes. In May 1801, William Nicholson and Sir Anthony Carlisle in London electrolyzed water producing hydrogen and oxygen.

In the 1830s, the British Chemist Sir William Robert Grove (1811–1896), a trained lawyer and judge and an amateur natural scientist, conducted a series of experiments on water electrolysis (Grove W., 1939). His device consisted of two platinum electrodes dipping into water acidified with sulfuric acid. He assumed that, if by the passage of current, water can be decomposed to hydrogen and oxygen, then there should be a possibility for the opposite reaction to occur. He saw, in fact, that after disconnecting the current, the electrodes at which hydrogen and oxygen had been

Electrochemical Power Sources: Batteries, Fuel Cells, and Supercapacitors, First Edition.
Vladimir S. Bagotsky, Alexander M. Skundin, and Yurij M. Volfkovich
© 2015 John Wiley & Sons, Inc. Published 2015 by John Wiley & Sons, Inc.

evolved as gases were "polarized," that is, a certain potential difference was preserved between them. When in this state they were linked by an external circuit, current was found to flow in this circuit. This invention was called a *gas voltaic battery* by Grove. His results were published in February 1839 in the *Philosophical Magazine*. This date is regarded as the date of creation of the first prototype of a fuel cell with gas reactants.

Grove himself did not regard his gas battery as a practical means for producing electrical energy (instead, he developed a battery involving zinc and nitric acid). He pointed out, though, that an improvement of such cells could be achieved by increasing the contact area between the gases, adsorbents, and electrolyte.

In 1894, the German physical chemist Wilhelm Ostwald came forward in the *Zeitschrift für Elektrochemie* with the proposal to build devices for a direct oxidation of natural kinds of fuel with air oxygen by an electrochemical mechanism without heat production (the so-called cold combustion of fuels) (Fig. 17.1). He wrote: "In the future, the production of electrical energy will be electrochemical, and not subject to the limitations of the second law of thermodynamics. The conversion efficiency thus will be higher than in heat engines." This paper of Ostwald was basic and marked the beginning of huge research into fuel cells.

Figure 17.1. Friedrich Wilhelm Ostwald (1853–1932); Nobel Prize in 1909.

17.2   THE PERIOD FROM 1894 TO 1960

Even the first experimental studies performed after the publication of Ostwald's paper showed that it is very difficult to build devices for the direct electrochemical oxidation of natural kinds of fuel.

Over the period from 1912 to 1939, a great deal of research into the electrochemical oxidation of coal and coal gasification products was done in Zürich (Switzerland) and Brunswick (Germany) by the Swiss scientist Emil Baur (1873–1944) and his coworkers. It was the basic assumption of these workers that the process will succeed only at those temperatures at which coal usually burns sufficiently rapidly. In their first studies, therefore (Baur et al., 1910, 1912, 1916, 1921, 1937, 1938) high-temperature molten electrolytes, such as a mixture of sodium and potassium carbonates or molten caustic soda, were used. Nickel, iron, and sometimes platinum were used as the anodes for the oxidation of gases. Molten silver was the cathode for the reduction of oxygen. With time, large difficulties due to corrosivity of the molten electrolytes and to instability of these fuel cells at high temperatures were encountered. Subsequent work was done with high-temperature solid electrolytes. Basis of these electrolytes was the "Nernst rod" consisting of 15% of yttria and 85% of zirconia.

Apart from its high cost, this electrolyte had the other large defect, of gradually losing conductivity during current flow. After numerous studies, a cell for the electrochemical oxidation of coal was built. In this cell, a tubular crucible consisting of a mixture of clay, cerium dioxide, and tungsten trioxide served as the electrolyte. The crucible was filled with powdered coke and placed into a clay vessel filled with annealed iron oxide Fe_3O_4. At temperatures above $1000°C$, the cell exhibited a voltage of 0.7 V and developed a power density (per unit of useful volume) of 1.33 W/dm^3. However, even this cell had a limited lifetime owing to gradual loss of conductivity and to mechanical destruction of the electrolyte, and also to the effect of salt impurities in the coal.

It is interesting to note that Walter Schottky, a prominent specialist in the field of solid electrolytes, in his 1936 paper thought fuel cells with solid electrolytes to be unpromising, and favored those with salt melts.

With the aim of producing a more highly conducting, mechanically and chemically stable electrolyte, the Armenian scientist Oganes Davtyan who worked at the Moscow G.M. Krzhizhanovsky Power Institute of the Russian Academy of Sciences, in the 1930s used a mixture of Urals' monazite sand (containing 3–4% of ThO_2 and up to 15–20% of rare earth elements), tungsten trioxide WO_3, calcium oxide, quartz, and clay. The cell developed by him was used to oxidize carbon monoxide CO with air oxygen at the temperature of $700°C$; at the voltage of 0.79 V and current density of 20 mA/cm², it could be operated for several tens of hours. Drastic temperature changes sometimes provoked cracking of the electrolyte. Davtyan also performed numerous studies on low-temperature hydrogen–oxygen cells. The results of his research were presented together with a review of earlier work in his book *The Problem of Direct Conversion of the Chemical Energy of Fuels to Electrical Energy* published 1947 in Moscow. This book is the first monograph in the world dedicated to fuel cells.

In 1932, Francis Thomas Bacon (1904–1992), an engineering professor at the Cambridge University in England, started work modifying the earlier fuel-cell proto-types. Instead of the highly corrosive acidic solutions, Bacon used an alkaline (KOH) electrolyte. He used the porous electrodes of the gas-diffusion type that had been suggested as early as 1923 by Schmid. To prevent gases from breaking through these electrodes, Bacon coated them on the side of the electrolyte with a gas-impermeable barrier layer with fine pores. The electrodes were made of nickel powder treated with hot lithium hydroxide solution in order to raise their corrosion resistance (lithi-ated nickel). His first alkaline fuel cell, the "Bacon cell," was patented in 1959. In 1960 Bacon gave a public demonstration of a fuel-cell battery producing a power of 5–6 kW. It contained concentrated alkali solution (37–50% KOH). High temper-atures (200°C and more) and high gas pressures (20–40 bar) were used to accelerate the electrode processes and attain high current densities (0.2–0.4 A/cm^2). Because of the high gas pressure, the cell was very massive and heavy.

17.3 THE PERIOD FROM 1960 TO THE 1990s

Bacon's battery demonstration drew great attention from the scientific and technical community, and research and design work in this field started on a large scale in many countries (the first "fuel-cell boom").

Large interest was also aroused in scientific circles by the work of Grubb and Niedrach (1960), who were able to show for the first time that at temperatures below 150°C hydrocarbons such as methane, ethane, and ethylene can be oxidized elec-trochemically at electrodes made of platinum metals. By comparison, the chemical process occurring at the same catalysts in the gas phase exhibits marked rates, only at temperatures higher than 250°C. This work showed that, despite the lack of suc-cess during the first half of the twentieth century, there was a basic chance for finding solutions for the direct conversion of the natural fuels' chemical energy to electrical energy.

Rising interest in fuel cells can be attributed to a considerable extent to the world-wide oil crisis that began to be felt in the late 1960s and came to a peak in 1973/1974. Large hopes were placed in fuel cells as a means for raising the efficiency of uti-lization of natural fuel resources. Another important factor stimulating fuel cells was the space race started between the United States and the Union of Soviet Socialist Republics in the early 1960s, and also the beginning of the Cold War between these countries. For these two countries fuel cells had a key role, with the effect of the mil-itary paying the cost of developments beneficial for later civil use. During the period examined in the present chapter almost all basic kinds of fuel cells known at present evolved.

17.3.1 Alkaline Fuel Cells (AFC)

Early in the 1960s, the aircraft and engine manufacturers Pratt and Whitney obtained a license to use Bacon's patents. With the aim of simplifying the initial design and lowering its weight, very high alkali concentrations (85% KOH) were introduced, so that the gas pressure could be lowered significantly. These P&W fuel cells were subsequently used in the Apollo program of space flights to the moon.

Other workers gradually went to less concentrated alkali (30–40% KOH) than found in Bacon's and P&W's batteries. For the space shuttle program, United Technology Corporation (UTC-Power) developed a battery of alkaline fuel cells where 35% KOH immobilized in an asbestos matrix was used as the electrolyte. The electrodes contained a relatively large amount of platinum catalysts, so that at a temperature of 250°C it was possible to work at very high current densities, of up to 1 A/cm^2.

In the laboratories of the agricultural equipment manufacturers Allis-Chalmers, a new version of fuel cell with immobilized alkaline electrolyte solution was developed. The company reequipped one of its tractors to electric traction with an electric motor powered by four batteries consisting of 252 alkaline fuel cells each. The traction was strong enough for a load of 3000 pounds. This tractor was a successful demonstration exhibited on different agricultural fairs in the United States.

In the 1950s already, the Austrian battery specialist Karl Kordesch built a rather efficient zinc air battery with a new type of carbon–air electrode. In later work at the American company Union Carbide, he developed a fuel cell with alkaline electrolyte using multilayer carbon electrodes with a small amount of platinum on the hydrogen side, and with cobalt oxide on the oxygen side. Kordesh put a battery of such fuel cells into his car and was the first person to regularly use an electric car with fuel cells (Kordesch, 1963).

A considerable contribution to the development of alkaline fuel cells was made toward the end of the 1950s by the German physicists Eduard Justi and his coworkers. They made electrodes with nonplatinum catalysts, the so-called Raney-type skeleton metals: nickel for the hydrogen side and silver for the oxygen side (Justi et al. 1954). The catalysts were included into a matrix of carbonyl nickel. These electrodes were named *Doppel-Skelett* (DSK) = double skeleton electrodes (Justi and Winsel, 1962).

17.3.2 Fuel cells with Proton-Exchange Membrane (PEMFC)

Toward the end of the 1950s, several groups of scientists and engineers headed by Grubb and Niedrach (1963) at General Electric worked on the development of fuel cells using a solid ion-exchange (proton-exchange) membrane as the electrolyte. After 1960, this "Grubb–Niedrach fuel cell" was commercialized and used for electric power generation in the Gemini spacecraft during its first flights in the early 1960s. When the flights of Gemini had ended, work on this version of fuel cell practically stopped. In its projects for subsequent space flights, NASA relied on the alkaline fuel cells, known at that time as being more efficient. Work on the proton-exchange membrane fuel cells (PEMFC) was taken up again only in the very late 1980s.

17.3.3 High-Temperature Molten Carbonate Fuel Cells (MCFC)

It is the great advantage of high-temperature fuel cells that not only hydrogen but also hydrocarbons as well as carbon monoxide (a coal processing product) can directly be used as fuel. High-temperature fuel cells are primarily intended for large power plants, in the megawatt range. It is important as well that the heat Q_{exh} evolved during the reaction is high-potential heat, that is, liberated at a high temperature, and can be used to generate additional electric energy in heat engines. In this way the original plant efficiency of 40–50% can be raised by another 10–20%. In the high-temperature molten carbonate fuel cells (MCFC), a molten mixture of sodium, potassium, and lithium carbonates is the electrolyte. The working temperature of these cells is around 650°C. Such an electrolyte had already been used by Emil Baur and his coworkers before 1920.

In about 1958–1960, a large program of studies into these fuel cells was conducted at the University of Amsterdam (Netherlands) by Broers and Ketelaar (Broers and Ketelaar, 1961). A little later, such studies were also initiated at the Institute of High-Temperature Electrochemistry in Yekaterinburg (Russia) (Stepanov, 1972–1974).

Starting in 1960, some work toward building real models of MCFC was done in the Institute of Gas Technology, Chicago, and in a number of companies in the United States, Europe, and Japan. The interest in MCFC gained a significant impetus with the realization that apart from hydrogen, one could use methane and carbon monoxide as fuels. In the fuel cell itself, these fuels undergo an "internal conversion" to hydrogen, making a prior reforming in an external plant unnecessary. In different countries, tens of power plants operating with such cells and having a power output of tens to hundreds of kilowatts were built. The largest plants were demonstration plants of 2 MW in Santa Clara, California, and in Japan.

17.3.4 High-Temperature Solid-Oxide Fuel Cells (SOFC)

Right after the work of Davtyan that we referred to earlier, vigorous studies into solid-oxide fuel cells (SOFC) with electrolytes on the basis of zirconium dioxide doped with oxides of yttrium and other metals were started in many places and particularly at the Institute of High-Temperature Electrochemistry already mentioned (Palguyev and Volchenkova, 1958; Chebotin et al., 1971). The working temperature in these cells was in the range of 800–1000°C.

In 1962, important research and development (R&D) work in this direction was started at Westinghouse. Analogous work was undertaken also in many research and manufacturing establishments in the United States, Europe, and Japan. Operating models with powers of tens of kilowatts were built.

17.3.5 Medium-Temperature Phosphoric-Acid Fuel Cells (PAFC)

The first information on medium-temperature phosphoric-acid fuel cells (PAFC) came out in 1961. Highly concentrated phosphoric acid (85–95%) is the electrolyte in these cells. The working temperature in these cells is in the range of 180–200°C. These fuel cells quickly aroused great interest and found wide distribution. On the basis of such cells, numerous power plants of up to 250 kW were built and used as an autonomous power supply for individual operating units such as hotels and hospitals. Also, in the United States, Japan, and other countries, megawatt plants were built on this basis and used to supply power to entire new city quarters. Compared to the low-temperature hydrogen–oxygen fuel cells, these phosphoric acid cells have the advantage of being able to work with less carefully purified technical hydrogen.

17.3.6 Fuel Cells Using Other Kinds of Fuel

Hydrogen is an electrochemically very active reducing agent (fuel). Thus, in phosphoric acid hydrogen–oxygen fuel cells, relatively high values of current density and specific power per unit weight have been achieved at acceptable values of the discharge or operating voltage. However, hydrogen as a fuel is very complicated in its handling, storage, and transport. This is a problem primarily for relatively small-size, low-power plants of the portable or mobile kind. For such plants, liquid fuels are much more realistic.

In the 1960s, several versions of alkaline fuel cells using liquid hydrazine as a fuel were built.

Hydrazine is also very active electrochemically and yields fuel cells with a high performance. However, apart from its high cost, hydrazine has another large defect: it is highly toxic. Therefore, the uses of hydrazine fuel cells have been limited to a few special areas (chiefly for the military).

Methanol is a much more realistic fuel for fuel cells. The specific energy content of methanol when electrochemically completely oxidized to CO_2 is 0.84 Ah/g. For fuel cells with methanol as a fuel, acidic electrolyte solutions must be used. Alkaline solutions are inappropriate, as the alkali combines with CO_2 produced in the fuel cell to insoluble carbonates. In the early 1960s, first laboratory models of methanol–air fuel cells were built. As large amounts of expensive platinum catalysts were used in these fuel cells, work in this direction soon ended and was not taken up again for many years.

Most types and models of fuel cells that had become known by the 1960s were described in monographs by Justi and Winsel (1962) and Vielstich (1965), as well as in handbooks on fuel cells edited by Young (1960, 1963) and Mitchell (1963).

During the next two decades, some decline of interest in fuel cells can be noticed, and fewer studies appeared in this field. Technical and design improvements were introduced into models of the alkaline fuel cell, molten carbonate, and sold oxide systems, and some power plants of large size were built. The basic structure of the fuel cells themselves (composition of electrodes and electrolyte) and also the specific performance figures (per unit surface area of the electrodes) changed little during this time. To a certain extent, the decline of interest in fuel cells was due to difficulties in commercial realization of earlier achievements. Despite the demonstration that low-temperature electrochemical oxidation of hydrocarbons is basically possible, reaction rates realized in practice were too low, and the amounts of platinum–metal catalyst required to achieve them were so large that the economic prospects of fuel cells using these reactions were very poor. Platinum catalysts were used in most of the fuel-cell types built by different manufacturers, despite the fact that in many studies it had been shown that nonplatinum catalysts could be useful for hydrogen and oxygen electrodes.

For economic reasons, the number of potentially interested users gradually decreased. The financial support of work on fuel cells decreased correspondingly. Work of more academic character concerned with the processes occurring in the different versions of fuel cells still continued through the entire period mentioned. It was the aim of this work to find ways to raise the rate and efficiency of the electrochemical reactions taking place at fuel-cell electrodes, mainly by raising their catalytic activity. These studies examined reactions, both at the oxygen (air) electrode and at the hydrogen electrode (see the monograph of Breiter, 1969). In many countries, the special features of electrochemical oxidation of methanol at platinum catalysts were studied, too.

All these studies as well as the work of Leonard Niedrach and Thomas Grubb, mentioned earlier, concerning the possibility of low-temperature hydrocarbon oxidation, led to the development of a new branch of electrochemical science, "electrocatalysis."

17.4 THE PERIOD AFTER THE 1990s

Approximately in the mid-1980s, apart from the work toward the primordial aim of building powerful plants for a more efficient grid-power generation, a second aim or application of fuel-cell work was discerned, namely, that of building autonomous power sources having intermediate or small capacity. Such power sources are intended for applications where grid energy is inaccessible, such as means of transport, portable devices, remote users, and so on. The expression "fuel cell" began to lose its original meaning of an electrochemical power source operated with natural fuels and acquired the meaning of an electrochemical power source, which in contrast to ordinary batteries works continuously so long as reactants (an oxidizer and a reducing agent—the fuel) are supplied. The first application of this type of work was in the hydrogen–oxygen fuel cells used to power the equipments of Apollo and Gemini manned spacecrafts.

During the 1960s, an American company du Pont de Nemours started to develop a new polymeric ion-exchange membrane marketed as "Nafion®." This had a drastic effect on the further development of fuel cells, particularly for applications of the second type. It was soon obvious that this membrane could help to strongly enhance the characteristics and lifetime of relatively small fuel cells. At the same time, success was achieved toward substantially lower platinum catalyst outlays in such fuel cells. These developments aroused the interest of potential fuel-cell users.

From the mid-1980s, published work on membrane fuel cells rapidly increased. This was driven by several reasons. One of the basic reasons was the air pollution caused by rising automobile population in large cities worldwide. This situation induced legislation in some places, such as in California, requiring the introduction of a certain quota of zero-pollution vehicles. All major car makers worldwide started a serious effort to develop different versions of electric cars and associated power plant based on storage batteries and fuel cells. Another factor contributing to the rising interest in fuel cells was the strong increase in the selection and number of different kinds of portable electronic devices (mobile phones, portable PC, etc.). The need to drastically extend the time of continued operation of these devices and to replace the ordinary batteries by power sources of higher capacity came to the fore.

All these factors led to a very strong increase in the number of studies performed in this area, and one may safely speak about the start of a new, second boom of fuel-cell R&D.

The largest progress was made in the area of membrane fuel cells. Today's membrane fuel cells differ widely from the prototypes of the 1960s in their design and characteristics. Improved polymer electrolyte membrane fuel cells of medium-size power output are widely used and produced on a large commercial scale.

Progress in the area of membrane hydrogen–oxygen fuel cells led to the development of fuel cells of a new type, the direct methanol fuel cells (DMFC) in which methanol is oxidized directly. The more convenient liquid fuel replaces hydrogen as a fuel inconvenient in its handling, storage, and transport. At present, methanol is regarded as a very promising kind of fuel for future electric cars. Two ways of using methanol in them are discussed: (i) its direct oxidation in DMFC, (ii) its prior conversion to technical hydrogen and use of this hydrogen in hydrogen–oxygen PEMFC. Each has its shortcomings and problems. The specific performance figures of DMFC are still too low for their use in an electric car. Power plants with prior conversion need complex, bulky additional equipment. Additional difficulties arise, as technical hydrogen always contains traces of carbon monoxide strongly affecting the platinum catalysts used in PEMFC.

The current development of DMFC is largely based on results obtained in the development of PEMFC. Similar problems must be solved, particularly those associated with the proton-exchange membrane and with the catalysts for the electrodes. In addition, each type has its own specific problems.

Apart from the large volume of research and design work for PEMFC and DMFC, many studies of improved high-temperature fuel cells, SOFC and MCFC, have been conducted since 1990. A marked rise in the number of power plants based on MCFC was seen between 2003 and 2005. The volume of work concerned with alkaline fuel cells has strongly declined as the late 1980s. As to PAFC, the literature of recent years has offered only a few indications of research in this area.

REFERENCES

Bacon FT. Ind Eng Chem 1960;52:301; Young GJ, editor. *Fuel Cells*. Vol. 1. 1960. p 51.

Baur E et al. Z Elektrochem 1910;16:286304; 1912;18:1002; 1916;22:409; 1921;27:409; 1937;43:727; 1938;44:695.

Broers GHJ, Ketelaar JAA. In: Young GJ, editor. *Fuel Cells*. Vol. 1. 1961. p 78.

Chebotin VN, Glumov MV, Palguev SF, Neuimin AD. Elektrokhimiya 1971;7:196.

Davtyan OK. *The problem of Direct Conversion of the Chemical Energy of Fuels into Electrical Energy* [in Russian]. Moscow: Publishing House of the USSR Academy of Sciences; 1947.

Grove W. Phil Mag 1839;14:447.

Grubb WT, Niedrach LW. J Electrochem Soc 1960;107:131.

Grubb WT, Niedrach LW. J Electrochem Soc 1963;110:1086.

Mitchell W. *Fuel Cells*. 1963a. p 253.

Justi E, Scheibe W, Winsel A, inventors. German Patent 1,019,361. 1954.

Kordesch K. Proceedings of IEEE, **51**(5); 1963. p. 806; Mitchell W. *Fuel Cells*. 1963. p 329.

Ostwald W. Z Elektrochem 1894;1:122.

Palguyev SF, Volchenkova ZS. Proceedings of Institute of High-Temperature Electrochemistry, No. 2; 1958. p. 183.

Schmid A. *Die Diffusions-Elektrode*. Stuttgart: Enke; 1923.

Stepanov GK. Proceedings of Institute of High-Temperature Electrochemistry, No. 18; 1972. p. 129; No. 20; 1973. p. 95; No. 21; 1974. p. 88.

MONOGRAPHS AND REVIEWS

Breiter MW. *Electrochemical Processes in Fuel Cells*. Springer-Verlag: New York; 1969.

Gottesfeld S. J Power Sources, 171, 37 (2007).

Justi EW, Winsel A. *Fuel Cells. Kalte Verbrennung*. Wiesbaden: Franz Steiner Verlag GmbH; 1962.

Mitchell W, editor. *Fuel Cells*. New York, London: Academic Press; 1963b.

Perry ML, Fuller TE. J Electrochem Soc 2002;149:859.

Vielstich W. *Brennstoffelemente*. Verlag Chemie: Weinheim, Germany; 1965.

Wand G. *Fuel Cell History*, Part 1: Fuel Cell Today. 2006. Part 2: Fuel Cell Today, April 2006.

Young GJ, editor. *Fuel Cells*. Vol. 1. New York: Reinhold Publishing Corporation; 1960. Volume 2, 1963.

18

PROTON-EXCHANGE MEMBRANE FUEL CELLS (PEMFC*)

The membranes in polymeric proton-exchange membrane fuel cells (PEMFC) serve as a solid electrolyte. The membrane's conductivity comes about because in the presence of water it swells, a process leading to the dissociation of the acidic functional groups and formation of protons free to move about throughout the membrane.

The electrochemical reactions occurring at the electrodes of polymer electrolyte membrane fuel cells, as well as the overall current-producing reaction are the same as in the hydrogen-oxygen fuel cells with liquid acidic electrolyte discussed in Section 16.4 (Eqs. 16.2 and 16.3).

It will become obvious from the following outline that an important fact to remember when looking at polymer membrane is the hydration of protons in the aqueous medium. The protons always exist in the form of species $H^+ \cdot nH_2O$. During discharge of the cell, when current flows, the hydrated protons migrate in the membrane from the anode toward the cathode, each proton dragging along n water molecules in the same direction.

18.1 THE HISTORY OF PEMFC

A first version of polymer electrolyte membrane fuel cells, which had a power of 1 kW, was built in the early 1960s by the General Electric for the Gemini spacecraft.

*The abbreviation PEMFC stands also for the equivalent term of polymer electrolyte membrane fuel cell.

Electrochemical Power Sources: Batteries, Fuel Cells, and Supercapacitors, First Edition.
Vladimir S. Bagotsky, Alexander M. Skundin, and Yurij M. Volfkovich.
© 2015 John Wiley & Sons, Inc. Published 2015 by John Wiley & Sons, Inc.

A sulfated polystyrene ion-exchange membrane was used as the electrolyte in these fuel cells. The electrodes contained about 4 mg/cm^2 of a platinum catalyst. Because of the marked ohmic resistance of the membrane, the current density was below 100 mA/cm^2, with a voltage of about 0.6 V for an individual cell. This corresponds to a specific power of the fuel cell of about 60 mW/cm^2. Because of the insufficient chemical stability of the membrane used, the total lifetime of the battery was below 2000 h. The high cost of such a battery excluded uses in fields other than space flight.

After a dormant period lasting almost three decades, striking improvements in polymer electrolyte membrane fuel cells' properties were achieved after 1990. The specific power of their current prototypes went up to 600–800 mW/cm^2 while less than 0.4 mg/cm^2 of platinum catalysts were used, and the lifetime is now several tens of thousands of hours. On the basis of the strength of these achievements, Costamagna and Srinivasan were able to title their comprehensive analytical review of 2001: "Quantum jumps in polymer electrolyte membrane fuel cell science and technology from the 1960s to the year 2000." This breakthrough, as well as the need to develop zero-emission vehicles after enactment of the corresponding laws in California in 1993, led to a brisk pace of scientific and engineering work on polymer electrolyte membrane fuel cells.

Basically a combination of three factors was responsible for this important progress:

(a) introduction of the new Nafion®-type proton-exchange membranes,
(b) the development of new methods yielding a drastic efficiency increase of platinum catalyst utilization in the electrodes,
(c) the development of unique membrane-electrode assemblies (MEAs).

18.1.1   The Nafion Membranes

Between 1960 and 1980, American chemical company du Pont de Nemours began the delivery of ion-exchange membranes of a new kind tradenamed Nafion. They are made of a perfluorinated sulfonic acid polymer (PSAP) consisting of a continuous skeleton of $-(CF_2)_n-$ groups to which a certain number of hydrophilic segments containing sulfonic acid groups $-SO_3H$ are attached. The hydrophobic skeleton is responsible for the very high chemical stability, many times exceeding that of membranes known previously. When the membrane is appropriately wetted, the sulfonic acid groups dissociate and at a temperature of 80°C provide a relatively high protonic conductivity of the membrane (about 0.1 S/cm), three to four times higher than for previously used membranes.

Initially, Nafion-type membranes had been developed for the needs of the chlorine industry. Chlorine electrolyzers with such membranes were widely introduced. Apart from chlorine, they yield alkali that is very pure as the second major electrolysis product. Soon after the advent of these membranes, work on fuel cells that would include them started. Fuel cells with such membranes had lifetimes two to three orders of magnitude longer.

At high current densities, the ohmic resistance of the membrane still affects the properties of the fuel cell. For improved performance, thinner membranes of Nafion 112, 50 μm thick, were introduced in the place of the standard Nafion 115 membranes (100 μm). Even thinner membranes have been developed (Nafion 111, 25 μm), but fuel cells with such membranes occasionally failed because of small membrane defects allowing the gases to mix.

Fuel cells with Nafion-type membranes usually are operated at a temperature of 80–90°C. A sufficient water content of the membranes must be secured by continued moistening, which is done by saturating the reactant gases supplied to the cell with water vapor, by passing them through water having a temperature somewhat higher (by 5–10°C) than the fuel cell's operating temperature. The optimum moistening conditions depend on membrane thickness and on the current drawn from the fuel cell.

A general disadvantage of all membranes of this type is the complicated manufacturing process and associated high cost. A Nafion membrane costs about 700 USD/m^2 or 120 USD/kW of designated power (at an average power density of 600 mW/cm^2).

18.1.2 More Efficient Utilization of the Platinum Catalysts

Platinum is the basic catalyst for the electrochemical reactions at the hydrogen and oxygen electrodes of polymer electrolyte membrane fuel cells. In the first models of such fuel cells, the dispersed platinum catalyst was used in a pure form, so that large amounts of platinum were applied. A considerably higher utilization efficiency and at the same time much lower metal input were attained when platinum was deposited onto highly dispersed carbon supports. The best results were attained with Vulcan XC-72 furnace carbon black as support, possibly, because of small amounts of sulfur contained in it that gives rise to special surface properties. At low platinum contents, its specific surface area on this support can be as high as 100 m^2/g (with a size of 0.3 nm of the platinum crystallites), while for unsupported dispersed platinum the specific surface area usually is not larger than 15 m^2/g (with a size of 1–2 nm of platinum crystallites). At low platinum contents in the carbon black, thicker catalyst layers must be used, which interferes with reactant transport to the reaction sites. The degree of dispersion of the platinum and its catalytic activity depend not only on the amount deposited on the support but also on the method of chemical or electrochemical deposition of platinum onto the support.

18.1.3 Membrane-Electrode Assemblies (MEAs)

In polymer electrolyte membrane fuel cells, like in many other kinds of fuel cells, gas-diffusion electrodes are used. They consist of a porous, hydrophobic gas-diffusion layer (GDL) and of a catalytically active layer. The diffusion layers (often called *backing layers*) usually consist of a mixture of carbon black and about 35% by mass of polytetrafluoroethylene (PTFE) applied to a conducting base (most often a thin graphitized cloth). The GDLs yield a uniform supply of reactant gas to all segments of the active layer. The porosity of the diffusion layer is 45–60%. This high porosity is particularly important when air oxygen rather than pure oxygen is used as the oxidizing agent.

The electrode reactions in polymer electrolyte membrane fuel cells proceed within the active layer along a highly developed catalyst-electrolyte-gas three-phase boundary. The active layer is supported either by a special support (carbon cloth or carbon paper made hydrophobic) or by the membrane itself.

An appreciable decrease in relative platinum input without any departure from the high electric parameters of polymer electrolyte membrane fuel cells was attained when adding a certain amount (30–40% by mass) of a proton-conducting ionomer to the catalytically active layer. To this end, an ink of platinized carbon black and PSAP ionomer (or a solution of low-equivalent-weight Nafion in alcohol) is homogenized by sonication and applied to the diffusion layer, the solvent is evaporated, and the whole mixture is treated at a temperature of about 100°C. The ionomer added to the active layer leads to a considerable increase in the contact area between catalyst and electrolyte (which, in the case of solid electrolytes, is actually quite small). For further improvement of the catalyst-membrane contact, the membrane is sandwiched between the electrodes, with the active layers facing the membrane, and subjected to hot pressing at a temperature of 130–155°C, pressures up to 100 bar, and pressing times of 1–5 min. This operation yields unique MEAs.

The basic development work for MEAs of the current type was done in the United States by Gottesfeld and Zawodzinski (1997) at Los Alamos National Laboratory and by Srinivasan et al. (1991) at the Center for Electrochemical Systems and Hydrogen Research, Texas A&M University. With all these developments, it was possible, without sacrificing performance, to lower the use of platinum metals toward the mid-1990s to 0.4 mg/cm^2 (one order of magnitude less than in the first models), and by 2001 even further to 0.05 mg/cm^2 for the hydrogen electrode, and to 0.1 mg/cm^2 for the oxygen electrode.

By the late 1990s, the US company E-Tek (now affiliated to BASF) started the commercial production of several versions of complete MEA used by other companies to manufacture different kinds of fuel-cell stacks.

18.2   STANDARD PEMFC VERSION OF THE 1990s

Membrane-type hydrogen–oxygen fuel cells underwent numerous changes in design and manufacturing technology in the course of their development. In the present section, we describe the structure of a polymer electrolyte membrane fuel cell version that had become kind of a standard in the mid-1990s, and will further be called the *standard version of the 1990s.*

18.2.1   Individual PEMFC

Figure 18.1 is a schematic drawing of the structure of an individual polymer electrolyte membrane fuel cell. Each cell is separated from a neighboring cell by an

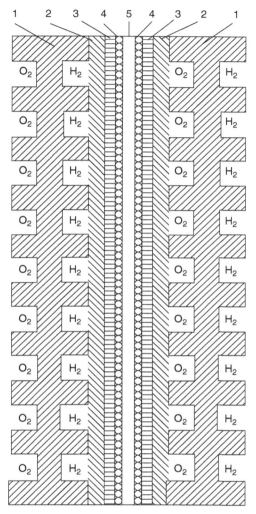

Figure 18.1. Schematic of a single polymer electrolyte fuel cell, (1) bipolar plates; (2) current collectors; (3) gas-diffusion layers; (4) catalytic layers; (5) membrane.

electronically conducting bipolar plate. This plate is in contact on one side with the positive electrode of a given cell and on the other side with the negative electrode of the neighboring cell. Such plates, thus, function as interconnectors for cells connected in series in a battery. The plate has channels cut out on either side that via manifolds are supplied with the reactant gases (hydrogen and oxygen). The gases are carried to the electrodes by these channels. Water vapor as a reaction product is eliminated via such channels from the oxygen side. For uniform gas supply, the channels are laid out in spirals or serpentines yielding a ramified flow field.

Between any two bipolar plates providing reactant access, a multilayer MEA is set up that consists of a positive electrode and a negative electrode pressed into the two sides of a proton-conducting membrane.

For reasons that are discussed in Section 19.4, the catalyst for the hydrogen electrode in polymer electrolyte membrane fuel cells is a mixed platinum–ruthenium catalyst applied to carbon black, rather than pure platinum. The overall thickness of modern MEA is about 0.5–0.6 mm (of which ~0.1 mm for the membrane, for each of the two GDLs, and for each of the two active layers). The bipolar plates have a thickness of about 1.5 mm, the channels on both sides having a depth of about 0.5 mm.

18.2.2   PEMFC Stack (Battery)

The discharge voltage of an individual polymer electrolyte membrane fuel cell is low, between 0.65 and 0.75 V, depending on the current density. For a stack to have a given voltage, therefore, a certain number of cells are connected in series. For a 30-V stack, for example, about 40 individual cells must be put together. As the polymer electrolyte membrane fuel cells have a flat configuration, a battery of filter-press design is built. The required number of cells is clamped as a block between two thick end plates also serving as the terminals for drawing the current. The end plates must be sufficiently massive in order not to become deformed when clamping the cells together and provide a compressing force that is uniform over the entire cell surface.

The membrane in MEA, as a rule, extends about 2 cm beyond the active layers of the electrodes on all sides of the assembly. This part of the membrane serves as a peripheral seal of the individual cells clamped together into the battery stack. The seal is needed to prevent the mixing of the reactant gases, hydrogen and oxygen. Additional O-rings may be provided when needed.

In addition to the individual polymer electrolyte membrane fuel cells and their bipolar plates, special heat-exchanger plates must also be included in the battery stack. Cooling fluid is circulated through these plates in order to eliminate the heat produced during battery operation. At least one such plate must be provided for any two cells when the battery is to be operated with high current densities. These plates could also be used to warm up the battery for a cold start-up.

18.3   OPERATING CONDITIONS OF PEMFC

Modern polymer electrolyte membrane fuel cell stacks are basically intended for high energy densities at the electrodes (up to 0.6 W/cm^2). For this reason, and also because of the compact design, the maximum values of the stacks' specific power per unit volume and weight are higher for them, than for all other batteries of conventional type. Often, polymer electrolyte membrane fuel cells are used as well for operation at lower energy densities.

The working temperature of a polymer electrolyte membrane fuel cell is about 80°C. The reactant gases are supplied to the battery at pressures of 2–5 bar. This relatively high pressure is needed because, in order to prevent drying out of the membrane, the gases entering the cells must be presaturated with water vapor, so that the resulting partial pressure of the reactant gas in the gas–vapor mixture is lower. For work at high current densities, however, a sufficiently high partial pressure must be maintained. This is attained by raising the overall pressure of the mixture.

The thermodynamic electromotive force of a polymer electrolyte membrane fuel cell at a temperature of 25°C is given by $\varepsilon = 1.229$ V. The open-circuit voltage (OCV) of a hydrogen–oxygen polymer electrolyte membrane fuel cell has values between 0.95 and 1.02 V, depending on the temperature and gas pressures.

Hydrogen–oxygen polymer electrolyte membrane fuel cells containing the new MEAs, are convenient insofar as they operate highly efficiently (with a relatively low reactant consumption) over a wide range of discharge current densities. In many applications, current densities of $0.8–1.0$ A/cm^2 can be drawn at cell voltages U_i of about 0.75 V, corresponding to power densities of $0.6–0.75$ W/cm^2. Under these conditions, the total voltage loss of $\varepsilon - U_i$ amounts to about 0.48 V. Of this loss, about 0.4 V are because of polarization of the oxygen electrode, about 20 mV are because of polarization of the hydrogen electrode, the remainder is because of ohmic resistances within the MEA.

18.4   SPECIAL FEATURES OF PEMFC OPERATION

18.4.1   Water Management

In polymer electrolyte membrane fuel cell operation, water is formed as a reaction product at the positive (oxygen) electrode. It can be seen from Equation 16.8 that water reaches the side of the membrane that is in contact with the positive electrode also by being dragged along in the form of hydrated protons. As a result of Equation 16.7, on the other hand, water leaves the side of the membrane that is in contact with the hydrogen electrode. In part, this unilateral membrane water transport is compensated by the back diffusion of water on account of the concentration gradient that develops, but it is not completely compensated. Asymmetry of the hydraulic pressure on both sides of the membrane may also contribute to membrane water transport. All this leads to excess water at the oxygen electrode during fuel cell operation, while the water content of the membrane next to the hydrogen electrode decreases. Both of these effects have negative consequences for fuel-cell operation. Excess water at the positive electrode may lead to flooding of the active layers' pores through which oxygen reaches the catalyst. This will affect the performance of the positive electrode and may in the end lead to a complete cessation of oxygen access to the active layer, and thus, to a cessation of fuel-cell operation. Dehydration of the membrane next to the hydrogen electrode, on the other hand, raises the ohmic resistance and leads to a lower discharge voltage of the fuel cell. To avoid these situations, usually hydrogen that is supplied is saturated with water vapor and oxygen is circulated. Passing next to the electrode, oxygen becomes saturated with water vapor. It then reaches a chamber with a lower temperature, where it condenses, and dry oxygen is returned to the electrodes. Sometimes, a periodic release of excess oxygen saturated with water vapor is practiced. The liquid water that is produced may be used for other purposes, such as drinking water.

All these processes must be controlled quite carefully. If water withdrawal is too fast, then there is a risk of water loss from the swollen membrane, which not only leads to a drastic raise in resistance but also to fragilization. Cracks may then develop across which the gases may mix, yielding an explosive mixture, with all the catastrophic consequences that ensue.

A large problem in polymer electrolyte membrane fuel cell operation is a possible partial condensation of water vapor when temperature gradients are present in the fuel cell and a dual-phase water system develops.

The liquid water forming within the metal-electrode assemblies or in the channels of the bipolar plates interferes with the access of reactant gases to the catalytically active layer, thus causing an additional polarization of the electrodes and a drop in the fuel cell's discharge voltage. For this reason, selecting optimum operating parameters for the system of water withdrawal from operating low-temperature fuel cells is a very important condition for stable operation of polymer electrolyte membrane fuel cells. Many studies have been done on this topic. The problem is complicated by the fact that the rate of water formation by Reaction 16.4 is directly proportional to the current. Most often, fuel cells operate under conditions of constantly changing discharge currents (dynamic operating mode). Yet the parameters of the systems for water withdrawal (such as the rate of oxygen supply and circulation) are usually conditioned to a particular value of the water formation rate. As a result, under real operating conditions of the fuel cell, transient situations may arise, where temporarily the electrodes are flooded or the membrane partly dries out. Optimum water management is a central concern of polymer electrolyte membrane fuel cell research and development (R&D) up to the present.

18.4.2   Heat Management

In the high-power-density operation of polymer electrolyte membrane fuel cells, large amounts of heat are generated. It follows from what was said in Section 1.3.2 that at a discharge voltage of 0.75 V, the thermal energy generated is approximately just as large as the electrical energy generated. At lower voltages of 0.6 V, for instance, the thermal energy generated is half as much as the electrical energy generated. Under these conditions, an efficient heat management is possible only by cooling with a liquid heat-transfer medium (water or other fluid). For efficient heat transfer, the temperature of the medium must be at least 10°C lower than the battery's operating temperature. The heat-exchange cooling plates, mentioned above, usually are designed like the usual cell units with a metallic membrane, but without electrodes. The gases supplied pass along one side of the membrane and the heat-transfer fluid passes along the other side of the membrane. After leaving these plates, the heat-transfer fluid reaches the definitive heat exchanger, usually equipped with cooling ribs radiating excess heat to the surroundings. The hot heat-transfer fluid can also be used for heating purposes (heat cogeneration).

18.4.3   Partial Pressure of the Reactant Gases

An important problem for polymer electrolyte membrane fuel cells is that of replacing pure oxygen (requiring special equipment for storage and transport) by oxygen from the ambient air. The partial pressure of oxygen in the air is low (about 0.2 bar). With passive air supply (air-breathing electrodes), where the air is not precompressed or pumped, the performance drop is rather significant. In this case, difficulties also arise with the elimination of water accumulating at the cathode. These difficulties can be overcome in part when using active air supply, by compressing the air to a few bars and circulating it. Air breathing is rarely used, not only in polymer electrolyte membrane fuel cells but also in other fuel-cell plants of relatively high power. For low-power fuel cells intended for power supply to portable equipment, passive air supply is the only possibility for the attainment of sufficiently low weight and volume.

18.4.4   Start-Up of a Cold Battery

Major problems arise during the start-up of a cold battery, which, for instance, has been warehoused or had a long idle period. When the battery's temperature is higher than the freezing point of water, then slow heating by drawing current through an external circuit of low or zero resistance is feasible. When its temperature is still lower, then ice incrustations may have formed in the gas-diffusion and catalyst layers from condensed and frozen water, partly or completely blocking gas access to the catalyst. Freezing water, increasing in volume, may destroy the structure of the electrode. Prior to shutting down a hot operating battery, therefore, one must use a dry gas to remove all residues of water vapor. In this dried state, a fuel cell will withstand tens of freeze-thaw cycles without any marked performance loss.

18.5   PLATINUM CATALYST POISONING BY TRACES OF CO IN THE HYDROGEN

For fuel-cell operation, most often technical hydrogen obtained by the conversion of primary fuels such as methanol or petroleum products is used, rather than pure hydrogen obtained by electrolysis. Technical hydrogen always contains carbon monoxide and a number of other impurities, even after an initial purification. In the first experiments conducted in the mid-1980s it was shown that traces of CO in hydrogen used for the operation of fuel cells with phosphoric acid electrolyte lead to a marked increase in the hydrogen electrode polarization.

An increase in polarization is noticeable already in the presence of traces of carbon monoxide (less than 10 ppm). With 25 ppm of carbon monoxide and a current density of $600 \, mA/cm^2$, the polarization of the electrode increases by 0.2–0.3 V, implying a loss of about 30–40% of electrical power. A more thorough elimination of carbon monoxide from hydrogen is difficult to attain. This electrode polarization is because of the fact that the platinum catalyst is a good adsorbent for carbon monoxide. On carbon monoxide adsorption, the fraction of the surface that is available for the adsorption of hydrogen and its subsequent electrochemical oxidation decreases drastically (the catalyst is "poisoned" by the catalytic "poison"). At least 10% of the surface of platinum must remain accessible for hydrogen if its polarization is to be kept within reasonable limits (10–20 mV), so that the degree of surface coverage by adsorbed carbon monoxide should not be higher than 80–90%.

18.5.1   The Use of Platinum–Ruthenium Catalysts

The most reliable and promising way of fighting the poisoning of the platinum catalyst by carbon monoxide impurities in the hydrogen is by modifying the catalyst itself, that is, by adding alloying elements. It was found in the late 1960s when studying the anodic oxidation of methanol at platinum catalysts that the reaction is much faster at mixed platinum–ruthenium catalysts than at pure platinum catalysts. It could be shown in a number of studies (e.g. Schmidt et al. (2002)) that in polymer electrolyte membrane fuel cells such platinum–ruthenium catalysts are appreciably less sensitive toward carbon monoxide poisoning than pure platinum. The reasons are not exactly clear. Possibly, the energy of adsorption of carbon monoxide decreases owing to changes in crystal lattice structure of the alloy relative to pure platinum (or changes in its electronic state). It is also possible that excess oxygen adsorbed on ruthenium sites oxidizes the adsorbed carbon monoxide species, thus eliminating them from the platinum surface. Many studies have been performed in recent years to elucidate the operating mechanism of Pt–Ru catalysts for hydrogen oxidation in the presence of carbon monoxide. It could be shown in particular that the effectiveness of the Pt–Ru catalyst depends on the moisture content of technical hydrogen, supplied to the anode.

At present, anodes with platinum–ruthenium catalyst (with about 50 atom% of ruthenium) are used in the majority of polymer electrolyte membrane fuel cells, designed to work with technical hydrogen. It, however, follows from experiments with pure hydrogen (free of carbon monoxide traces) that the catalytic activity of the mixed platinum–ruthenium catalyst is somewhat lower than that of pure platinum. In this connection, it is worth noting the suggestion of Chinese workers to use electrodes having a special structure with two catalytically active layers. The first layer closer to the diffusion layer from which the gas is supplied contains the platinum–ruthenium catalyst at which carbon monoxide is oxidized. It thus acts like a filter that does not pass this gas. The hydrogen that has been freed of carbon monoxide reaches the following layer containing the more active, pure platinum catalyst.

18.5.2   Higher Working Temperatures

Still another way of fighting hydrogen catalyst poisoning by carbon monoxide impurities is that of raising the operating temperature.

Raising the temperature, the adsorption equilibrium between hydrogen and carbon monoxide, jointly adsorbing on platinum, shifts in favor of hydrogen adsorption. This raises the highest admissible threshold concentration of carbon monoxide. The effect could be seen in fuel cells with phosphoric acid electrolyte, which work at temperatures of about 180–200°C and admit carbon monoxide concentrations in hydrogen as high as 100 ppm, despite the fact that platinum catalysts are used.

Raising the temperature of polymer electrolyte membrane fuel cells, beyond 100°C, brings a number of difficulties. Works addressing these problems are described in Section 18.8.

18.6 COMMERCIAL ACTIVITIES IN RELATION TO PEMFC

At present, polymer electrolyte membrane fuel cells and power plants based on such fuel cells are produced on commercial scale by a number of companies in many countries. As a rule, the standard battery version of the 1990s is used in these batteries, though in certain cases different ways of eliminating water and regulating the water balance (water management) have been adopted.

A Canadian company Ballard Power Systems started research in polymer electrolyte membrane fuel cell manufacturing technology in 1989, and between 1992 and in 1994 delivered a few prototypes of power units in different sizes. The first commercial unit having a power of 1.2 kW was made in 2001. At present, this company produces different kinds of power units between 4 and 21 kW for different applications: electric cars, power back-up, and plants for heat and power cogeneration—combined heat and power (CHP) Systems. In these plants, water is eliminated by supplying to the anodes hydrogen with a low water vapor content, so that in the membrane, water is transferred by diffusion from the cathode to the anode. In this way oxygen circulation along the cathode surface can be lowered. In 2006, Ballard reported delivery of the new power unit Mark 1020 ACS applying for the first time air cooling and a simplified membrane humidification system.

The American UTC Power company (a United Technologies Company affiliate) produces power units with polymer electrolyte membrane fuel cells for different military and civil applications. In 2002, a regular electric bus service using fuel-cell batteries developed by this company was started. The Pure Cell™ model 200M Power Solution power plant delivers 200 kW of electric power and about 900 btu/h (about 950 kJ/h) of thermal power. An American company, Plug Power, founded in 1997, delivers emergency power plants since 2000 on the basis of polymer electrolyte fuel cells providing uninterruptible power supply for hospitals and other vitally important objects in cases of sudden losses of grid power.

At present, apart from the United States, polymer electrolyte membrane fuel cells and power plants on their basis are developed in many other countries including China, France, Germany, South Korea, and the United Kingdom, and so on. The major part of the power plants delivered in 2006 (about 60%) was for power supply to portable equipment. The second place (about 26%) was taken by small stationary power plants for uninterruptible power supply.

Approximately 75% of the work on polymer electrolyte membrane fuel cells is conducted in industrial organizations, the remaining 25% in academic and government organizations. This proves that the initial research and engineering stage has already been terminated and the commercial development of these fuel cells is under way. A detailed analysis of the different ways, to make polymer electrolyte membrane fuel cells, can be found in the review of Mehta and Cooper (2003).

18.7   FUTURE DEVELOPMENT OF PEMFCs

At the time of writing this book (in 2013), polymer electrolyte membrane fuel cells and power plants built on their basis have attained a high degree of perfection. They work reliably, exhibit rather good electrical characteristics, are convenient to handle. Such plants have found practical applications in different areas, for example, for securing uninterruptible power supply in the case of grid breakdowns to strategically important objects (hospitals, command stations, water works, gas suppliers, telecommunication centers, etc.). They are also used for the combined power and heat supply in individual residences and office buildings. In a number of places, regular bus service with electric traction provided by such power plants has been organized.

Yet these uses are not on a mass scale, and the commercial success, deriving from the production of such power plants, is still very limited. A wider use of this kind of fuel cells can only be expected, when they have conquered two new areas of application, light electric vehicles, and portable electronic equipment. For success in this direction, a number of important and rather complex problems must be solved first:

(a) longer lifetime of the power plants and a higher stability of the catalysts and membranes associated with this longer lifetime;

(b) a lower cost of production, both of the polymer electrolyte membrane fuel cells as such and of the entire power plant, and the development of catalysts without platinum and of cheaper membranes associated with this lower cost;

(c) a higher tolerance of polymer electrolyte membrane fuel cells for carbon monoxide impurities in the hydrogen, particularly by developing versions of these fuel cells operating at higher temperatures (Section 18.8);

As power supply for a variety of portable devices is one of the more important future applications of polymer electrolyte membrane fuel cells, great efforts are made at present to reduce the dimensions and weight, and to even miniaturize both the fuel-cell stack and all auxiliary equipment needed for a power plant.

18.7.1   Longer Lifetime

There are a few different aspects to the concept of lifetime. They cover values of the following parameters of a power plant with fuel cells that have either been attained or are guaranteed or expected:

1. the time of smooth, uninterrupted operation in a given operating mode;
2. the number of (admissible) on−off cycles;
3. the number of (admissible) temperature cycles between ambient and operating temperatures and back.

By definition, fuel cells should work without interruption so long as reactants are supplied and reaction products eliminated. Actually almost all varieties of fuel cells exhibit some time-dependent decline of their characteristics during long-term discharge. When operating at constant current, for instance, the voltage will gradually decrease. The rate of voltage decrease of an individual fuel cell is stated in microwatt per hour or, where currents vary, as a function of time in microwatt per ampere-hour.

Figures for the time required for a smooth operation of polymer electrolyte membrane fuel cells (and other fuel cells used in the same applications) are given variously as 2000–3000 h for the power plants in portable devices, as up to 3000 h over a period of 5–6 years for the power plants in electric cars, and as 5–10 years for stationary power plants. Much time will, of course, be required to collect statistical data for the potential lifetime of different kinds of fuel cells. Research efforts, therefore, concentrate on finding the reasons for the gradual decline of performance indicators and for premature failure of fuel cells. In recent years, many studies have been conducted in this area.

The decline of polymer electrolyte membrane fuel cells parameters may be reversible or irreversible. When it is reversible, then better performance can be reestablished by the operators, for instance, by changing over to a lower operating power. When it is irreversible, then there is no way for bringing them back to the original level. A reversible voltage drop in most cases will be caused by an upset water balance within the MEA, leading to flooding of the catalytically active layer of the cathode and/or some dehydration of the membrane layer next to the anode. Flooding of cathodic layers leads to hindrance of oxygen transport to the catalytically active sites and decrease in its concentration (partial pressure) at these sites, an effect described as oxygen transport difficulties or oxygen concentration polarization. Membrane dehydration causes higher ohmic resistance of the membrane. When changing to lower power, often the original optimum distribution of water within the MEA is reestablished.

Recrystallization of the Catalyst

Irreversible changes have many reasons. A basic reason is recrystallization of the very highly disperse platinum catalyst on its carbon support. At the positive potentials of the oxygen electrode, platinum undergoes perceptible dissolution producing platinum ions. This has two consequences: (i) part of these ions settle in the membrane, so that the total amount of catalyst in the catalytic layer decreases; (ii) most of them diffuse to nearby platinum grains, deposit on them, and cause them to grow larger. This leads to a smaller true surface area of the catalyst and higher true current densities (higher fraction of the discharge current per unit of true surface area). Polarization of the electrode increases in response to the higher true current density, and the discharge voltage of the fuel cell as a whole decreases. At the potentials existing at the hydrogen electrode, the rate of platinum dissolution is much lower, yet, on account of solid-phase reactions, the highly disperse platinum present in the catalytically active layer of this electrode may also become coarser by recrystallization, the true surface area, thus, becoming smaller.

Corrosion of the Carbonaceous Support

Another reason for higher polarization of the oxygen electrode is corrosion (oxidation) of the carbon material serving as support of the platinum catalyst. This causes loss of catalyst's contact with the support (Cai et al., 2006). That is a de facto exclusion of part of the catalyst. The loss of Pt–C bonding also favors recrystallization of the platinum catalyst.

Degradation of the Nafion Membrane

Membranes of the Nafion type are chemically highly stable, yet under the conditions existing in an operating fuel cell, slow degradation is seen. In part, this is because of oxygen-containing free radicals forming during side reactions at the oxygen electrode. These radicals are strong oxidizers and may attack the membrane, the attending degradation leading to higher membrane resistance and sometimes even brittleness and mechanical defects.

A shortened lifetime may be caused, not only by changes in the MEA (its catalytically active layers or the membrane), but also by problems arising in other components of the fuel cell. Bipolar graphite plates have some porosity, and a perceptible permeation of gases through them gives rise not only to irreproducible reactant losses but also to certain irreversible changes. Bipolar metal plates are subject to corrosion, and heavy-metal ions produced by corrosion markedly depress the activity of the catalysts, when depositing on them. The sealing materials may also be the reason for the degradation of the fuel cells—it has, in fact, been known for a long time that polymer oxidation and decomposition products affect the active catalysts. A marked decrease in catalyst activity was noticed when seals made of organic silicon material were used and traces of silicon deposited on the catalyst surface.

The influence of a variety of contaminants in reactants and in the fuel cells themselves on the polymer electrolyte membrane fuel cell lifetime, as well as the mechanisms of this influence, have been examined in a review by Cheng et al. (2007).

Nonuniformity of Current Distribution

Nonuniformity in the distribution of the current-producing reaction over the electrode surface has a large effect on the degradation rate of the fuel cell's performance. Such nonuniformity may have a technological origin (such as fluctuations in the thickness of individual MEA components), or arise during fuel-cell operation. The latter include local formation and accumulation of water drops in the channels of the bipolar plates, causing a nonuniform supply of reactant gases across the electrode surface.

A rather dangerous situation arises when individual fuel cells of a multicell fuel-cell stack deviate. Such nonuniformity is most often because of problems in reactant supply. Two systems of gas supply exist: parallel and series. In parallel supply, the gas reaches each fuel cell through a narrow channel coming from a common manifold. The pressure in these channels is the same for all elements where they leave the manifold, but on account of differences in gas flow resistance, the amounts of gas (or the pressure) reaching each fuel cell may differ. In series supply, gas is fed to a first individual cell, flows through it, and continues to the next cell, and so on. In each fuel cell in series the amount of gas needed for the reaction is consumed leaving a lower pressure of reactant gas for the next cell. The fuel cells thus operate at different working pressures of the gas, which constitutes nonuniformity.

Most often parallel gas supply is practiced. In fact, for the oxygen that should be circulated for the purposes of water elimination, this is the only possibility. If in a battery working with an external load, one of the fuel cells is in short supply with one of the gases, the corresponding electrode reaction will cease, but the closed circuit of the load and all series-connected cells implies that current is forced through the affected fuel cell. Assuming that hydrogen supply had stopped, then at the hydrogen electrode, anodic hydrogen oxidation is no longer possible and anodic oxygen evolution will instead be caused by the current. When oxygen supply has stopped, then, analogously, oxygen reduction at the oxygen electrode will be replaced by hydrogen evolution. De facto, the deviating fuel cell is forced into a mode of electrolysis by the normal fuel cells ahead of it and behind it. As a result, this fuel cell, instead of contributing its voltage (of, e.g., 0.65 V) to the stack voltage, draws voltage equivalent to a "negative contribution" of about 2.0 V ("cell reversal"). This polarity inversion produces not only a marked voltage drop of the stack, but also irreversible changes in the membrane-electrolyte assembly of the affected fuel cell. More serious consequences include the formation of an explosive gas in the system.

18.7.2   Cost Estimation for PEMFC

At this time a generally accepted way of estimating the cost of items such as fuel cells that just now start to go into industrial mass production does not exist. The sales price of fuel cell power plants produced today is usually not made public and is determined by an arrangement between seller and buyer. The sales price of a new item usually depends not only on the manufacturing cost itself, but also on additional charges for research, development, and special design features. It is only after the start of mass production that the manufacturing price becomes determining for the price of a new item.

Few data on this topic can be found in the literature on polymer electrolyte membrane fuel cells. In addition, these data often are contradictory. The first data became known in 2002. That is, about a decade after the establishment of the "standard version of the 1990s." Schmidt et al. estimated the cost of the first polymer electrolyte membrane fuel cell as 20,000 $/kW (here and below, all prices are in US dollars). Of this figure, 90% was for labor, considering that these cells were handmade, and only 10% was for materials. In mass production, for instance, when making one million 50-kW power plants per year, labor cost could fall to 10 $/kW. In the opinion of the authors, it would be necessary to develop new membranes costing no more than 20 $/m^2 and lower the platinum content of the electrodes to 0.25 mg/cm^2, in order to bring the materials' cost down from 2000 $/kW to a desirable figure of 30 $/kW, or at least to a temporarily acceptable value of 100–200 $/kW (also, platinum should be recovered and recycled). A very important and difficult problem is that of making cheaper corrosion-resistant metallic bipolar plates. All auxiliary devices also must become much cheaper.

A detailed cost analysis for a polymer electrolyte membrane fuel cell power plant of 5 kW was provided in 2006 by Kamarudin et al. According to their data, the total cost of such a plant will be about $1200 of which $500 is for the actual fuel-cell stack and $700 for the auxiliary equipment (pumps, heat exchangers, etc.). The cost of the fuel-cell stack is derived from the components as 55 $/kW for the membranes, 52 $/kW for the platinum, 128 $/kW for the electrodes, and 148 $/kW for the bipolar plates.

By comparison, the cost of an internal combustion engine of the same power is between 500 and 1500 $/kW according to data from different authors, so that, in this respect, a power plant with fuel cells is hardly inferior to a combustion engine.

Other than monetary costs, one must also take into account the availability of raw materials needed for fuel-cell production. Assuming a total platinum content in both electrodes of 0.8 mg/cm^2 and an optimistic value of 1 W/cm^2 for the specific power, then one will need 0.8 g platinum/kW. With a price of platinum of 30 $/g, this gives 24 $/kW. Therefore, an electric car with a power of 50 kW would have a price tag of $1200 for only the platinum in the fuel cells. It must of course be taken into account here that with a production volume of 1 million electric cars per year, 40 tons of platinum metal were needed, representing about 20% of current world production. An even more difficult situation would arise in the mass production of fuel cells with mixed platinum–ruthenium catalysts. World reserves of ruthenium are very limited and would not admit mass production on such a scale. For this reason, it will be one of the prime tasks in further fuel-cell development to broadly search for ways to lower the platinum metal content of the catalytic layer and to find new nonplatinum catalysts.

18.7.3 Bipolar Plates

Bipolar plates are a very important component of fuel-cell batteries. They largely determine the efficiency of the battery and its possible lifetime. The bipolar plates take up considerably more than half of the total battery volume (sometimes as much as 80%), and have the corresponding share of the battery's weight. The cost of the bipolar plates is up to15% of the battery's total cost.

Bipolar plates have a number of functions in fuel-cell batteries. Mechanically, they are the backbone of the individual fuel cells and of the battery as a whole. They provide the electrical contact between individual fuel cells in the stack and channel the reactant supply to the entire working surface area of the electrodes. The plates must meet a number of requirements in order to fill all these functions: They must (i) be sufficiently sturdy, (ii) be electronically conducting, (iii) have a low surface resistance in contact with other conductors, (iv) be impermeable to gases, (v) be corrosion resistant under the operating conditions of the fuel cells.

Materials that can be considered for making bipolar plates are graphite, various carbon-polymer composites, and various metals. Graphite and the composites are sufficiently highly conducting and have a low surface resistance, but are not sufficiently strong; they are brittle and poorly withstand shock and vibration. Machining of the gas channels in graphite plates is laborious and expensive. Composites are more favorable in this respect, as the channels may be made by pressing and stamping. Both the composites and graphite are porous to some extent, and permeable to hydrogen. In view of their mechanical strength and impermeability to gases, metals such as stainless steel, titanium, and aluminum are very attractive. In metals, the channels are readily made by separate stamping of metal sheets for each side of the bipolar plate and subsequent welding of the two halves. An important defect of metallic bipolar plates is their inadequate corrosion resistance under the operating conditions of the fuel cells (high temperature and high humidity, contact with the acidic proton-conducting membrane, oxygen atmosphere).

Corrosion of the plates not only detracts from their mechanical properties but also gives rise to undesirable corrosion products, namely, heavy-metal ions, which, when depositing on the catalysts, strongly depress their activity. The corrosion processes also give rise to superficial oxide films on the metal parts, and these cause contact resistance of the surfaces. For a lower contact resistance, metallic bipolar plates sometimes have a surface layer of a more stable metal. Thus, in the first polymer electrolyte membrane fuel cell, developed by General Electric for the Gemini spacecraft, the bipolar plates consisted of niobium and tantalum coated with a thin layer of gold. A bipolar plate could also be coated with a layer of carbide or nitride.

Metals as the material for bipolar plates have the additional advantage that thinner, lighter plates can be made. Future developments in this area probably will yield new ways to protect the metal surfaces from corrosive attack by the medium and from formation of the superficial oxide films that lead to contact resistance. All the problems associated with bipolar plates have been discussed in detail in a review by Tawfik et al. (2007).

18.8   ELEVATED-TEMPERATURE PEMFCs (ET-PEMFCs)

Fuel cells of this variety sometimes are called high-temperature or mid-temperature polymer electrolyte membrane fuel cells, but is preferable to use the designation given in the section heading, as in their application to high-temperature fuel cells, high-temperature and mid-temperature refer to different temperature ranges.

The primary interest of studies of polymer electrolyte membrane fuel cells working at the temperatures of 120–130°C rather than the usual 80°C was that of attaining a higher tolerance of the platinum catalysts for carbon monoxide present as an impurity in the hydrogen supply. In fact, if (as already mentioned in Section 18.4.4) at 80°C the admissible carbon monoxide concentration in hydrogen is not higher than 10 ppm at 130°C this value goes up to 1000 ppm, making prior purification of the hydrogen much simpler.

Apart from the tolerance to carbon monoxide that is highly important, the higher working temperature of polymer electrolyte membrane fuel cells has a favorable effect also on other processes occurring in these fuel cells:

(a) The electrode processes are greatly accelerated, the polarization of the electrodes decreases accordingly, and the discharge voltage of the fuel cell rises.

This is seen particularly distinctly at the oxygen electrode where, as mentioned earlier, the energy loss because of the lack of equilibrium and polarization of the electrode is particularly large.

(b) Cooling of the battery is greatly simplified, utilization of the heat set free at higher temperature becomes possible (for instance, for the steam reforming of methanol or for indoor heating). Another fact to consider is the inability of car radiators existing today to cope with dissipating heat that is set free at a temperature of only 80°C.

(c) The elimination of water from the operating fuel cell is greatly facilitated. At a temperature of 80°C part of the water accumulates in liquid form in the pores of the diffusion layer and in the gas supply channels, hampering gas transport to the catalyst. In winter, this residual water may even freeze when the battery has been turned off and damage the MEA. In ordinary polymer electrolyte membrane fuel cells a careful purge with dry gas is needed to eliminate these problems. At temperatures higher than 100°C they no longer exist.

In view of all these advantages, most of the research into polymer electrolyte membrane fuel cells concentrates on the elevated-temperature variant. At higher temperatures, the thermodynamic EMF value of a hydrogen–oxygen fuel cell is somewhat lower, but the cell's OCV is practically unaffected:

Temperature, °C	25	80	120
EMF, ε, V	1.229	1.19	1.14
OCV, U_0, V	1.05	1.05	1.05

The cell's operating voltage U_i increases considerably with increasing temperature because of the factors mentioned in the item (a) above.

A number of challenges also arise at higher operating temperatures of polymer electrolyte membrane fuel cells:

(a) At temperatures of 120–130°C, the vapor pressure of water is about 2.5 bars. The total pressure in the system must then be at least 3–4 bars so that the partial pressures of the reactant gases will retain acceptable values of about 0.5 bar.

(b) At higher temperatures, the rate of hydrogen diffusion through the membrane (crossover) increases. At 80°C, no more than 3% of the hydrogen is lost by this crossover, which is equivalent to a current loss of about 3 mA/cm². These losses rise markedly with temperature. The hydrogen diffusing to the oxygen electrode is oxidized there by a chemical mechanism. Then free oxygen radicals of the type of OH⁻ or OOH⁻ may form, which attack the membrane, accelerating its degradation.

(c) Degradation processes of fuel-cell components are accelerated at higher temperatures: catalyst recrystallization and the associated decrease in working surface area of the catalyst, superficial oxidation of the catalyst's carbon support, and the associated loss of contact between the support and the catalyst, oxidative attack of the membrane (see above), superficial oxidation and corrosion of the bipolar plates, attack of various structural materials (seals, etc.).

The data available up to now as to the effect of higher temperatures on all these phenomena of degradation and destruction are very limited. Research into these problems needs considerable time. Various accelerated degradation tests that are conducted often lead to ambiguous results.

(d) The main problem in elevated-temperature-polymer electrolyte fuel cell operation is degradation of the membrane at the higher temperature. Marked water loss raises the ohmic resistance of the membrane, causes brittleness, and may give rise to crack formation. For this reason, most polymer electrolyte fuel cells research at present addresses the question of how to maintain the membrane in good working condition in an elevated-temperature-polymer electrolyte fuel cell.

Different approaches to this problem have been suggested:

(a) *Nafion-Type Membranes with Hydrophilic Additives.* When inorganic compounds, such as SiO_2, TiO_2, $Zr(HPO_4)_2$ or heteropoly acids of the type of phosphotungstic acid $H_3PW_{12}O_{40}$ or silicotungstic acid $H_4SiW_{12}O_{40}$ are introduced into the membrane, they will combine with water present in the membrane and make its escape at elevated temperatures more difficult. Such additives may be introduced both during membrane manufacture and posttreatment of finished membranes.

(b) *Solvents Other Than Water.* The water present in the membrane makes possible the dissociation of the sulfonic acid groups providing protonic conductivity. Water could be replaced by other solvents having the same effect, but at the same time having a higher boiling point, and hence would be less easily lost from the membrane. Phosphoric acid and imidazoles [e.g., poly(2,5)benzimidazole (PBI)] could be used in this capacity. Sometimes they are used together with added heteropolyacids as mentioned above, which by themselves are proton conducting.

(c) *Membranes Made from Other Polymers.* It has been suggested instead of the membranes made from perfluorinated sulfonic acids (PFSA) to use membranes made from other, heat-resistant polymers containing sulfonic acid groups. One could think here of sulfonated polyimides (SPI) and sulfonated polyether ether ketones (SPEEK). Membranes made from such polymers have a sufficiently high protonic conductivity at elevated temperatures but are less sensitive to lower humidity and water loss.

Membranes that display protonic conductivity in the complete absence of water are of interest as well. A material of this kind is cesium hydrogen phosphate CsH_2PO_4 described in 2001 by Norby. At temperatures above 230°C where a "superprotonic state" is attained, the protonic conductivity of this material is fairly high.

In their review of 2007, Shao et al. called attention to the fact that in classical versions of polymer electrolyte membrane fuel cells, Nafion-type membranes were used while adding a polymer of the same type as a proton-conducting additive to the catalytically active layers. If, now, for the membrane one uses other materials than for the proton-conducting additive, complications may arise during humidity changes in the system, which may lead to differences in the degrees of swelling or compression and to a loss of contact between individual MEA components. It had taken about 5 years as new samples of Nafion membranes had become available, until an optimum structure of the MEA and an optimum technology for its manufacture was established. With the advent of new membrane materials, existing experience can be used only in part, and considerable efforts must be expended to arrive at said optimization.

From all that has been said above, it can be concluded that polymer electrolyte membrane fuel cells, working at elevated temperatures, are highly promising. Many difficulties must still be overcome in order to develop models, which will function in a stable and reliable manner, and for extended periods of time. At present, about 90% of all publications on fuel cells are concerned precisely with the attempts to overcome these difficulties. Most of the publications deal with research into new varieties of membrane materials. Some results of these works are described in the reviews on elevated-temperature-polymer electrolyte fuel cells (Zhang et al., 2006; Shao et al., 2007).

REFERENCES

Cai M, Ruthkosky MS, Merzougui B, et al. J Power Sources 2006;160:977.

Kamarudin SR, Daud WRW, Som AM, et al. J Power Sources 2006;157:641.

Norby T. Nature(London) 2001;410:877.

Schmidt H, Buchner P, Datz A, et al. J Power Sources 2002;105:243.

REVIEWS

Costamagna P, Srinivasan S. J Power Sources 2001;102:242253.

Cheng X, Shi Z, Glass N, et al. J Power Sources 2007;165:739.

Gottesfeld S, Zawodzinski TA. In: Alkire RC, Gerischer H, Kolb DM, Tobias CW, editors. *Advances in Electrochemical Science and Engineering*. Vol. 5. Weinheim, Germany: Wiley-VCH Verlag GmbH; 1997. p 195.

Mehta V, Cooper JS. J Power Sources 2003;114:32.

Shao Y, Yin G, Wang Z, Gao Y. J Power Sources 2007;167:235.

Srinivasan S, Yelev OA, Parthasarthy A, Manko DJ, Appleby AJ. J Power Sources 1991;36:299.

Tawfik H, Hung Y, Mahajan D. J Power Sources 2007;163:755.

Zhang J, Xie Z, Zhang J, Tang Y, et al. J Power Sources 2006;160:872.

19

DIRECT LIQUID FUEL CELLS WITH GASEOUS, LIQUID, AND/OR SOLID REAGENTS

Methanol is a highly promising type of fuel for fuel cells. It is considerably more convenient and less dangerous than gaseous hydrogen. These advantages of methanol are important for applications in mobile power plants and, particularly, in small-size fuel cells of low power, intended to power-portable electronic devices (personal computers, cellular phones, and the like). In contrast to petroleum products and other organic kinds of fuel, methanol has a rather high electrochemical activity. Its specific energy content of about 6 kWh/kg, although lower than that of gasoline (about 10 kWh/kg), is still quite satisfactory. For this reason, even its use in fuel cells for medium-sized power plants in electric vehicles is widely discussed today.

Two possibilities exist for the use of methanol as a fuel: (i) its prior catalytic or oxidizing conversion to technical hydrogen, (ii) its direct anodic oxidation at the electrodes in the fuel cell. The former possibility implies that additional unwieldy equipment for the conversion of methanol to technical hydrogen and for subsequent purification of this hydrogen is needed. The second possibility is more attractive but involves certain difficulties related to the relatively slow anodic oxidation of methanol even at highly active platinum electrodes. In the view of these difficulties, much attention is given at present to the development of such direct methanol fuel cells.

Electrochemical Power Sources: Batteries, Fuel Cells, and Supercapacitors, First Edition.
Vladimir S. Bagotsky, Alexander M. Skundin, and Yurij M. Volfkovich.
© 2015 John Wiley & Sons, Inc. Published 2015 by John Wiley & Sons, Inc.

19.1   CURRENT-PRODUCING REACTIONS AND THERMODYNAMIC PARAMETERS

The electrode reactions taking place at the electrodes of direct methanol fuel cells, the overall current-producing reactions, and the corresponding thermodynamic values of equilibrium electrode potentials E^0 and electromotive force (EMF) ϵ^0 of the direct methanol fuel cell are given as follows:

$$\text{anode} : \ CH_3OH + H_2O \rightarrow CO_2 + 6H^+ + 6e^- \quad E^0 = 0.02\,V \quad (19.1)$$

$$\text{cathode} : \ (3/2)O_2 + 6H^+ + 6e^- \rightarrow 3H_2O \qquad E^0 = 1.23\,V \quad (19.2)$$

$$\text{overall} : \ CH_3OH + (3/2)O_2 \rightarrow CO_2 + 2H_2O \qquad \epsilon^0 = 1.21\,V \qquad (19.3)$$

The thermodynamic parameters of the reaction (19.3) are

$$\text{Reaction enthalpy(or heat } q_{react})\Delta H = -726\,kJ/mol = 1.25\,eV$$

$$\text{Gibbs reaction energy(or maximum work } W_e)\Delta G = -702\,kJ/mol = 1.21\,eV$$

19.2   ANODIC OXIDATION OF METHANOL

The first studies of methanol oxidation's special features and of the kinetics and mechanism of anodic methanol oxidation at platinum electrodes began in the early 1960s, in the period known as *the first boom of work in fuel cells*. In the years after that, this reaction was the subject of countless studies by many groups in different countries. In summary one can say of all this work that, by now, the mechanism of this reaction has been established rather reliably (for reviews see Bagotsky et al., 1977; Iwasita and Vielstich, 1990; Kauranen et al., 1996), while conflicting views persist on certain detailed aspects. Work on these questions is continuing even now.

The major reaction product is carbon dioxide, but in certain cases, the transient production of small amounts of other oxidation products, such as formaldehyde, formic acid, and so on, is seen. Six electrons are given off in the complete oxidation of methanol to carbon dioxide, so that the specific capacity of methanol is close to 0.84 Ah/g.

Methanol oxidation is a reaction with several consecutive stages. At the first stage, the methanol molecules undergo dehydrogenation:

$$CH_3OH \rightarrow COH_{ads} + 3H_{ads} \qquad (19.4)$$

forming chemisorbed species COH_{ads}.

At the next stage, these species are oxidized by way of chemical interaction with oxygen-containing species OH_{ads} adsorbed on neighboring sites of the platinum surface:

$$COH_{ads} + 3OH_{ads} \rightarrow CO_2 + 2H_2O \qquad (19.5)$$

Ionization of adsorbed hydrogen atoms and the anodic formation of OH_{ads} species from water molecules are the steps actually producing the current:

$$H_{ads} \rightarrow H^+ + e^- \tag{19.6}$$

$$H_2O \rightarrow OH_{ads} + H^+ + e^- \tag{19.7}$$

Under certain conditions, as when the cell is temporarily at open circuit, when the formation of species OH_{ads} is not possible, species COH_{ads} "age" and change to species CO_{ads} that are hard to oxidize and capable of inhibiting further methanol oxidation.

The thermodynamic potential of the methanol electrode is +0.02 V, a value that is rather close to the hydrogen electrode potential. The steady-state potential of a platinum electrode in a methanol solution is about +0.3 V. The working potential of a steady-state methanol oxidation depends on the current density and varies within the range of 0.35–0.65 V. This means that the working voltage of a methanol–oxygen fuel cell will have values within the range of 0.4–0.7 V.

19.3 USE OF PLATINUM–RUTHENIUM CATALYSTS FOR METHANOL OXIDATION

Interest in anodic methanol oxidation reaction received a marked boost when it was shown that the platinum–ruthenium system has a significantly higher catalytic activity for this reaction than pure platinum (while pure ruthenium has no activity at all for this reaction). This synergy effect was the subject of many studies that continue to this day. Some authors tend to believe that this effect is following changes in the electronic structure of platinum when alloyed with ruthenium. Now, the most widely accepted concept is that of a so-called bifunctional mechanism, according to which the organic species primarily adsorb on platinum sites while ruthenium sites facilitate adsorption of the OH_{ads} species needed for the oxidation of the adsorbed organic species.

19.4 MILESTONES IN DMFC DEVELOPMENT

In the beginning of the 1960s, repeated attempts were made to build test models of methanol/oxygen (air) fuel cells. In the first of these attempts, an alkaline electrolyte solution was used. On account of undesirable carbonate formation in this electrolyte, for later studies of anodic methanol oxidation solutions of methanol in aqueous sulfuric acid were used, and the same solutions were used when building the first models of methanol fuel cells, more particularly, by Shell in England and Hitachi in Japan.

Owing to the relatively low rate of methanol oxidation at platinum, considerable amounts of platinum catalyst had to be used in these models (up to $10\,mg/cm^2$). Still, the specific power attained was quite small (about $20\,mW/cm^2$). For this reason, interest gradually subsided. Just a few papers have been published since in the scientific and technical literature on this topic. This period ended only in the mid-1990s when great success was achieved with membrane hydrogen–oxygen (air) fuel cells, and when it had become possible to transfer these achievements to other kinds of fuel cells, namely, to direct methanol fuel cells. Lamy et al. (2002) titled their 2000 review: "Direct methanol fuel cells: from a 20th century electrochemists' dream to a 21st century emerging technology."

At present, most of the work toward building methanol fuel cells relies on technical and design principles, developed previously for polymer electrolyte membrane fuel cells. In both kinds of fuel cells, it is common to use platinum–ruthenium catalysts at the anode and a catalyst of pure platinum at the cathode. In the direct methanol fuel cells, the membrane commonly used is of the same type as in the hydrogen–oxygen fuel cells. The basic differences between these versions are discussed in Section 19.7.

19.5   MEMBRANE PENETRATION BY METHANOL (METHANOL CROSSOVER)

A basic problem associated with the operation of direct methanol fuel cells is the gradual penetration of methanol to the oxygen electrode by diffusion through the membrane. This "crossover" has two undesirable consequences: (i) a part of methanol is unproductively lost; (ii) the potential of the oxygen electrode moves to more negative values (a "mixed oxygen–methanol potential" is becoming established at it), that is, the working voltage of the fuel cell decreases.

The rate of methanol penetration i_{cross} is often given in electrical units of current density, equivalent to the oxidation of the amount of methanol involved. Many factors influence this rate, including the nature of the membrane, its thickness, and the temperature. The rate increases with increasing methanol concentration in the aqueous methanol solution. The crossover rate decreases with increasing working current density: the following relation between the values of i_{cross} and i_{load} at a temperature of 80°C and two methanol concentrations (0.5 M and 1 M) were measured:

i_{load}, mA/cm^2	0	100	300	600
i_{cross}, mA/cm^2 ($c = 1$ M)	100	85	60	40
i_{cross}, mA/cm^2 ($c = 0.5$ M)	60	40	20	(0)

It can be seen from these data that the methanol concentration should be below 0.5 M when working at low current densities, in order to avoid large losses of methanol (i.e., realize a high efficiency of methanol utilization). From the same point of view, somewhat higher concentrations could be admitted when working at high current densities. However, to realize really high current densities (i.e., raise the reaction rate of methanol), an important increase in methanol concentration is needed. This leads to an antagonism between the conditions needed for working at high current densities (higher methanol concentrations) and the conditions needed for working with a high efficiency of methanol utilization (lower methanol concentrations). In recent years, a large amount of research was done in order to overcome this antagonism.

Negative consequences of methanol crossover were first detected for Nafion-type membranes. Three possibilities can avoid them:

(a) finding ways to treat the membrane that will lower its permeability for methanol, or finding new membrane materials, where permeation is not at all observed. Some success has been achieved in lowering the permeability, but the results (so far) are not sufficient for solving this problem;

(b) replacing methanol in the solution with another substance, readily undergoing electrochemical oxidation, but not penetrating the membrane. Work in this direction is discussed in Section 19.10;

(c) attempting to act on the crossover rate of methanol by improving the membrane–electrode assembly design and the design of the cell as a whole.

The effect, mentioned earlier, of crossover rate decreasing with increasing working current density is because of the drastic decrease in methanol concentration in the anode's catalytic layer that occurs when methanol undergoes rapid consumption at high current densities. This, in turn, leads also to a decrease in methanol concentration in the membrane's surface layer adjacent to the catalytic layer. As methanol diffusion into the membrane starts precisely in this surface layer, the diffusion rate also decreases.

Some workers have attempted to attain the same effect artificially, delaying access of the methanol solution to the metal–electrode assembly by placing some barrier between the bipolar plate and the anodic part of the metal–electrode assembly that should slow down access of the concentrated methanol solution to the electrode boundary.

At present, as a practical way of overcoming excessive methanol penetration through the membrane, most direct methanol fuel cell types use aqueous solutions, having methanol concentrations of no more than 2 M.

Some success was attained in overcoming the second of the negative consequences of methanol crossover, that is, poorer performance of the oxygen electrode with a platinum catalyst. It was shown by a number of researchers that if a dispersed alloy of platinum with a metal of the iron group (rather than pure platinum) is used as the catalyst for the oxygen electrode or, if platinum deposited on a carbon base (that contains FeTPP—iron tetramethoxyphenylporphyrin) is used, then the influence of methanol on the cathodic catalyst decreases while the catalyst's activity for the oxygen reaction increases.

Methanol crossover, apart from the undesirable consequences mentioned above, also has some useful influence (small, to be honest) on the operation of direct methanol fuel cells. Methanol that has diffused to the cathode will be chemically oxidized by oxygen to carbon dioxide (i.e., a reaction without generating current), under the influence of the platinum catalyst. This reaction produces additional heat, and this heat may serve to accelerate the start-up of a cold fuel cell battery.

19.6   VARIETIES OF DMFC

19.6.1   Varieties with Regard to Reactant Supply

Not only methanol must be supplied to the anode space of direct methanol fuel cells, but also water, in order to secure a sufficiently high conductivity of the proton exchange membrane. This can be done in two ways:

(a) by supplying a mixture of methanol vapor and water vapor produced by evaporation of an aqueous methanol solution in a special evaporator;

(b) by directly feeding a liquid aqueous methanol solution.

The first of these possibilities has the advantage of keeping the fuel cell at a higher temperature and thus realizing a better electrical performance. Shukla et al. (1995) have described such a direct methanol fuel cell. A temperature of 200°C was maintained in an evaporator for a 2.5 M aqueous methanol solution. The fuel cell's working temperature was lower. At a temperature of 100°C current density of 200 mA/cm^2 could be realized at a voltage of about 0.5 V (the total platinum content of both electrodes was 5 mg/cm^2).

The electrical parameters of varieties supplied with a liquid solution are lower than those of the varieties supplied with gases, but research in direct methanol fuel cells was done mainly with liquid supply. This type of fuel cell is much simpler to design and operate, inasmuch as neither a special evaporator nor dual temperature control (for the evaporator and for the reaction zone of the fuel cell) is needed. With the supply of a liquid methanol solution, all risk of the membrane drying out close to the anode side is eliminated. The elimination of heat is also easier with cells having liquid solution supply. All subsequent information on direct methanol fuel cells in the present chapter refers to the variety supplied with liquid water-methanol solution.

One distinguishes fuel cells with active and passive reactants (fuel and/or oxygen) supply. In passive direct methanol fuel cells, methanol gets into the cell either under the effect of gravitational forces (where the vessel with methanol solution is situated above the fuel cell itself), or under the effect of capillary forces through special wicks. A third way of passive methanol supply is hydraulic, under the effect of pressure of the gases evolved in the reaction. In completely passive cells, the surrounding air having direct access to the cathode's gas diffusion layer is used as source of oxygen (air-breathing cells).

In active direct methanol fuel cells, special pumps are used to supply the methanol solution. They yield a higher rate of methanol flowing into the fuel cell, and, when needed, a better control of this rate. Moreover, forced methanol supply secures a more uniform access of methanol to the different cells of a multicell battery. Active oxygen supply can be accomplished from high-pressure cylinders across regulating valves. When air is used as the oxidizing agent, fans or precompression to a relatively high pressure can be used for forced air supply.

The electrical performance figures of active direct methanol fuel cells are much higher than those of passive ones. Yet owing to the need to use many auxiliary pieces of equipment, such as pumps, valves, and controllers, power plants with active direct fuel cells are much more complex and, in a number of cases, are less reliable in their operation. Also, part of the electrical energy generated by them is getting used in the auxiliary equipment.

As a rule, passive fuel cell systems are simpler to operate, more reliable, and cheaper than active systems. They are used mainly for power plants with low output, such as those supplying power for portable devices. Active systems are more appropriate for power plants with higher power output.

19.6.2 Varieties by Application

Basically, all variants of direct methanol fuel cells are built according to one and the same principle, but, depending on the intended use and way of use, different requirements may have to be met, which would be reflected in special design and operating features. In contrast to plants built up from other kinds of fuel cells, direct methanol fuel cells, on economic grounds, are not meant to be used for relatively large fixed-site plants having a power of hundreds of kilowatts or even a few megawatts for centralized or decentralized power supply of isolated settlements or towns.

The main applications of direct methanol fuel cells in the near future will be the power supply of various portable electronic devices with civil, commercial, or military uses. In the future, power plants for electric vehicles and other transportation will join the above-mentioned applications. All these applications have specific, different features:

(a) Power plants for portable devices ranging in power from a few watts to 20 W must have a low working temperature (for instance, not higher than 60°C), should work with a passive supply of air oxygen, and must be easy to handle. These power plants replace conventional batteries, either disposable or rechargeable ones, and should secure long uninterrupted operation of the portable devices powered by them. The basic criterion reflecting this requirement is the specific energy output per unit mass (Wh/kg) or unit volume (Wh/l) of the fuel cell stack. For the conventional rechargeable batteries, which today have the largest power density (lithium ion batteries), the specific energy output is about 150 Wh/kg or 250 Wh/l. In the case of power plants with fuel cells, this criterion should account, as well, for the weight and volume of the reactants needed for uninterrupted work during a given period of time (including their storage containers).

(b) Power plants for electric vehicles having a power of tens of kilowatts can have working temperatures of up to 100°C, may use precompressed air, and their specific cost (per kilowatt of power output) should be lower than in the case of less powerful plants for portable devices. These power plants replace ordinary internal combustion engines. The basic criterion for the operating efficiency is the average or maximum available power and the reactant consumption per unit electrical energy output (or per unit driving distance, the analog of miles per gallon).

19.7   SPECIAL OPERATING FEATURES OF DMFC

The peripheral equipment needed for direct methanol fuel cells is largely analogous to that of polymer electrolyte membrane fuel cells. The mechanical basis of fuel cells and stacks on the whole consists of bipolar plates between which the sandwiched membrane–electrode assemblies are arranged. For the venting of heat, cooling plates with a circulating heat transfer agent are set up in a particular order between individual fuel cells in the stack.

The major distinguishing features of direct methanol fuel cells relative to polymer electrolyte membrane fuel cells are

(a) use of an aqueous methanol solution as reactant for the anode, rather than hydrogen gas;
(b) liberation of the reaction product, carbon dioxide, not at the cathode, but at the anode;
(c) a relatively slow anodic reaction (oxidation of methanol), leading to a considerable polarization of the negative electrode and, thus, to a lower working voltage of the fuel cell;
(d) supply of a large amount of water to the membrane–electrode assembly inasmuch as the feed consists of an aqueous solution; hence, water elimination must be just as intense.

Different provisions have been suggested to account for these special features.

19.7.1   Reactant Supply and Product Elimination

Considerable changes are needed in the anodic part of the membrane–electrode assemblies in order to accommodate the first two of the above-mentioned points. Instead of the porous gas diffusion layer that in polymer electrolyte membrane fuel cells ensures a uniform distribution of hydrogen across the surface, a gas–liquid diffusion layer that contains a set of hydrophilic as well as a set of hydrophobic pores is needed here. Through the hydrophilic pores, this layer must secure the unobstructed access of the aqueous methanol solution to the reaction zone and its uniform distribution. Through the hydrophobic pores, this layer must secure the unobstructed elimination of carbon dioxide, as the gaseous reaction product, from the reaction zone. Analogous changes must be made in the catalytically active anode layer of the membrane–electrode assemblies, where the gas is actually formed, and must be removed toward the gas–liquid diffusion layer.

By introducing carefully determined amounts of Nafion and polytetrafluoroethy-lene into the gas–liquid diffusion layer and into the catalytically active layer, the ratio needed between hydrophilic and hydrophobic micropores is achieved.

19.7.2   Water Management in DMFC

The operation of direct methanol fuel cells is based on principles that, at first sight, look absurd: water, which, together with carbon dioxide, is the major product of the current-producing reaction (see Eq. 19.1), is supplied to the fuel cell in large amounts as an aqueous methanol solution. Certainly this water is useful for moistening the membrane and possibly for the elimination of the heat of the reaction, but, when considering the electrochemical reaction, this water is merely inert ballast, supplied to and withdrawn from the fuel cell. The aqueous solution needs much larger and heavier containers for reactant storage than would be needed for pure methanol.

This raises the question of whether water could be recirculated. One makes a distinction between outer and inner water recirculation. In the former, water vapor coming out of the anode compartments (and consisting of the two components: water as a reaction product and evaporated solution water) is condensed and redirected into a special mixer. In this mixer, pure methanol is admixed, as needed for maintaining the required solution concentration, while the reaction proceeds. The methanol concentration can be controlled with different sensors.

Inner water circulation is much fancier, but also harder to realize. Under ordinary operating conditions, both in direct methanol fuel cells and in polymer electrolyte membrane fuel cells, a flow of water from the anode toward the cathode is observed. This flow has two origins: water molecules are dragged along by hydrated hydrogen ions moving in the electric field from the anode to the cathode, and water diffuses under the influence of its own concentration gradient.

To arrange an inverse flow of water from the cathode to the anode, it was suggested to arrange several layers of paper treated with a mixture of polytetrafluorethylene and carbon black, on each side of the cathode's current collector. The thickness of such an assembly is 0.5 mm. These hydrophobic layers do not allow liquid water to pass through and redirects the water flow aiming for the cathode to the opposite side. Water vapor that has gone through this barrier layer is eliminated together with the flow of dry oxygen. The exhaust contains about two molecules of water per molecule of methanol that had reacted. This indicates that only that part of the water that had formed in the reaction quits the fuel cell, while there is no change in the overall water content within the membrane–electrode assembly. This would eliminate any need for introducing water together with methanol into the fuel cell. The cell that operates under these conditions is termed a *water-neutral direct methanol fuel cell.*

19.7.3 CO_2 Buildup in the Bipolar Plate Channels

A further problem in the operation of direct methanol fuel cells is because of the evolution of gaseous carbon dioxide at the anode. The gas bubbles that can locally interfere with the flow of the aqueous methanol solution may form in the flow field on the anodic side of the bipolar plates. This leads to a nonuniform reaction distribution (and, thus, of the current) across the surface of the membrane–electrode assembly. This effect is particularly noticeable when the solution is passively supplied (e.g., by free flow from a tank above). To overcome it, one should use active reactant supply at flow rates several times in excess of the stoichiometric requirements.

19.8   PRACTICAL PROTOTYPES OF DMFC AND THEIR FEATURES

In parallel with the large amount of work done to study the mechanism and operating features of methanol fuel cells with proton-conducting membranes, operating models of such fuel cells started to appear in the mid-1990s, first as laboratory-type small single-element fuel cells, then, finally, in the form of multicell stacks of relatively large power.

The electrical parameters (and particularly, the current–voltage relationship) were found to depend on many factors, including ones related to the technical parameters, such as thickness and nature of the membrane, nature of the catalyst and its amount per unit surface area of the electrode, as well as factors related to the test and operating conditions: the temperature, methanol solution concentration, rate of flow of the solution through the fuel cell, use of pure oxygen or air, the pressure of these gases, and so on. In published data on the development and testing of direct methanol fuel cell prototypes, the above-mentioned factors exhibit a wide variation. For this reason, it is impossible to compare the performance parameters obtained by different researchers and in different laboratories. It is merely possible with some approximation, to select an optimum technical solution (for particular operating conditions) or select optimum operating conditions (for particular technical solutions). To provide examples, later we report some selected data on the development and testing of experimental DMFC prototypes.

In the Los Alamos National Laboratory test, fuel cells were studied at temperature above 100°C with air and pure oxygen. A fuel cell with a Nafion 112 membranes at a temperature of 130°C, and pure oxygen, at a pressure of 5 bar, and at a voltage of 0.5 V, attained a current density of 670 mA/cm^2, which corresponds to a power density of 400 mW/cm^2. With the same fuel cell at a temperature of 110°C, using air at a pressure of 3 bar, a maximum power density of 250 mW/cm^2 was achieved.

In the Jülich Forschungszentrum in Germany, a fuel cell stack, having a power of 500 W, was built. This battery consisted of 71 fuel cells, having a surface area of 144 cm^2 each. The battery was intended to work with air oxygen under a pressure of 1.5–3 bar. In the publication about this experiment, mainly data on the testing of individual fuel cells or groups of three fuel cells were reported. Using a 1 M methanol solution and air oxygen at a pressure of 3 bar, the current density was about 320 mA/cm^2 at a temperature of 80°C and a working voltage of 0.3 V.

At the University of Hong Kong, China, a direct methanol fuel cell with passive supply of the methanol solution was built. The solution was supplied to the fuel cell by natural convection from a vessel, situated above the fuel cell. Carbon dioxide evolved from the fuel cell was fed into this vessel, which was tightly closed off. Increasing current density led to a higher rate of gas evolution and to a higher pressure in the vessel. This also produced a larger flow of methanol solution into the cell. Thus, the rate of methanol supply was controlled by the load on the cell. At current densities up to $150 \, mA/cm^2$, this cell had approximately the same characteristics as the same fuel cell when operated with an active supply of methanol solution via pumps. At higher current densities, natural convection proved inadequate. Difficulties also were seen at low current densities (below $50 \, mA/cm^2$), when operation became unstable. At current densities below $5 \, mA/cm^2$, gas evolution was too insignificant, and the system ceased to function.

At the Korea Institute for Science and Technology, the operation of a six-fuel-cell stack, having a total power of 50 W, was investigated. The stack was fed with a 2 M methanol solution. The fuel cell was supposed to work at ambient temperature, but internal heat production raised its temperature to above 80°C. Using oxygen, a maximum power of $254 \, mW/cm^2$ was attained. With air this value was $85 \, mW/cm^2$.

19.9 THE PROBLEMS TO BE SOLVED IN FUTURE DMFC

Quite in contrast to polymer electrolyte fuel cells, and despite the large volume of research performed, direct methanol-type fuel cells are still not in commercial production or wide practical use. Indeed, the true performance of direct methanol fuel cells is hard to assess. The experience gathered in the tests of individual samples, performed under a variety of conditions, does not give a clear reliable picture of fuel cell's performance when used for different needs. For such an assessment, one would need statistical data obtained from tests on a sufficiently large number of cells of any single type, and accounting for all cell parameters and test conditions.

Yet even now one potential area of application can be recognized distinctively. This is for relatively low-power energy sources (no more than 20 W) in electronic equipments such as notebooks, cameras, video cameras, DVD players, some medical devices, and so on. In these areas, direct methanol fuel cells, taking up little space, should replace the disposable or rechargeable conventional batteries used at present.

Another potential field of application for direct methanol fuel cells, that of power sources for electric vehicles, is, so far, very remote. A large amount of research and engineering work still has to be done to master this application, mainly, work aiming at improving their technical and economic parameters to solve the particular problems described later.

19.9.1 A Longer Lifetime

The production volume of direct methanol fuel cells and their long-term testing results are yet insufficient for an estimate of the lifetime of such fuel cells. Even so, the researchers mainly look at all the major reasons giving rise to the gradual performance drop and/or premature failure of these fuel cells. The reasons for performance drop and failure considered in the corresponding section dealing with polymer electrolyte membrane fuel cells, in the case of methanol fuel cells, must be supplemented by two more, as follows.

Crossover of Ruthenium Ions

At the working potential of an electrode, the platinum–rhuthenium catalyst is rather stable, but in long runs and following the slightest shift of potential in the positive direction (as may occur, following an arrest of methanol supply), ruthenium is seen to dissolve selectively from the catalyst and move in the form of ruthenium ions into the solution or membrane. Different optical techniques have been used for a rather detailed examination of this phenomenon. The dissolution of ruthenium not only leads to a gradual drop in the activity of the platinum–ruthenium catalyst, but also, when these ions pass the membrane and reach the cathode, to a significant inhibition of the cathodic reduction of oxygen. This may lead to a decrease in the working voltage of the fuel cell by almost 0.2 V.

Aging of Methanol Adsorption Products on Platinum

During the electrochemical oxidation of methanol, products of the –COH type become adsorbed on the platinum catalyst surface. They are the product of dehydrogenation of the original methanol molecules. Under normal conditions, these species are rapidly oxidized by –OH species adsorbed on neighboring segments of platinum or ruthenium. Under certain conditions, such as long current breaks, these particles may age, that is, change into species of the –CO type and be unwilling to leave the surface. They then cause an important slowdown of the continued oxidation of new quantities of methanol.

19.9.2   A Higher Efficiency

The operating efficiency of direct methanol fuel cells is greatly lowered because of methanol crossover. This effect leads to unproductive methanol consumption and to a marked decrease in working voltage, caused by the action of methanol on the oxygen electrode potential. So far, only two options can be seen to lessen or completely eliminate this effect:

(a) developing ways to treat membranes of the Nafion type so as to lower their permeability for methanol, or developing new kinds of proton-conducting membrane materials altogether impermeable to methanol.

(b) replacing methanol in fuel cells of the direct methanol fuel cell type, by other liquid reducing agents not penetrating through the membrane, that is, changing over from direct methanol fuel cells to other direct liquid fuel cells (DLFCs). Some varieties of this type are considered in Section 19.10.

19.9.3   A Lower Cost

Reasonable estimates for the possible production cost of direct methanol fuel cells are almost nonexistent. Even today it can be seen, however, that, apart from technical problems, their large-scale use in such an area as future electric vehicles will be possible only if two basic problems are solved:

(a) development of new membrane materials costing at least an order of magnitude less than Nafion;

(b) development of new kinds of catalysts with a much lower platinum metal content and, yet, with an activity higher than that of catalysts in current use. A special concern is ruthenium world production that would hardly be able to supply the needs arising in the production of many millions of electric vehicles with the current ruthenium content of the catalyst.

19.10 DIRECT LIQUID FUEL CELLS (DLFC)

19.10.1 The Problem of Replacing Methanol

The basic advantage of methanol–oxygen fuel cells with proton-conducting membranes over analogous hydrogen–oxygen fuel cells with proton-conducting membranes resides in a much more convenient liquid reactant replacing hydrogen a gas that has drawbacks in the transport, storage, and manipulation. Undoubtedly, the specific energy content per unit weight is much larger for hydrogen than for methanol, but, if we include all equipment needed for reactant storage (cylinders, containers for metal hydrides, etc.), then the energy content per unit weight turns out to be three times higher for methanol than for hydrogen.

Yet, using methanol in fuel cells is associated with certain difficulties and inconvenient features:

(a) relatively low rate of the electrochemical reaction;

(b) the need to eliminate from the system large amounts of water introduced together with methanol in an aqueous methanol solution;

(c) the need to eliminate from the system carbon dioxide formed as a gaseous reaction product;

(d) considerable amount of heat evolving because of the relatively low values of working voltage;

(e) as the boiling point of methanol being around 65°C, raising the temperature in order to accelerate the electrode reactions will be possible only when working under higher pressures;

(f) methanol crossover through the membrane and the need to deal with all the consequences of this effect;

(g) toxicity of methanol requires certain precautions be taken against the leakage of methanol and its vapors from the system as a whole and from the storage vessels.

Methanol is not the only possible liquid reactant for anodes in fuel cells. During the past two decades, parallel to very active work on methanol fuel cells, numerous studies were made to examine possibilities of using other liquid reactants with reducing properties (both organic and inorganic).

A number of factors are considered when selecting a new reducing agent: (i) the new compounds' equilibrium potential of its anodic oxidation reaction will influence the EMF and open circuit voltage (OCV) of the fuel cell; (ii) the relative rate of this reaction, which will influence the polarization of the electrode and, thus, the working voltage; (iii) the price and availability of the substance; (iv) the ecological cleanliness and harmlessness of the reactant itself and of its direct and potential secondary oxidation products.

19.10.2   The Problem of Using Ethanol in Fuel Cells

In its chemical properties, ethanol is similar to methanol. A considerable advantage over methanol is the much lower toxicity. It must be pointed out that from an ecological viewpoint, ethanol is exceptional among all other kinds of fuel. In the oxidation of all kinds of organic fuels (natural gas, oil-derived products, and coal), carbon dioxide is formed, which leads to the well-known global warming effect in the atmosphere of the earth, to a global temperature raise. This is equally true for chemical uses of these fuels in heat engines by "hot combustion," as for electrochemical uses in fuel cells by "cold combustion."

Ethanol can be obtained by fermentation of different agricultural biomasses, which, in turn, are formed by photosynthesis, involving that same carbon dioxide and solar energy. This implies that the combustion, hot or cold, of ethanol will not lead to the accumulation of excess carbon dioxide, and will not upset the overall balance of this gas in the atmosphere. Using ethanol as an energy vector in essence is one of the practical ways of using solar energy. Ethanol is the only kind of chemical fuel in renewable supply. In Brazil, mass production of ethanol from biomass has already been started, the corresponding infrastructure was created, and large part of automotive transport is transformed so as to work with ethanol, rather than with gasoline. In the European Union, the question is discussed of making ethanol a compulsory fuel additive. Attention should be drawn to current discussions of the ecological (and other) implications. The overall energy content of ethanol (about 8 kWh/kg) is rather close to that of gasoline (about 10 kWh/kg).

It is quite natural that all these considerations have led to enhanced interest in the uses of ethanol in fuel cells. A lot of research went into the development of direct ethanol fuel cells during the past decade.

The reaction of anodic oxidation of ethanol in an ethanol–oxygen fuel cell ideally is a multistep 12-electron reaction:

$$C_2H_5OH + 3H_2O \rightarrow 2CO_2 + 12H^+ + 12e^- \qquad (19.8)$$

In this reaction, the overall energy content of ethanol is about 8 kWh/kg a value rather close to that of gasoline (about 10 kWh/kg).

Unfortunately, the use of ethanol in fuel cells is connected with a problem not yet solved. Experimental data show that during anodic oxidation of ethanol on platinum catalysts, carbon dioxide is not the only reaction product, but also appreciable amounts of acetaldehyde and acetic acid are formed. The formation of acetaldehyde and acetic acid is because of the fact that rupture of the C$-$C bond in the original ethanol molecule would be required for carbon dioxide formation. In the chemical oxidation at high temperatures (combustion) of ethanol, this bond is readily ruptured in the hot flame, and the only reaction product (in addition to water) is carbon dioxide. However, in the electrochemical oxidation, occurring at temperatures below 200°C, this bond is not ruptured and the ethanol molecule is only partially oxidized.

The formation of acetic acid by ethanol oxidation is a four-electron process, following the equation:

$$C_2H_5OH + H_2O \rightarrow CH_3COOH + 4H^+ + 4e^- \tag{19.9}$$

while the formation of acetaldehyde is a two-electron process and follows the equation:

$$C_2H_5OH + H_2O \rightarrow CH_3CHO + 2H^+ + 2e^- \tag{19.10}$$

The smaller number of electrons involved in the reaction leads to a considerable decrease in the de facto energy content of ethanol from 8 kWh/kg for the 12-electron process, to 2.6 kWh/kg for a four-electron process, and to as little as 1.3 kWh/kg for a two-electron process.

The drop in energy content is not the only negative consequence of these side reactions. In contrast to carbon dioxide gas, which is readily vented from the system into the atmosphere, considerable problems of removal and disposal arise when aldehyde and acetic acid are formed in the reaction.

The very idea of building ethanol fuel cells is very attractive and promising. But, to this day, the development of these fuel cells is still in its initial stages. A large amount of research in electrocatalysis is required to find ways of making fully adequate ethanol fuel cells. This work should lead to the development of basically new polyfunctional catalysts that would secure a high rate for organic compound's low-temperature oxidation via C$-$H bond rupture. Only then it will be possible to think of a widespread application for future highly efficient ethanol fuel cells in ecologically harmless electric vehicles.

19.10.3 Direct Formic Acid Fuel Cell (DFAFC)

Formic acid HCOOH is a somewhat unusual kind of "fuel" for fuel cells, and in many of its properties it differs from other substances used as fuels. On the one side, the theoretical energy content of formic acid (1.6 kWh/kg) is much lower than for all other reactants considered in this chapter. On the other side, the equilibrium electrode potential for the oxidation of formic acid (-0.171 V) is more negative than that for the other organic fuels, that is, in a thermodynamic sense, formic acid is a very strong organic reducing agent. The thermodynamic EMF of a formic acid$-$oxygen cell is 1.45 V.

Formic acid has a number of properties making its use in fuel cells very attractive. First of all, this substance is ecologically absolutely harmless (the US Food and Drug Administration allows it to be used as a food additive). Its only oxidation products are carbon dioxide and water. It is practically impossible that side products or intermediates be formed. Formic acid is a liquid. In aqueous solutions it dissociates yielding $HCOO^-$ ions. This is of basic significance for its use in fuel cells with proton-conducting membranes. These membranes have a skeleton containing negatively charged ionic groups (as in the case of Nafion with sulfonic acid groups SO_3^-). On account of the electrostatic repulsion, these groups hinder (or at least strongly retard) penetration of formiate ions into the membrane. Thus, in the case of formic acid, the effect that constituted the major difficulty for the development of methanol fuel cells (crossover of the anodic reactant from the anodic region through the membrane to the cathodic region) is practically absent.

Formic acid is used in membrane-type fuel cells as an aqueous solution. A 20 M HCOOH solution contains about 75% of formic acid. Owing to the low water content, the membrane is not sufficiently moistened in such a concentrated solution, and its resistance increases. In solutions with concentrations less than 5 M, the current densities that can be realized are low because of slow reactant supply by diffusion to the catalyst surface. An optimum concentration for fuel cell operation is 10–15 M. In contrast to direct methanol fuel cells, an increase in reactant concentration for direct formic acid fuel cells does not produce complications related to reactant crossover.

In direct formic acid fuel cell, at a temperature of 70°C and a working voltage of 0.4 V, a power density of about $50 \, mW/cm^2$ was attained in 12 M formic acid. For comparison, the power density in a typical methanol fuel cell under the same conditions is about $30 \, mW/cm^2$.

With the experience gathered in the development of direct methanol fuel cells, platinum–ruthenium catalysts were used for the anodic process in the first studies on direct formic acid fuel cells. Then, it was shown that much better electrical characteristics can be obtained with palladium black as the catalyst. Importantly, with this catalyst, one can work at much lower temperatures. In particular, at a temperature of 30°C power densities of $300 \, mW/cm^2$ were obtained with a voltage of 0.46 V, and about $120 \, mW/cm^2$ with a voltage of 0.7 V. Considering all these special features, it will be very convenient to use formic acid as a reactant in fuel cells of small size, for power supply in portable equipment, ordinarily operated at ambient temperature.

19.10.4   Direct Borohydride Fuel Cells (DBHFC)

Sodium borohydride $NaBH_4$ is a substance with a relatively high hydrogen content (10.6 wt%). Besides, it is a strong reducing agent. In strong alkaline aqueous solutions, sodium borohydride is stable, but it undergoes hydrolysis, forming $NaBO_2$ and evolving hydrogen gas, in neutral and acidic solutions:

$$NaBH_4 + 2H_2O \rightarrow NaBO_2 + 4H_2 \tag{19.10}$$

On account of this feature, this compound sometimes is used as a convenient hydrogen source for small-size polymer electrolyte membrane fuel cells.

The first data concerning a possible direct use of this compound as a reducing agent (fuel) in fuel cells appeared in the early 1960s. In the first prototypes of such fuel cells an anion-conducting membrane was used as electrolyte. The work in this direction was not followed up, apparently, because of the fact that, so far, anion-exchange membranes are not stable enough in concentrated alkaline solutions. After 2003 it was suggested that a proton-conducting membrane may be used to this end. This variant is the one generally adopted at present.

Usually, a solution containing 10–30 wt% of $NaBH_4$ and 10–40% NaOH is supplied to the anode compartment of membrane-type borohydride–oxygen fuel cells.

The reaction of anodic oxidation of sodium borohydride can be written as

$$NaBH_4 + 8OH^- \rightarrow NaBO_2 + 6H_2O + 8e^- \tag{19.11}$$

Thus, theoretically, sodium borohydride yields eight electrons on oxidation, that is, it is a reactant with very high energy content (9.3 kWh/kg). As a matter of fact, fewer electrons are involved in the current-producing reaction. This is because the electrode potential of borohydride is more negative than the potential of the hydrogen electrode in the same alkaline solution. When the borohydride solution is getting into the fuel cell, a cathodic hydrogen evolution is possible at the anodic metal catalyst. Because of the negative potential this is accompanied by the coupled (but not current-producing) reaction of borohydride oxidation. This is analogous to the corrosion of an electronegative metal under the effect of "local elements" on which hydrogen evolution is facilitated. As a result, part of the borohydride is unproductively lost. At nickel catalysts for the anodic reaction, the effective number of electrons in the current-producing reaction is about 4. At platinum catalysts, this number is even lower. High values (6.9) were observed for gold. Different intermetallic compounds, including some containing rare earth elements, have also been suggested as anodic catalysts.

Even when taking into account the lower number of electrons, the specific energy content of borohydride remains rather high. Under the assumption of six electrons, it is close to 7 kWh/kg. This substance, therefore, is rather promising for the development of small-size fuel cells as power supply for portable equipment.

So far, it is not clear, how to solve the following problem: on contact with NaOH solution, H^+ ions in the ion-exchange membrane are exchanged (at least in part) for Na^+ ions. During current flow, these sodium ions start to migrate from the anode toward the cathode. Hydrogen ions when reaching the cathodic zone become involved in oxygen reduction, and then are eliminated as water vapor. However, the sodium ions will accumulate in the cathodic zone as NaOH, which must then be eliminated from the fuel cell.

19.10.5 Direct Hydrazine Fuel Cells (DHFC)

In the 1970s, a number of studies dealing with the design of hydrazine–oxygen fuel cells with alkaline electrolyte were published. Several prototypes of such fuel cells were actually built for portable devices and for various military objects, but, owing to the high toxicity and also to the high price of hydrazine, this work was not further developed. Hydrazine is a strong reducing agent, both in a thermodynamic sense (the equilibrium potential of its oxidation reaction is rather negative) and under the viewpoint of an easy electrochemical reaction. The theoretical value of the EMF of a hydrazine–oxygen fuel cell is 1.56 V.

In 2003, it was suggested to use this substance in a fuel cell with proton-conducting (i.e., an acid) membrane. Hydrazine was used as a 10% aqueous solution of hydrazine hydrate ($N_2H_4 \cdot H_2O$). In the aqueous solution, hydrazine, on account of its strong alkaline properties, dissociates into the ions $N_2H_5^+$ and OH^-. The anodic oxidation of hydrazine can be written in terms of the equation:

$$N_2H_4 \cdot H_2O \rightarrow N_2 + 4H^+ + H_2O + 4e^- \qquad (19.12)$$

the only reaction products being nitrogen and water. Disperse powders of platinum and platinum–ruthenium were used as catalysts for the anodic reaction in the experimental fuel cell. When working at a temperature of 80°C this DHFC had the following parameters (the parameters for a DMFC of analogous design working at 100°C are given in parentheses) OCV 1.2 V (0.7 V), voltages at current densities of 50, 100, and 150 mA/cm^2: 0.95 V (0.58 V), 0.57 V (0.50 V), and 0.48 V (0.46 V). It can be seen that under these conditions, at current densities below 50 mA/cm^2, the hydrazine fuel cell owing to its high voltage has definite advantages with respect to voltages at higher current densities, and these advantages disappear. In gases, evolving during the operation of the cell, noticeable quantities of ammonia have been detected. This is because of the fact that apart from the current-producing reaction, catalytic decomposition of hydrazine takes place.

The basic problem, arising in fuel cells of this type, is the free penetration of $N_2H_5^+$ ions through the proton-conducting membrane, to the cell's cathodic zone, which leads to consequences that had been described earlier with respect to methanol crossover.

REFERENCE

Shukla AK, Christensen PA, Hamnett A, Hogarth MP. J Power Sources 1995;55:87.

REVIEWS

Bagotsky VS, Vassiliev YB, Khazova OA. J Electroanal Chem 1977;81:229.
Demirci UB. J Power Sources 2007;169:239.
Dillon R, Srinivasan S, Arico AS, Antonucci V. J Power Sources 2004;127:112.
Iwasita T, Vielstich W. In: Gerischer H, Tobias CW, editors, Vol. 1. *Advances in Electrochemical Science and Engineering*. New York: VCH; 1990. p 127.
Kauranen P, Skou E, Munk J. J Electroanal Chem 1996;404:1.

Lamy C, Lima A, LeRhun V, Delime F, Coutanceau C, Léger J-M. J Power Sources 2002;105:283.

de Leon CP, Walsh FC, Pletcher D, Browning DJ, Lakeman JB. J Power Sources 2006;155:172.

Qn W, Wilkinson DP, Shen J, Wang H, Zhang J. J Power Sources 2006;154:202.

Wee J-H. J Power Sources 2006;161:1.

20

MOLTEN CARBONATE FUEL CELLS (MCFC)

20.1 SPECIAL FEATURES OF HIGH-TEMPERATURE FUEL CELLS

Together with the solid oxide fuel cells, which is considered in Chapter 21, molten carbonate fuel cells (MCFCs) are representatives of the class of high-temperature fuel cells that have a working temperature of over 600°C.

High-temperature fuel cells have a number of advantages in comparison to other fuel cell types:

(a) the possibility to attain higher efficiency using the reaction heat for generating additional electrical energy;
(b) high rate of the electrode reactions and relatively little electrode polarization, hence, no need to use platinum catalysts;
(c) the possibility of using technical hydrogen with a large concentration of carbon monoxide and other impurities;
(d) the possibility of directly using carbon monoxide, natural gas, and a number of petroleum products via an internal conversion of these fuels to hydrogen within the fuel cell itself.

Yet, changing over to higher temperatures implies a certain diminution of thermo-dynamic indices of the fuel cells. The Gibbs free energy $-\Delta G$ of hydrogen oxidation by oxygen decreases with increasing temperature. It amounts to 1.23 eV at 25°C but

Electrochemical Power Sources: Batteries, Fuel Cells, and Supercapacitors, First Edition.
Vladimir S. Bagotsky, Alexander M. Skundin, and Yurij M. Volfkovich
© 2015 John Wiley & Sons, Inc. Published 2015 by John Wiley & Sons, Inc.

decreases to 1.06 eV at 600°C and to 0.85 eV at 1000°C. (Identical figures in volts would be the numerical values for the thermodynamic EMF of hydrogen–oxygen fuel cells at these temperatures.) On the other side, the reaction enthalpy, $-\Delta H$, which at 25°C has a value of 1.48 eV, will not change significantly with increasing temperature. For this reason the thermodynamic efficiency of the reaction, $\eta_{thermod} = -\Delta G / -\Delta H$, also decreases with increasing temperature. At temperatures of 25, 600, and 1000°C, it has values of 0.83, 0.72, and 0.57, respectively.

In fuel cells operated at higher temperatures, numerous problems associated with the limits of chemical and mechanical stabilities of various materials that are used in them also become important.

20.2   THE STRUCTURE OF HYDROGEN–OXYGEN MCFC

The working temperature of molten carbonate fuel cells is around 600–650°C. Mixed carbonate melts containing 62–70 mol% of lithium carbonate and 30–38 mol% of potassium carbonate, with compositions close to the eutectic point, are used in molten carbonate fuel cells as an electrolyte. Sometimes, sodium carbonate and other salts are added to the melts. This liquid melt is immobilized in the pores of a ceramic fine-pore matrix, made of sintered magnesium oxide or lithium aluminate powders.

Porous metallic gas diffusion electrodes are used in these fuel cells. The anode consists of a nickel alloy with 2% of chromium. Chromium that is added prevents recrystallization and sintering of the porous nickel though it works as an electrode. This action is based on chromium forming a thin layer of chromium oxide at the nickel grain boundaries, which interferes with the surface diffusion of the nickel atoms.

The cathode consists of lithiated nickel oxide. Nickel oxide is a p-type semi-conductor, having a rather low conductivity. When doped with lithium oxide, its conductivity increases tens of times, owing to a partial change of Ni^{2+} to Ni^{3+} ions. The lithiation is accomplished by treating the porous nickel electrode with a lithium hydroxide solution in the presence of air oxygen. The compound produced has a composition given as $Li_x^+Ni_{1-x}^{2+}Ni_x^{3+}O$. This lithiation of nickel oxide was first applied in 1960 by Bacon in his alkaline fuel cell.

The matrix with its fine pores filled by the carbonate melt is a reliable protection against gases bubbling through and getting into the "wrong" electrode compartment. Therefore, there is no need to provide the gas diffusion electrodes with a special gas barrier layer.

The fuel cell components have thicknesses as follows: the anode is 0.8–1.5 mm thick; the cathode, 0.4–1.5 mm, the matrix, 0.5–1 mm. In a fuel cell of the filter-press type, the individual cells are separated by bipolar plates made of nickel-plated stainless steel, contacting the anode with their nickel side, and the cathode with their steel side. All structural parts are made of nickel or nickel-plated steel. In a working fuel cell, the temperature of the outer part of the matrix electrolyte is lower than that of the inner part, so that in the outer part the electrolyte is solidified. This provides for tight sealing around the periphery of the individual fuel cells.

In a hydrogen–oxygen molten carbonate fuel cell, the following reactions take place:

$$\text{Anode } H_2 + CO_3{}^{2-} \rightarrow H_2O + CO_2 + 2e^-, \quad E^0 = 0\,V \tag{20.1}$$

$$\text{Cathode } \tfrac{1}{2}O_2 + CO_2 + 2e^- \rightarrow CO_3{}^{2-}, \quad E^0 = 1.06\,V \tag{20.2}$$

$$\text{Overall } H_2 + \tfrac{1}{2}O_2 \rightarrow H_2O \quad \epsilon^0 = 1.06\,V \tag{20.3}$$

It is a special feature of the electrode reactions in molten carbonate fuel cells that, unlike most other versions of fuel cells, the cathodic reaction consumes not only oxygen (or air) but also carbon dioxide. In the anodic reaction, carbon dioxide is evolved at the anode. It is imperative, therefore, in the design of molten carbonate fuel cells, to provide for the possibility of carbon dioxide evolved at the anode to return to the cathode (Fig. 20.1).

The values of electrode potential given in Equations (20.1) and (20.2) refer to the condition where all gases involved in the reactions (H_2, O_2, CO_2, and water vapor) are in their standard states, that is, they have partial pressures of 1 bar. During the reaction, in fact, their amounts and, hence, their partial pressures constantly change. This implies that the equilibrium potentials of the electrodes also change. According to the Nernst equation, the potential of the hydrogen anode is given by

$$E_a = E_a^0 + \left(\frac{RT}{2F}\right) \ln \left(\frac{p_{H_2O}\, p_{CO_2}}{p_{H_2}}\right) \tag{20.4}$$

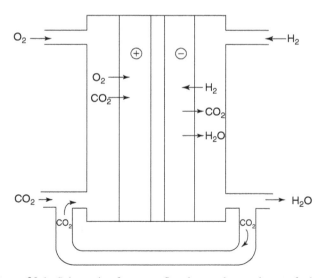

Figure 20.1. Schematic of reactant flow in a molten carbonate fuel cell.

and the potential of the oxygen cathode is given by

$$E_c = E_c^0 + \left(\frac{RT}{2F}\right) \ln \left(p_{O_2}^{1/2} p_{CO_2}\right) \tag{20.5}$$

If the reactant gases were enclosed, then the potential of the hydrogen electrode would shift in the positive direction with the progress of hydrogen consumption, while the potential of the oxygen electrode would shift in the negative direction with the progress of oxygen and carbon dioxide. The resulting power losses are called Nernst losses. As a result of these losses, the overall value of thermodynamic EMF ($\epsilon = E_c - E_a$) would fall below the standard value of 1.06 V.

Reacting gases are continuously supplied into an operating fuel cell. Therefore, their partial pressures do not change, the Nernst losses are mitigated, and the fuel cell's EMF retains its standard value.

Depending on the partial pressures of the reactant gases and the open circuit voltage of hydrogen–oxygen molten carbonate fuel cells have values of 1.00–1.06 V. A practically linear relation between cell voltage U_i and current density i is a special feature of these fuel cells. This implies a practically constant value of the apparent internal resistance R_{app} and leads to a more important voltage decrease with increasing cell discharge current than in other fuel cell types. At a reactant gases' pressure of 1 bar, and a current density of $100\,mA/cm^2$, the voltage is about 0.85 V. At a current density of $150\,mA/cm^2$, the voltage is only about 0.6 V. Current densities above $150\,mA/cm^2$ are practically not used in molten carbonate fuel cells.

The linear relation between voltage and current density is not a result of the internal resistance of the fuel cell, being a purely ohmic resistance. Apart from the ohmic voltage drop across the electrolyte, kept in the pores of the matrix, a marked contribution to the voltage decrease comes from the polarization of the oxygen electrode. At a current density of $100\,mA/cm^2$, this amounts to 0.5 V. The polarization of the hydrogen electrode is considerably lower.

20.3   MCFC WITH INTERNAL FUEL REFORMING

From the very outset of molten carbonate fuel cell development, research workers were attracted by the fact that not only hydrogen but also carbon monoxide could be used as a reactant fuel (reducing agent). Carbon monoxide (the so-called water gas, a mixture of CO and H_2) is readily obtained by the steam gasification of coal:

$$C + H_2O \rightarrow CO + H_2 \tag{20.6}$$

This opens up possibilities for an indirect electrochemical utilization of huge coal reserves. The possibility of direct electrochemical oxidation of carbon monoxide was soon questioned. It was suggested that hydrogen, rather than the carbon monoxide, is involved in the electrochemical reaction after being formed from CO by the Boudouard reaction:

$$CO + H_2O \rightarrow CO_2 + H_2 \tag{20.7}$$

which readily occurs under the operating conditions of the molten carbonate fuel cell. Later it was shown that pure carbon monoxide actually can be oxidized electrochemically, according to the reaction:

$$CO + CO_3^{2-} \rightarrow 2CO_2 + 2e^- \tag{20.8}$$

though in fact, the rate of this reaction is about 20 times lower than that of the hydrogen oxidation, reaction (20.1). The exchange current density (as an index of reaction rates) is about $0.7 \, mA/cm^2$ for the oxidation of water gas (a mixture of 56% $H_2 + 8\% \, CO + 28\% \, H_2O$), and is about $1 \, mA/cm^2$ for the oxidation of pure hydrogen. This implies that the use ("combustion") of carbon monoxide in molten carbonate fuel cells occurs mainly, but not exclusively, via the intermediate formation of hydrogen.

The possibilities of using the products of gasification of biomass or residential waste that have a large carbon monoxide content but include a variety of contaminants have been examined by different authors.

The rates of direct electrochemical oxidation of hydrocarbons (and, particularly, of methane) in molten carbonate fuel cells are negligibly small. Therefore, only natural kinds of fuel can be used in these fuel cells after their prior conversion (reforming) to hydrogen. As the reforming process is highly endothermic and requires a large heat supply, the idea sprang up that this process should be carried out within the fuel cell itself where the heat of reaction, evolving in the fuel cell, could be utilized for the reforming (by analogy with the example of carbon monoxide conversion to hydrogen reported earlier).

Two kinds of internal-reforming fuel cells (IRFCs) are distinguished: the direct internal-reforming fuel cells (DIRFCs), where the reforming process takes place within the fuel cell at its anode catalyst, and the indirect internal-reforming fuel cells (IIRFCs), where plates with special reforming catalysts are included within the fuel cell's stacks.

The most active company working in this field is the American company FuelCell Energy, which, starting from the late 1970s developed and delivered a large series of power plants designated as Direct Fuel Cells®. These plants include a combined internal-reforming fuel cell system. In the stacks, special plates for prior fuel reforming are placed between groups of 8–10 direct internal-reforming fuel cells. With this system one can achieve a more uniform temperature distribution within the stack; heat is evolved in the fuel cells and consumed at the reforming plates.

20.4   THE DEVELOPMENT OF MCFC WORK

The work of Baur et al. (1916, 1921) must be recognized as the first work in the field of molten carbonate fuel cells. It is important to note that the "solid electrolyte" (a mixture of monazite sand and other components including alkali metal carbonates), with which O. Davtyan worked in the 1930s, actually represented a carbonate melt immobilized in a solid skeleton of silicates.

Work on molten carbonate fuel cells was resumed after 1960 in many places: in the Institute of High-Temperature Electrochemistry in Yekaterinburg, Russia (Stepanov, 1972–1974), and in the Institute of Gas Technology in Chicago. A large contribution in this field was made by Broers and Ketelaar (1958–1960) in Amsterdam (Broers, Ketelaar (1961)).

Industrial organizations were first to become interested in molten carbonate fuel cells, in about the mid-1960s. At present, a large number of companies, both in the United States and in other countries (Germany, Japan, Korea, etc.), are working in this field. Large-scale production of power plants operating with such fuel cells has been initiated. A particularly vigorous growth of the production and use of molten carbonate power plants is seen since 2002. In the power plants, Direct Fuel Cells (DFC®), described in the previous section of this chapter, electrodes having a surface area of about 1 m^2 are used. In total, a stack includes some 350–400 individual fuel cells. The average voltage given out by an individual fuel cell is 0.8 V. The overall efficiency of energy conversion in such a power plant (referred to the lower heating value) is about 51%. Starting from 2000, more than 40 power plants in sizes from 250 kW to 1 MW have been delivered and installed in various countries. According to a statistical report, including data up to April 2005, these plants had produced more than 70 million kilowatt-hours of electrical energy.

Work to build power plants on the basis of molten carbonate fuel cells has also been done at UTC and a number of companies in different countries: in Germany by MTU CFC Solutions GmbH, Munich, jointly with FCE, United States, in Italy by Ansaldo S.p.a., in Japan by Ishikawajima Harima Heavy Industries Co., and others.

The largest power plant with molten carbonate fuel cells was a 2-MW installation, built in Santa Clara, California, as a joint project of five utilities, the US Federal government, and five research organizations. The total cost of the project was 46 million US dollars. The plant consisted of 16 stacks, each having a power of 125 kW, laid out in four modules. Construction started in April 1994 and was finished in June 1995. After 720 h of operation, problems came up that had to do with thermal decomposition of insulating materials, leading to deposition of carbon particles on the electrode surfaces and connecting buses. In 1996 the operation of this plant was definitely terminated. In all, this plant has worked for 5290 h (partly at half power) and produced 2500 MWh of electrical energy.

20.5 THE LIFETIME OF MCFCs

In view of their ability to work with different kinds of fuel, MCFCs are of great interest. The electrical and operating parameters of these fuel cells are quite sufficient for building economically justified stationary power plants with a relatively large power output. The only point that, so far, that poses a problem is an insufficiently long trouble-free operation. The minimum length of time a large (and expensive) power plant should work, until it would be replaced, is 40,000 hours (4.5–5 years).

First laboratory prototypes of molten carbonate fuel cells, built in the 1960s, were in the best case operative for just a few months. At present, intense research and engineering efforts have made it possible to build individual units that have worked several hundreds and thousands of hours. Yet, the road to a guaranteed 5-year operation is still long. There are many reasons that lead to a gradual decline in the performance of such power plants or even to a premature failure. The most important reasons associated with the fuel cells themselves (rather than with extraneous reasons rooted in auxiliary equipment or operating errors) are stated later.

20.5.1 Gradual Dissolution of Nickel Oxide from the Oxygen Electrode

Nickel oxide dissolves in the carbonate melt according to the equation:

$$NiO + CO_2 \rightarrow Ni^{2+} + CO_3^{2-} \tag{20.9}$$

Owing to this dissolution process, the weight and thickness of the cathode will decrease by 3% after thousand operating hours of the fuel cell. The concentration of Ni^{2+} ions in the melt can attain values of $10-15$ ppm. These ions spread by diffusion all across the electrolyte-filled matrix. The nickel sites are reduced to metallic nickel by diffusing hydrogen according to the equation:

$$Ni^{2+} + H_2 + CO_3^{2-} \rightarrow Ni + CO_2 + H_2O \tag{20.10}$$

Deposits of metallic nickel have been discovered in different parts of operating fuel cells. Experimental results concerning the influence of various factors on the rate of nickel oxide dissolution and nickel ion transport that were obtained by different researchers are contradictory. The so-called basicity of the melt has a distinct effect. In carbonate melts, the basicity is not determined by the concentration of OH^- ions, as in aqueous solutions, but by the concentration of the O^{2-} ions, formed in the equilibrium reaction:

$$CO_3^{2-} \rightleftarrows O^{2-} + CO_2 \tag{20.11}$$

According to this equation, the basicity of the melt (or concentration of O^{2-} ions) is a function of carbon dioxide partial pressure. Under the conditions of molten carbonate fuel cell operation, the basicity of the melt is low (the acidity is relatively high), and nickel oxide will dissolve with a relatively high rate. This rate can be reduced by raising the basicity of the melt via the lowering of the carbon dioxide partial pressure. This, though, will lower the rate of the cathode reaction and the overall operating efficiency of the fuel cell. Thus, it will be necessary to select an intermediate value of carbon dioxide partial pressure that represents a compromise. An acceptable value is $0.15-0.20$ bar. Another way to achieve a certain increase in melt basicity is by raising the fraction of lithium carbonate in the melt. A similar effect is observed when adding a small amount of alkaline earth metal ions (Mg^{2+}, Ca^{2+}, Sr^{2+}, and Ba^{2+}) to the melt.

20.5.2 Anode Creep

When assembling a filter-press battery from individual fuel cells of any type, considerable compression must be applied in order to minimize the contact resistance between individual fuel cells. When the stack is in a vertical position, the fuel cells in the lower part of the stack experience an additional load because of the weight of the fuel cells above. In molten carbonate fuel cells, where the working temperature (in Kelvin) is higher than 50% of the nickel's melting point, a prolonged action of mechanical load and temperature leads to a plastic deformation of the nickel anodes that has been named "anode creep." As a result of this creep, the anode's thickness may decrease by up to $1-3\%$.

This compression leads to a smaller pore volume in the anode, reflected in the degree of the electrode's gas filling and, thus, in the electrode characteristics. In addition, anode creep may disturb the contact between anode and matrix. Both factors are limiting factors for the lifetime of molten carbonate fuel cells. Nickel is alloyed with chromium and/or aluminum in order to reduce the plasticity of the nickel anode and make it more rigid.

20.5.3   Corrosion of Metal Parts

Another factor limiting the lifetime of molten carbonate fuel cells is corrosion of various structural metal parts of the fuel cells. Usually, nickel-plated stainless steel is used for making these parts. The temperature in molten carbonate fuel cells leads to a high corrosive activity of the melt. This activity is particularly high near the anode space with its reducing properties. Here a carburetion of steel is possible, which leads to inferior mechanical properties. In addition to this, corrosion leads to a higher contact resistance between bipolar plates and electrodes. Earlier on, the bipolar plates were nickel-plated only on the side of the anode. It has now been established that both sides of these plates must be nickel-plated. During prolonged operation of the fuel cells, individual components of the stainless steel may diffuse through the nickel layer thus diminishing the protective power of the layer.

REFERENCES

Baur E, Peterson A, Füllemann Q. Z Elektrochem 1916;22:409.

Baur E, Treadwell G, Trümpler G. Z Elektrochem 1921;27:409.

Broers GHJ, Ketelaar JAA. In: Young GC, editor. *Fuel Cells.* New York: Reinhold, Vol. 1; 1961. p 78.

REVIEWS AND MONOGRAPHS

Bischoff M. J Power Sources 2006;160:842.

Davtyan OK. *The Problem of Direct Conversion of the Chemical Energy of Fuels to Electrical Energy*[in Russian]. Moscow: Publishing House of the USSR Academy of Sciences; 1947.

Farooque M, Maru HC. J Power Sources 2006;160:827.

Joon K. J Power Sources 1996;61:129.

Selman JR. J Power Sources 2006;160:852.

Tomczyk P. J Power Sources 2006;60:858.

21

SOLID OXIDE FUEL CELLS (SOFCs)

Fuel cells of this class are built with solid electrolytes having unipolar ion conductivity. Best known among these electrolytes is yttria-stabilized zirconia YSZ, that is, zirconium dioxide doped with the oxide of trivalent yttrium: $ZrO_2 + 10\%$ Y_2O_3 or $(ZrO_2)_{0.92}(Y_2O_3)_{0.08}$. This compound was the basis of Nernst's (glower) lamp of 1897, which became known as Nernst mass, and was already regarded as a candidate electrolyte for fuel cells by Baur (1937) and Davtyan (1938). It is commonly found in the oxygen sensors (lambda probes) used to optimize the operation of internal combustion engines.

The Y^{3+} ions introduced into the crystal lattice of zirconium dioxide act as dopant ions by giving rise to the formation of oxygen vacancies in the lattice, and electrical conductivity comes about by O^{2-} ions jumping from a current position into a neighboring vacancy, thus filling this vacancy while leaving a vacancy behind (one could also visualize this process as vacancy migration, which occurs in a direction opposite to that of the O^{2-} ion migration).

The conductivity of the prototype zirconia-type electrolytes becomes acceptable (with values of about 0.15 S/cm), only at temperatures above 900°C. For this reason the working temperature of fuel cells having such an electrolyte is between 900 and 1000°C. Such fuel cells are called *conventional solid oxide fuel cells* in the following chapters.

Electrochemical Power Sources: Batteries, Fuel Cells, and Supercapacitors, First Edition.
Vladimir S. Bagotsky, Alexander M. Skundin, and Yurij M. Volfkovich
© 2015 John Wiley & Sons, Inc. Published 2015 by John Wiley & Sons, Inc.

21.1 SCHEMATIC DESIGN OF A CONVENTIONAL SOFC

Conventional solid oxide fuel cells exist in several design variants. The basic variants are tubular and planar cells. So-called monolithic cells joined the first two variants around 1990. The specific design and operating features of these and other variants are described in the subsequent sections of this chapter. Those factors that are common for all variants of conventional solid oxide fuel cells are described in this section. The anodes of these fuel cells consist of a cermet (ceramic–metal composite) of nickel and zirconia electrolyte. This material is made from a mixture of nickel oxide (NiO) and the YSZ electrolyte. The nickel oxide is *in situ* reduced to metallic nickel forming highly disperse particles that serve as the catalyst for anodic fuel–gas oxidation reactions. These particles are uniformly distributed in the solid electrolyte, and are prevented by this electrolyte from agglomerating during fuel cell operation, thus, retaining their catalytic activity. The prototype zirconia-doped material present in the anode also improves the contact between the nickel catalyst and the fuel cell's electrolyte layer.

The cathodes consist of manganites or cobaltites of lanthanum doped with divalent metal ions, for example, $La_{1-x}Sr_xMnO_3$ (LSM) or $La_{1-x}Sr_xCoO_3$ (LSC), where $0.15 < x < 0.25$. Apart from their O^{2-} ion conductivity, these cathode materials also have some electronic conductivity that secures a uniform current distribution over the entire electrode. The doped lanthanum cobaltite has a higher ionic conductivity than the doped lanthanum manganite, but is more expensive and leads to problems, in particular on account of a possible chemical interaction with the electrolyte.

The three components of the fuel cell, anode, cathode, and electrolyte form a membrane–electrolyte assembly, as, by analogy with polymer electrolyte fuel cells, one may regard the thin layer of solid electrolyte as a membrane. Any one of the three membrane–electrode assembly components can be selected as the entire fuel cell's support and made relatively thick (up to 2 mm) in order to provide mechanical stability. The other two components are then applied to this support in a different way: as thin layers (tenths of a millimeter). Accordingly, one has anode-supported, electrolyte-supported, and cathode-supported fuel cells. Sometimes though an independent metal or ceramic substrate is used to which, then, the three functional layers are applied.

The important components of solid oxide fuel cells, which to a large extent govern their operating reliability (and sometimes also their manufacturing cost), are the interconnectors needed for combining the individual fuel cells to a stack. These must be purely electronically conducting, chemically sufficiently stable toward the oxidizing and reducing atmospheres within the fuel cells, and free of chemical interactions with the active materials on the electrodes.

Hydrogen and carbon monoxide can be used as the reactive fuels in solid oxide fuel cells. The anodic oxidation of carbon monoxide can be represented by the equation:

$$CO + O^{2-} \rightarrow CO_2 + 2e^- \tag{21.1}$$

The open cell voltage U_0 in hydrogen–oxygen solid oxide fuel cells, at a temperature of 900°C, has a value of about 0.9 V, the exact value depending on the reactant composition. The relation between current density and voltage of an operating cell U_i versus i is practically linear, that is, the fuel cell's apparent internal resistance, R_{app}, has a constant value that does not depend on current density. This does not constitute evidence for the voltage drop (that is, the electrolyte's ohmic resistance) having merely ohmic origins. Rather, this parameter is due, in part, to special features of electrodes' polarization. At a current density of 200 mA/cm^2, the cell voltage typically has a value of about 0.7 V, that is, the, parameter R_{app} has a value of the order of 1 Ω cm^2 (its exact value will depend on the thickness of the electrolyte layer).

21.2 TUBULAR SOFCs

21.2.1 Tubular Cells of Siemens–Westinghouse

Right at the start of the new upswing in fuel cell development, at the American Westinghouse Electric Corporation, a hydrogen–oxygen fuel cell was built around a tube of ion-conducting electrolyte, $(ZrO_2)_{0.85}(CaO)_{0.15}$, that is, a material somewhat different from yttrium-doped zirconia. Like a test tube, it was closed off on one end. In the bottom part of this fuel cell, thin platinum electrodes (less than 25 µm thick) were disposed on the inside and outside. Oxygen was fed into the tube; from outside it was bathed in a hydrogen stream. At a temperature of 1000°C and gas pressures of about 1 bar, the cell had an OCV of about 1.15 V; at a current density of 110 mA/cm^2, the cell voltage was about 0.55 V. The current–voltage relation was strictly linear. The authors attributed the linear voltage drop with increasing current density to the ohmic resistance of the rather thick electrolyte layer between the electrodes. Not long after building this first prototype of a tubular solid oxide fuel cell of electrolyte-supported design that was associated with large ohmic losses, Westinghouse switched to a new cathode-supported design admitting a much thinner electrolyte layer and, thus, much lower ohmic losses. Also, the yttria-stabilized zirconia electrolyte was used for all subsequent work. In 1998, Westinghouse joined forces with the German company Siemens that, up to then, had worked on planar solid oxide fuel cells, which gave rise to a new enterprise called Siemens–Westinghouse (S–W). This enterprise, then, specialized in the further development and commercialization of tubular solid oxide fuel cells and soon became the world leader in this field.

In the S–W cells, ceramic tubes produced by extrusion of the cathode material, lanthanum manganite (with some added alkaline earth metal oxides), are used. They have a porosity of 30%, lengths between 50 and 150 cm, and a diameter of 22 mm. From outside, a thin layer of a doped prototype electrolyte (40 µm) is applied by chemical vapor deposition from the vapor of a mixed $ZrCl_4$ + YCl_3 + water vapors and oxygen. From inside, a layer of lanthanum manganite with a magnesium or strontium dopant is applied as cathode material by plasma spraying. Anode material is deposited on top of the electrolyte from a slurry of Ni (or NiO) and YSZ material, and then sintered. A narrow band, 85 µm thick, of lanthanum chromite doped with divalent cations (Ca^{2+}, Mg^{2+}, or Sr^{2+}), which is an electronically conducting semiconductor material, is plasma-sprayed along the tube's outside to serve as cell interconnector for series combination of fuel cells. In a power plant built with such fuel cells, several tubes are combined into a bundle.

A 100-kW power plant was built by S–W in Westervoort (Netherlands) from tubular cells. The fuel cell stacks used in this plant contained four bundles of said type, combined in series to form a row, 12 rows then being placed in parallel. Between the rows, units for conversion of natural gas were installed. The plant also included units for desulfurization and pre-reforming of the natural gas.

During the years 1998–2000, the power plant operated for 16,000 h, providing local grid power of 105–110 kW and an additional 65 kW equivalent as hot water to the district heating scheme. The system had an electrical efficiency of 46% and an overall efficiency of 75%. The operation of this plant gave proof of the high functional reliability of tubular solid oxide fuel cells under real-world conditions of a large power plant (including several temperature cycles caused by temporary stoppages, and the use of natural gas, containing sulfur). In March 2001 the system was moved from the Netherlands to a site in Essen, Germany, where it was operated by a German utility company RWE for an additional 3700 h, which amounts to a total of over 20,000 h. Following the Essen experience, the system was brought to GTT-Turbo Care in Turin, Italy. So far the system has reached an operating time of about 37,000 h, with a minimal degradation.

21.2.2   Other Versions with Tubular Electrodes

In order to achieve higher specific power, it was proposed to use tubular cells of very small diameter (microtubular or submillimeter tubular solid oxide fuel cells). Tubes of smaller diameter have another important advantage. In fact, the mechanical stresses experienced by all ceramic parts under conditions of drastic temperature changes (such as switching the fuel cell on or off) will lead to cracks when these changes occur repeatedly. These stresses, however, will be less significant, the smaller a linear dimension (here, the diameter) of the ceramic part is. With the aim of drawing lower currents from the electrode's unit surface area of the electrodes, and bringing the fuel cell design closer to that with flat electrodes, a new shape of tubes, the so-called flat-tube with an oblong cross-section, was introduced. Cells of this shape yield a higher specific power of the stack per unit weight and per unit volume. Flat tubular electrodes are used up to now by Siemens for building large, SOFC-based power plants.

21.3   PLANAR SOFCs

Planar solid oxide fuel cells are built analogously to other kinds of fuel cells, such as polymer electrolyte fuel cells. Usually, one of the electrodes (the fuel anode or the oxygen cathode) serves as support for the membrane–electrode assembly. To this end, it is relatively thick (up to 2 mm), and thin layers of the electrolyte and the second electrode are applied to it. In batteries of the filter-press design, the membrane–electrolyte assemblies alternate with bipolar plates, having systems of channels through which reactant gases are supplied to the electrodes and reaction products are eliminated from them. In addition, the bipolar plates act as intercell connectors in the battery by passing the current from a given cell to its neighbor. Alternating with groups consisting of several cells each, special heat exchangers are installed for cooling the operating battery and heating an idle battery during start-up.

The development of planar solid oxide fuel cells started later than that of tubular fuel cells. They have several advantages over the latter. In stacks of planar design, higher values of specific power per unit weight and volume can be realized than in stacks with tubular cells. This is because of the fact that in planar design stacks, the current's path from all electrodes' surface segments to the current collector is shorter. The current path between the individual fuel cells is also shorter. All this leads to an important decrease of the ohmic losses in the stack. Another advantage of the planar version is a much simpler, less expensive manufacturing technology. The technological processes used in making planar solid oxide fuel cells are more flexible and allow different types of materials to be used for electrodes and electrolyte. For this reason, the basic work on entirely new solid oxide fuel cell versions and, particularly, those, able to work at intermediate and low temperatures has all been done with fuel cells of the planar type.

However, in the development of planar solid oxide fuel cells considerable difficulties turned up that, so far, have not been resolved definitively, thus preventing a broad commercialization of these fuel cells. The difficulties are chiefly related to the fact that the selection of materials having sufficient chemical and mechanical strengths for operation at temperatures of $900-1000°C$ in the presence of oxygen and/or hydrogen is rather restricted. This holds true both for the materials of electrodes and electrolyte and for the different structural materials. The major problems arising in the development and manufacturing of planar-type solid oxide fuel cells are the following.

21.3.1 Sealing

A large problem in high-temperature fuel cells is careful sealing. The gas compartments of the fuel and oxygen (or air) electrodes should be protected against the entry of the "wrong" gas. The joints and welds of all inner channels and gas manifolds should be free of gas leaks. For planar cells this is a much harder problem than for tubular cells, as their entire perimeter must be sealed. Also, planar cells usually have a larger number of gas feeds.

Two kinds of sealing materials, rigid and compressive, are in use. The elastic materials require constant compression during fuel cell operation. Rigid materials, on the other hand, must meet certain requirements of adhesion (wetting) and of compatible thermal expansion coefficients. Glass or glass ceramics were used initially as the basic sealant. By changes in glass composition and in the conditions of crystallization of glass ceramics, it was possible to adapt the properties of these materials to the operating conditions prevalent in the fuel cells, but these materials have the basic defect of brittleness. In recent years, therefore, the development of rigid and elastic sealants on the basis of metals and ceramics was initiated. By using multiphase materials, one can adjust the elastic properties and wetting of surfaces in contact. More details concerning the various sealing materials, discussed today, can be found in the review of Fergus (2005).

21.3.2 Bipolar Plates

The bipolar plates, functioning as intercell connectors, are the most expensive and at the same time the most vulnerable components of planar solid oxide fuel cells. The basic requirements that must be met by these plates are a high chemical stability under the fuel cell operating conditions, a high electronic conductivity, and complete impermeability to gases. Usually, two kinds of materials are used to make the bipolar plates: ceramics and thermally stable high-alloy steels. The ceramics mainly used for bipolar plates are oxides on the basis of lanthanum chromite $LaCrO_3$ doped with MgO, CaO, or SrO. It is a defect of this kind of material that at elevated temperatures structural changes occur in it, causing internal stresses in the plates. Another common defect of ceramic materials is their brittleness and insufficient mechanical strength (high sensitivity) under mechanical and heat shocks.

Chemically and thermally resistant steels contain considerable quantities of chromium (more than 20%). On the side of the oxygen, cathode metallic chromium is oxidized to Cr_2O_3. Depending on the temperature and oxygen partial pressure, this oxide may further oxidize to volatile compounds CrO_3 and $CrO_2(OH)_2$, which could then settle at the cathode/electrolyte interface and hinder oxygen reduction. Such an "evaporation" of chromium is the basic difficulty in the use of metallic bipolar plates.

21.3.3 Stresses in Planar SOFC

Considerable internal stresses often develop when making and using ceramic parts. Strong temperature changes as well as temperature gradients within a given part are the most important reasons for the development of these stresses. Exactly these factors distinctly arise in solid oxide fuel cells operated at temperatures going up to 1000°C. When such fuel cells are started or stopped, the temperature may change at a rate of hundreds of degrees per minute. Local temperature gradients arise when colder reactant gases enter a hot fuel cell. The risk that such stresses develop is particularly high in ceramic parts consisting of a number of different material layers that have a mismatch of their thermal expansion coefficients. The membrane–electrode assemblies in solid oxide fuel cells are exactly in this category. For this reason all materials used to make high-temperature fuel cells should have identical or at least very close thermal expansion coefficients. For many of the materials used this coefficient has values between 9 and $14\,K^{-1}$. A somewhat lower probability of internal stress development in ceramic parts can be achieved when making them from mixtures of coarse and fine powders of given materials.

All the problems and difficulties listed will arise precisely at the high working temperatures of conventional solid oxide fuel cells. For this reason, many research groups in different countries have tried over the past two decades to develop new fuel cell variants, working at lower temperatures.

21.4 VARIETIES OF SOFCs

21.4.1 Single-Chamber SOFC

In the design variant with a single chamber, the input streams of fuel and oxidant gases are not separate but enter the common reaction chamber together. A potential difference between the electrodes can develop, as highly selective catalysts are used in them. The catalyst in the negative electrode reacts only with the reducing agent (reformed hydrocarbon, syngas), and oxidizes it electrochemically to the final product, while the catalyst in the positive electrode reacts only with oxygen (usually, air oxygen), reducing it to water. This design variant is much simpler, than the usually adopted design with separate reactant supply. It is essential that the need for reliable, heat-resistant sealing, encountered with the usual designs, does not exist here. Yet, this variant has the important defect of a lower fuel utilization efficiency, as parasitic side reactions, not related to the generation of electrical energy, may occur because of a direct chemical interaction between the reactants or because of a reaction of a reactant at the "wrong" electrode. In many applications, and particularly so in small power plants for mobile and portable equipment, a simple design and simple gas management are more important than a high fuel utilization efficiency. To a certain extent, the ratio of the rates of the current-generating and parasitic reactions can be adjusted by changing the amount of oxygen in the gas mixture entering the fuel cell compartment.

21.4.2 Ammonia SOFC

In recent years, ammonia was suggested as a fuel for solid oxide fuel cells by a number of researchers. Ammonia is a large-scale product of chemical industry. At temperatures above 500°C, ammonia is readily decomposed to nitrogen and hydrogen at nickel catalysts. Therefore, when ammonia is introduced into such a cell, it is completely converted to nitrogen and hydrogen at the nickel-containing anode; the hydrogen then undergoes electrochemical oxidation. This direct ammonia fuel cell is actually a direct internal ammonia-reforming fuel cell. Relative to hydrogen, ammonia has many advantages as a primary fuel for fuel cells. It is much simpler to handle, store, and ship. It poses practically no explosion hazard. At room temperature, it can be kept and transported as a liquid under a pressure of 8.6 bar. Unlike methanol, it does not need added steam for steam reforming. Unlike natural liquid fuels (petroleum products), its conversion to hydrogen is not attended by the formation of harmful contaminants (sulfur compounds), and charring of the catalyst will not occur in the fuel cells. The other product of ammonia decomposition (nitrogen) is completely harmless for the environment and for the operation of the fuel cell. The specific energy content of ammonia is rather (high 1.45 kWh/l). Ammonia is cheaper than many other kinds of fuel. The cost of ammonia producing 1 kWh of electrical energy is about $1.20, while for methanol and for pure (electrolytic) hydrogen, the cost figures are $3.80 and $25.40, respectively.

21.5   THE UTILIZATION OF NATURAL FUELS IN SOFCs

An important difference exists between solid oxide fuel cells and other kinds of fuel cells in that various kinds of natural fuels or products of a relatively simple processing of such fuels may also be directly utilized. As we know, the original aim of all work on fuel cells has actually been precisely the direct transformation of the chemical energy of natural fuels to electrical energy. In seeking solutions to this problem, researchers have encountered numerous difficulties, which in many cases could not be overcome practically. These difficulties were associated with the very low rates of electrochemical oxidation of these fuels and, also, with the presence of various contaminants hindering and sometimes completely blocking these reactions.

21.5.1   Preliminary Catalytic Conversion of Natural Fuels

For this reason, the most realistic way of utilizing these natural fuels in a fuel cell includes their prior chemical (catalytic) conversion to other substances, primarily hydrogen, that are more readily oxidized electrochemically. In addition to conversion, the final processing product must also be carefully freed from all contaminants that could hinder the electrochemical reaction. In view of the main goal of attaining a highly efficient utilization of the fuel's chemical energy, it will be obvious that installations combining conversion units, purification units, and fuel cells will be advantageous, apart from their not very large space requirements. Also, fuel conversion processes usually are highly endothermic, so that appreciable thermal energies must be supplied, usually, by burning part of the fuel. On the other hand, in fuel cell operation a large amount of heat is evolved, which must be rejected to the environment via a system of heat exchangers.

21.5.2   Internal Reforming Plants

The conversion processes proceed at elevated temperatures, close to those of solid oxide fuel cells' operation temperatures. Therefore, an important aspect of the work on solid oxide fuel cells has been the attempt to build unified plants, combining conversion processes and fuel cell operation. In this way the heat from the fuel cells could be transferred directly to the converters, with a much lower loss of thermal energy. This combination has been called "internal reforming." It had already been pointed out in Chapter 20 that there are two ways of conducting internal reforming in high-temperature fuel cells—direct internal reforming, when reforming occurs directly at the fuel cell electrodes, and indirect internal reforming, when reforming is conducted in a separate unit located within the fuel cell stack. In solid oxide fuel cells, a direct reforming is possible, as their nickel anodes are good reforming catalysts at the cell's operating temperature. However, two problems arise in the operation of such direct solid oxide fuel cells: (i) coking of the catalyst may occur, and (ii) the catalyst may become poisoned by sulfur contaminants of the natural fuel. These two problems are considered in the following two sections.

21.5.3   The Problem of Carbon Formation

The formation of carbon deposits (coking) is a serious problem in solid oxide fuel cell operation when natural hydrocarbon fuels are used. Carbon deposits on the surface of nickel-containing anode catalysts will drastically lower their activity. Two mechanisms for the formation of such deposits were found. According to the first mechanism, the organic molecule that is the carbon source becomes adsorbed on the catalyst surface. While interacting with the metal, carbon atoms penetrate into the metal and then evaporate from there to form fibrous deposits on the surface. This mechanism produces a dry (pitting-type) corrosion or "dusting" of the metals in contact with the hydrocarbons at high temperatures. According to the second mechanism, the organic molecules undergo pyrolysis in the gas phase, producing free radicals that polymerize to resinous substances. These substances block the pores of the anode and, in this way, interfere with reactant access to the reaction sites. Coking of the catalyst is particularly pronounced when dry hydrocarbons are supplied to the fuel cells. Simultaneous supply of steam, mixed with the hydrocarbons, for instance, in a ratio of $H_2O : C = 1.2$, is the best way of fighting coking. Coking can also be prevented by using copper instead of nickel in the anodes. A copper surface is practically insensitive, to all phenomena producing coking, but copper is also not sufficiently active as a catalyst for the electrochemical oxidation of hydrogen and carbon monoxide.

21.5.4   The Problem of Sulfur Compounds in Natural Fuels

All natural kinds of fuel contain a certain amount of sulfur compounds (sulfides, H_2S). The amount is relatively small in natural gas (less than 1 ppm), but in syngas and biogases, as well as in diesel fuel, there may be as many as $50-300$ ppm sulfur compounds, which will poison the catalysts consisting of nickel and many other materials. In conventional types of solid oxide fuel cells, operated at temperatures of about 900°C, this poisoning may be reversible, but at lower working temperatures the catalysts are poisoned irreversibly and completely lose their activity. In many cases, therefore, the gases entering a fuel cell must undergo desulfurization in special units. Electrode materials exist (for instance, sulfide-based materials), which are insensitive to any sulfur compounds, that may be present, but their catalytic activity for hydrogen oxidation is relatively low.

Aspects of the poisoning effects produced by sulfur compounds during the operation of solid oxide fuel cells have been considered in greater detail in a review by Gong et al. (2007).

21.6   INTERIM-TEMPERATURE SOFCs (ITSOFCs)

The high working temperatures of solid oxide fuel cells, between 900 and 1000°C, lead to numerous problems in the development, manufacture, and practical use of these fuel cells.

Various incidental processes leading to a gradual degradation of the fuel cells are accelerated as the temperature increases. This is true not only for corrosion processes, but also for the diffusion of individual components from the electrodes and electrolyte into phases with which they are in contact, which lowers the conductivity of these phases. In addition, the likelihood of thermal shocks and thermal gradients increases at high temperatures (for instance, when colder reactants hit hot segments of the electrodes) leading to mechanical stresses and, possibly, to the cracking of electrodes and electrolyte.

Many of the problems listed could be eliminated, if it is possible to lower the operating temperature of solid oxide fuel cells without any significant loss of performance and without losing the valuable possibility of direct internal hydrocarbon fuel reforming. Therefore, starting in the 1990s, many research groups initiated huge efforts to develop solid oxide fuel cells operating at lower temperatures. The first aim was that of lowering the operating temperature from 900°C to at least temperatures in the range from 600 to 700°C. Fuel cells having such an operating temperature were called "interim-temperature solid oxide fuel cells." Attempts to lower the operating temperature further to temperatures below 600°C were initiated a little later, approximately at the start of the twenty-first century. Cells having such an operating temperature were called "low-temperature solid oxide fuel cells." The temperature boundary between these two categories is conditional and may somewhat differ in the publications of different authors.

The task of lowering the working temperature of solid oxide fuel cells is made difficult by two facts: (i) the conductivity of the conventional yttria-doped zirconia-type electrolyte drops off sharply with decreasing temperature, (ii) the rates of the electrochemical reactions occurring at the electrodes used in conventional solid oxide fuel cells decrease with decreasing temperature, and the polarization of the electrodes (particularly, that of the cathode) increases accordingly. To some extent, the first factor could be overcome by using thinner electrolytes. The ohmic resistance of a layer of yttria-doped zirconia electrolyte less than 5 μm thick is not a limiting factor for designing fuel cells having a large specific power. However, it is a rather difficult task to make defect-free electrolyte membranes so thin, and yet sufficiently stable and reliable.

For this reason, the search for solid oxide fuel cells operable at lower temperatures has been primarily via the development of new types of materials for the electrolyte and the electrodes that could work at these temperatures.

21.6.1   New Types of Solid Electrolytes

Materials having relatively high ionic (oxide-ion) conductivity in the temperature range considered have been found in numerous studies in the field of

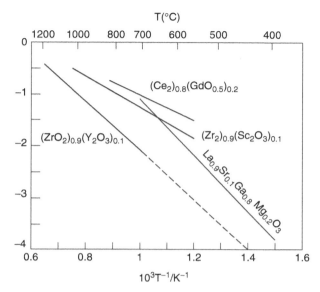

Figure 21.1. Temperature dependence of electrical conductivity for some solid oxide electrolytes: $ZrO_2 \cdot Y_2O_3$ (YSZ), $ZrO_2 \cdot Sc_2O_3$ (SDZ), $Ce_{0.8}Gd_{0.2}O_2$ (CGO), $La_{0.9}Sr_{0.1}$ $Ga_{0.8}Mg_{0.2}O_3$ (LSGM).

solid-state physics. Two such materials, doped ceria and doped lanthanum gallate, have been particularly attractive for work on solid oxide fuel cells development. Figure 21.1 shows plots of log σ against $1/T$ for the electrical conductivity σ as a function of temperature for four electrolyte materials: yttria-doped zirconia, scandia-doped ceria, gadolinia-doped ceria, and strontium-doped lanthanum magnesite gallate.

It can be seen from the figure that, over the temperature range from 1000°C to 400°C, the last two electrolytes have a markedly higher conductivity, than yttria-doped zirconia. It can also be seen from the figure that scandia-doped zirconia has a markedly higher conductivity than yttria-doped zirconia, but it is very rarely used on account of its high price.

For a higher conductivity, cerium dioxide can be doped either with gadolinium $Ce_{1-x}Gd_xO_2$ or with samarium $Ce_{1-x}Sm_xO_2$; doped cerias are quite stable chemically. In solid oxide fuel cells they lack the effect of interactions between the electrolyte and the cathode materials that would lead to the formation of poorly conducting compounds. However, doped cerias have an important defect in that at low oxygen partial pressures (such as those existing close to the anode) they develop a marked electronic conduction. This is entirely inadmissible for the electrolyte, as it leads to internal self-discharge currents and even to a complete internal short circuit. The electronic conduction comes about when Ce^{4+} ions in the lattice are partly reduced to Ce^{3+} ions, which creates the possibility for electrons to hop between ions of different valency.

Lanthanum gallate is doped with divalent metal ions, for instance, strontium and magnesium: $La_{1-x}Sr_xMg_yGa_{1-y}O_3$; usually, x and y have values between 1 and 0. Unlike the ceria-based electrolytes, this electrolyte can also be used at low oxygen partial pressures without the menace of developing an electronic conductivity. When in contact with cathodes of the LSM- or LSC type, some diffusion of manganese and cobalt from the cathode into the electrolyte is possible, but this has little effect on the fuel cell's performance. A certain defect of this electrolyte material is its complicated preparation procedure.

21.6.2   New Cathode Materials

The conductivity of strontium-doped lanthanum manganite, $La_{1-x}Sr_xMn_3$, used as cathode material in solid oxide fuel cells strongly decreases with decreasing temperature, the properties of the electrode, and of the fuel cell as a whole falling off accordingly. For these reasons, different new kinds of cathode materials were suggested for interim-temperature solid oxide fuel cells. A material based on strontium-doped barium cobaltite ferrite $BaFeO_3$: $Ba_{1-x}Sr_xCo_yFe_{1-y}O_3$ (where $0.1 \leq x \leq 0.5$, $0.2 \leq y \leq 0.8$) was suggested. Electrodes made from this material exhibited very good properties. However, it was shown that when using this material, difficulties arise because of the complex temperature dependence of the thermal expansion coefficients.

21.6.3   New Anode Materials

The anodes consisting of a nickel catalyst and of cermet mixed with yttria-doped zirconia electrolyte that are used in conventional solid oxide fuel cells also lose their ability to work at lower temperatures because of a loss of conductivity by the ceramic. This suggests that, for the ceramic in the anode, a material having a higher conductivity at intermediate temperatures should be used. It was in fact shown that an anode made with a nickel/samaria-doped ceria cermet has a much lower polarization than the conventional variant.

21.6.4   Prototypes of IT-SOFC

In 2006 at the Boston University, a prototype of an intermediate-temperature solid oxide fuel cell with a magnesium-doped lanthanum gallate electrolyte based on strontium was assembled. The thickness of the electrolyte was about 1 mm. The anode was a nickel/$Ce_{0.6}La_{0.4}O_2$ cermet with a thickness of 30–40 μm, and a porosity of 25–35%. Between the anode and the electrolyte, a thin layer (1.5 μm) of lanthanum-doped ceria was arranged in order to prevent a direct interaction between the nickel anode and the electrolyte, which would lead to the formation of a poorly conducting lanthanum nickelate phase. The fuel cell worked with air and moistened (about 3%) hydrogen. The cell's OCV at 800°C was 1.118 V, which is rather close to the theoretical EMF value. At 800°C and a current density of 400 mA/cm², the cell voltage was 0.45 V. The maximum energy density was 190 mW/cm². At a lower temperature of 600°C and a current density of 60 mA/cm², the cell voltage was 0.5 V, the maximum energy density was 30 mW/cm².

21.7 LOW-TEMPERATURE SOFCs (LT-SOFC)

Efforts to lower the operating temperature of SOFC to values below 600°C have continued right to the present time. Two directions are followed in this work: reducing the thickness of the electrolytes having higher conductivities at lower temperatures (such electrolytes were mentioned in Section 21.6.1), and developing new composite electrolytes having even higher conductivities.

21.7.1 Low-Temperature SOFCs with Composite Electrolytes

In collaboration between groups of scientists at the Royal Institute of Technology in Stockholm, Sweden, and the University of Science and Technology in Hefei, China, systematic investigations of composite electrolytes based on doped ceria and different nanosized inorganic additives were carried out. It was shown that certain properties of the ceria electrolyte undergo substantial change in such composites. First, electronic conduction that is harmful to fuel cell operation is suppressed; second, the conductivity increases dramatically at both high and low temperatures. In many cases stunning effects were seen.

The work on composite electrolytes is of great interest and may bring solid oxide fuel cells closer to commercialization.

21.8 FACTORS INFLUENCING THE LIFETIME OF SOFCs

It had been pointed out in Section 21.2 that a power plant with a battery of tubular solid oxide fuel cells had been more or less successfully operated for about 16,000 h. Apart from this communication, very few data can be found in the literature as to any results of long-term testing of solid oxide fuel cells. As this type of high-temperature fuel cells is primarily intended for large stationary power plants and, thus, has large investment needs, information as to potential lifetimes and reasons for gradual performance degradation or possible cases of sudden failure is extremely important.

In Section 21.3, the fundamental problems arising in the development of planar solid oxide fuel cells have been listed: sealing, corrosion of the bipolar plates, and the development of mechanical stresses in the numerous ceramic components found in such fuel cells. Solutions depend not only on the selection of suitable materials for the individual component parts of the fuel cells, but also on respecting certain principles of design and operation.

So-called thermal shocks have a highly negative effect on lifetime. These are sudden temperature changes in individual segments of the fuel cell. The development of considerable temperature gradients within given segments is also detrimental. In fuel cell operation, considerable heat is set free. In solid oxide fuel cells with internal fuel conversion, part of the heat evolved is absorbed by the endothermic reforming reaction. Yet, artificial cooling is usually required for the elimination of excess heat. In power plants with solid oxide fuel cells this is often achieved by blowing large amounts of air (large relative to the stoichiometric needs of the reaction). Owing to the temperature gradients that develop, the point of entry of the cold air into the hot fuel cell is very vulnerable. Various schemes to preheat the incoming air with heat from the exhaust gases have been developed.

The same problem arises when feeding fuel and the components (water or steam) needed for fuel reforming into the fuel cell. According to Equation (1.17), heat evolution is stronger the lower the operating voltage of the fuel cell. From this point of view, it will be preferable to work not with a cell voltage of 0.7 V, which is often used, but rather, with cell voltage not lower than 0.75–0.80 V and low currents. This will not only lower the influence of thermal shocks, but also lead to large savings in the energy needed to operate the cooling system (pumps, ventilators, etc.).

Another important aspect is that of thermal management during the start-up and shutdown of a power plant. Various processes taking place in the electrodes and in the electrolyte of a fuel cell also influence the lifetime of solid oxide fuel cells, in addition to the factors already mentioned earlier. As in other types of fuel cells, the highly disperse metal catalysts have a tendency that their active working surface area gradually shrinks with time. This decrease leads to a gradual increase in the polarization of the corresponding electrode.

This phenomenon is seen particularly in the highly disperse nickel used in the anodes of most solid oxide fuel cell variants.

A serious problem arising during solid oxide fuel cell's operation is the interaction between the materials of electrodes and of the electrolyte, by diffusion of individual components from a given phase to a neighboring phase in contact with it. This interaction often gives rise to the formation of new phases or compounds having a low conductivity.

Many of the factors influencing the solid oxide fuel cell's lifetime that are associated with ageing and degradation of these cells have been discussed in a review of Tu and Stimming (2004).

REFERENCES

Baur E et al. Z Elektrochem 1937;43:727.

Davtyan OK. *The problem of Direct Conversion of the Chemical Energy of Fuels into Electrical Energy [in Russian].* Moscow: Publishing House of the USSR Academy of Sciences; 1947.

MONOGRAPHS AND REVIEWS

Fergus JF. J Power Sources 2005;147:46.

Gong M, Liu X, Trembly J, Johnson C. J Power Sources 2007;168:289.

Riley B. J Power Sources 1990;29:223.

Singhal SC, Kendall K. *High Temperature Solid Oxide Fuel Cells.* Amsterdam: Elsevier; 2003.

Tu H, Stimming U. J Power Sources 2004;127:284.

Yamamoto O. Electrochim Acta 2000;45:2423.

22

OTHER TYPES OF FUEL CELLS

22.1 PHOSPHORIC ACID FUEL CELLS (PAFCs)

22.1.1 Early Work on Phosphoric Acid Fuel Cells

A hydrogen–oxygen fuel cell with a liquid acidic electrolyte was described as an example in Section 16.4. For that discussion it was not necessary to specify a particular acid, as the acid anion does not take part in the electrode reactions. In electrochemical studies solutions of sulfuric acid are usually used in different concentrations as an acid electrolyte.

However, throughout the history of fuel cell development, very few attempts were made to build a hydrogen–oxygen fuel cell with a sulfuric acid solution as an electrolyte. Sulfuric acid proved not to work well in fuel cells as an electrolyte.

In the early 1960s, concentrated solutions of phosphoric acid, with which the working temperature could be raised to 150°C, were introduced. Even at that temperature, and even in large quantities, the platinum catalysts proved not to be sufficiently active for a practical fuel cell working with natural fuels (hydrocarbons and other fuels). The experience gathered in these attempts later led to successful hydrogen–oxygen phosphoric acid fuel cells.

Electrochemical Power Sources: Batteries, Fuel Cells, and Supercapacitors, First Edition.
Vladimir S. Bagotsky, Alexander M. Skundin, and Yurij M. Volfkovich.
© 2015 John Wiley & Sons, Inc. Published 2015 by John Wiley & Sons, Inc.

22.1.2   The Special Features of Aqueous Phosphoric Acid Solutions

Like other acids, phosphoric acid in aqueous solutions dissociates into ions according to

$$H_3PO_4 \rightarrow H^+ + H_2PO_4^-. \tag{22.1}$$

Unlike other acids, in concentrated phosphoric acid solutions (where the water concentration is low) hydrogen ions exist not as hydrated ions $H^+ \cdot nH_2O$, but as ions solvated by phosphoric acid molecules $H^+ \cdot nH_3PO_4$. For this reason, the solution's conductivity is a complex, nonmonotonous function of concentration. As for the conductivity mechanism, these ions do not move like a spherical particle in a viscous medium when an electric field is applied to the solution (the Stokes mechanism), but rather the protons alone jump in the field direction from one acid molecule to another acid molecule (the Grotthuss mechanism, suggested in 1806 as an explanation for the conductivity behavior in aqueous solutions).

Another special feature of phosphoric acid is its dimerization in aqueous solutions at a concentration of about 85 wt% and at higher temperatures:

$$2\,H_3PO_4 \rightarrow H_4P_2O_7 + H_2O \tag{22.2}$$

This change is very important for fuel cell operation. Phosphate ions adsorb well on the platinum catalyst surface, displacing the electrochemical reactants, which leads to an appreciably slower reaction. The pyrophosphate ions adsorb much less and so, when they are present, the reaction happens much faster than in the presence of phosphate ions, reducing the polarization of the electrodes and increasing fuel cell's voltage at high current densities.

Another feature of concentrated phosphoric acid solutions that is very important for fuel cells is the water vapor pressure, which decreases drastically with increasing concentration of the acid. This feature allows the phosphoric acid solution to be immobilized in a porous solid matrix, greatly simplifying the elimination of water as a reaction product from the fuel cell's cathode space by gas (oxygen or air) circulation. It is safe to adjust this circulation to the maximal current load (maximum rate of water production), without the need to readjust it at lower loads, as there is no risk of excessive drying of the matrix on account of the said feature. In this way, water elimination in phosphoric acid fuel cells has a peculiar self-regulation. No such feature exists in sulfuric acid solution, where for water elimination the acid itself would have to be circulated, which would cause problems of sealing and corrosion.

22.1.3   Construction of PAFCs

Basically, the construction of phosphoric acid fuel cells differs little from what was said in Section 20.4 about fuel cells with a liquid acidic electrolyte. In the development of phosphoric acid fuel cells and, two decades later, in the development of polymer electrolyte membrane fuel cells many similar steps can be distinguished, such as the change from pure platinum catalysts to catalysts consisting of highly disperse platinum deposited on a carbon support with a simultaneous gradual reduction of platinum content in the catalyst from 4 to 0.4, and then to 0.25 mg/cm^2, and the change from pure platinum to platinum–ruthenium catalysts.

Bipolar graphite plates having special channels for reactant supply and distribution over the entire electrode surface, which are now widely used in polymer electrolyte membrane fuel cell stacks, were for the first time used in phosphoric acid fuel cells.

The concentrated acid solution in phosphoric acid fuel cells is absorbed in the pores of a porous matrix with fine pores and a total thickness of about 50 µm. From the outside, this matrix electrolyte behaves like a solid electrolyte, preventing the reactant gases, hydrogen and oxygen, from getting to the "foreign" electrode and preventing their mixing. Among new materials suggested for the porous electrolyte matrix in phosphoric acid fuel cells a mixture of silicon carbide and polytetrafluoroethylene can be mentioned. A suspension of the components is mixed in a ball mill for a long period of time and then spread onto the surfaces of the cathode and anode. This procures good contact between the electrodes and the electrolyte immobilized in the matrix.

The first hydrogen–oxygen phosphoric acid fuel cells in the mid-1960s had 85% phosphoric acid and were operated at temperatures not higher than 100°C. Relative to the results obtained with alkaline hydrogen–oxygen fuel cells, the performance of these fuel cells was poor. For this reason, the phosphoric acid concentration was subsequently gradually raised, first to 95% and then to 100%, and then the temperature was brought up to 200°C. During the decade between 1975 and 1985, research in this field was forceful, and large industrial organizations gradually joined the efforts.

22.1.4 Commercial Production of PAFC

On account of the special features mentioned earlier (of the partial water vapor pressure decreasing with increasing phosphoric acid concentration), the elimination of product water from phosphoric acid fuel cells, they became markedly simpler. Thus, it was possible to make electrodes as large as 1 m^2. In fuel cells of the polymer electrolyte membrane fuel cell type, electrodes of such a size would give rise to unmanageably large difficulties in water elimination and in maintaining the water balance within the membrane electrode assembly. Thus, the first prototypes of large-scale power plants, producing a power of hundreds of kilowatts and more and having an industrial importance, were built with phosphoric acid fuel cells.

United Technologies Corporation built a large plant for phosphoric acid fuel cell production in 1969. Together with the Japanese company Toshiba a special enterprise Fuel Cells International was created for mass production of these fuel cells and for their further improvement. A little later, a daughter company was set up. In these industries, mass production of power plants PC-25, having an output of 200 kW, was started. These power plants are designated for the combined on-site heat and power supply of individual residential and municipal structures, such as hospitals. They were operated autonomously with natural gas, and, in addition to the phosphoric acid stacks, included equipment for the conversion of natural gas to hydrogen and for subsequent hydrogen purification. Such a plant had a total weight of about 16 tons and occupied four square meters floor space. The electrical efficiency was 35–40%. Including the thermal energy produced, the total energy conversion efficiency was

85%. The thermal energy was produced in the form of hot water with a temperature around 80°C, or also as superheated steam at 120°C (the total thermal power was about 800 MJ/h). Unlike ordinary power plants of such a size, a PC-25 power plant operates without producing noise or vibrations, and without noticeable output of contaminants into the ambient air. These power plants had a relatively large commercial success. In the United States, Japan, and various European countries, about 300 plants of this type were installed.

22.1.5   Development of Large Stationary Power Plants

A big effort toward large-scale phosphoric acid fuel cell-based power plants was made in Japan, in part with the involvement of the United Technology Corporation. Within the framework of a national program, backed by Japanese gas companies, numerous prototypes of power plants having a size between 25 kW and 1 MW were built. In 1983, the United Technology Corporation developed and set up in Japan a power plant with an electrical output of 4.8 MW. The phosphoric acid fuel cells in this plant had electrodes with an area 0.34 m^2. The cells in this plant worked at a temperature 190°C, and at a current density 250 mA/cm^2 developed a voltage of 0.65 V. This plant was operative during the years 1983–1985 and produced a total of 5.4 GWh of electrical energy. In 1991 the same company built an even larger phosphoric acid based power plant in Japan, and it had an electrical output of 11 MW. In this plant, the working temperature of the fuel cells was 250°C.

22.1.6   The Future for PAFC

Toward the end of the 1990s, interest in power plants based on phosphoric acid fuel cell power plants gradually waned, despite the success that had been achieved: the relatively large number of intermediate-power PC-25 plants were built and the installation of several megawatt-sized power plants took place in a number of places.

On the one hand, this had strict economic reasons—the high cost of such plants. On the other hand, there were purely technical problems, that is, an insufficient long-term operating reliability.

The Cost of PAFC-Based Power Plants

It was reported by an official of the United Technology Corporation and of Toshiba that the joint company together with several other companies and a number of government agencies spent about 200 million US dollars worth of research and technology work to develop and manufacture the PC-25 plants. The production cost of each of the early units was slightly over 1 million dollars, that is, more than $5000/kW. It should be pointed out for comparison that the cost of alternative power plants in the same class, for example, of wind turbines, is about $500–700/kW. A further decrease in the production cost of the fuel cell-based power plants would be possible by lowering labor cost in higher production volumes. One had hoped for a "virtuous cycle," but experienced a "vicious cycle," seeing that an increase in production volume and sales volume was not possible without lower prices, but lowering the prices without raising the production volume was equally impossible. Of course, measures have been taken to get through this impasse, but, so far, substantial results have not yet been achieved.

Reliability and Long-Term Operation of PAFC-Based Power Plants

Phosphoric acid fuel cell-based power plants of intermediate and large power output are designed to work for 50,000 h, which is approximately 6 years. During this period, mandatory checking and tune-ups are performed. Many of the large number of PC-25 type plants installed have worked for the designated period and continue to work. Yet, in individual units malfunctions and also a gradual fuel cell performance decline were seen, mainly because of processes on the positive oxygen electrode, particularly, to strong corrosion (electrochemical oxidation) of the platinum catalysts' carbon support in the cathodes' catalytic layer. This corrosion leads to a contact loss between the support and the catalyst particles. In addition, the highly disperse platinum particles recrystallize, and thus, the catalyst's working surface area decreases. No such effects occurred at the hydrogen electrode, even after prolonged operation. The corrosion processes have to do with a rather highly positive potential of the oxygen electrode. At lower current densities, and more so at open circuit and at low current loads, the potential shifts even further to the positive side, leading to a further corrosion rate increase.

22.1.7 Importance of PAFC for Fuel Cell Development

The first data about fuel cells based on concentrated phosphoric acid solutions date from the mid-1960s. After three decades, since about 1995, very few papers on phosphoric acid fuel cells have appeared in the scientific literature. Of course, many of the numerous intermediate and large-size power plants that had been built in prior years still function, but work toward their further development and improvement has practically ceased.

Despite the fact that the period during which these fuel cells were receiving prime interest lasted only about three decades, they had an important role in the development of the fuel cells as such. For the first time, a relatively large-scale industrial production of fuel cell-based power plants was initiated, and these plants were widely spread among many users. As a result, many scientific and business circles worldwide have recognized that cells, rather than science fiction, are an entirely real possibility for applications, useful to mankind. During this period of prolonged phosphoric acid fuel cell development, the first large government schemes and projects for fuel cell research and engineering were implemented in many countries worldwide (United States, Japan, Russia, and others).

During phosphoric acid fuel cell research and development, technical solutions were found, which later were adopted successfully in the development of other fuel cell types. This is true, in particular, for the use of platinum catalysts not in a pure form, but as deposits on carbonaceous supports (for instance, carbon black), leading to a considerable drop in the amounts of platinum needed to manufacture fuel cells. During this same period, the influence of carbon monoxide traces in hydrogen on the platinum catalysts' performance was first investigated. It was shown that platinum–ruthenium catalysts could be used to reduce the influence of carbon monoxide. Also during this period it was shown for the first time that the performance of oxygen electrodes could be improved by using catalysts of platinum alloyed with iron-group metals.

22.2   REDOX FLOW FUEL CELLS

By definition, redox flow fuel cells are a kind of fuel cells, set up in loops of circulating electrolyte solutions, containing reversible redox systems. By design these fuel cells are intended for a temporary accumulation of electrical energy that they later can give off to a user. In this way, they accomplish the function of a conventional storage battery stack. Unlike other types of fuel cells, they produce the "fuel" themselves by accepting power (charging up their redox systems). Like other fuel cells, they produce power so long as their redox systems are pumped through and give off electric charges to the electrodes. As in an ordinary fuel cell, the fuel (the redox solutions) is stored outside, and the fuel reserve can be as large as would be feasible in the overall context. The redox reactions occur at suitable inert electrodes.

22.2.1   Iron–Chromium Redox Flow Fuel Cells

Redox flow fuel cells working with redox systems consisting of iron and chromium ions are the simplest representatives of redox-type fuel cells and constitute a convenient example for describing redox flow fuel cells' working principles. A fuel cell consists of two halves separated by an ion-exchange membrane. The positive electrode (or positive half-cell) resides in one, and the negative electrode (or negative half-cell) resides in the other half-cell. An electrolyte solution, containing divalent and trivalent iron ions, flows through the positive half-cell. When the fuel cell delivers electric charges, the cathodic reaction occurring at the positive electrode is

$$Fe^{3+} + e^- \rightarrow Fe^{2+} \quad E^0 = 0.77 \, V \tag{22.3}$$

When the fuel cell is recharged, this reaction occurs in the opposite (anodic) direction.

An electrolyte solution containing divalent and trivalent chromium ions flows through the negative half-cell. When the fuel cell delivers electric charges, the anodic reaction occurring at the negative electrode is

$$Cr^{2+} \rightarrow Cr^{3+} + e^- \quad E^0 = -0.41 \, V \tag{22.4}$$

When the fuel cell is recharged, this reaction takes place in the opposite (cathodic) direction. The two reactions occur simultaneously, when the external circuit is closed. The overall current-generating reaction that occurs when the fuel cell delivers electrical charges is given by the equation:

$$Fe^{3+} + Cr^{2+} \rightarrow Fe^{2+} + Cr^{3+} \quad \varepsilon^0 = 1.18\,V \quad (22.5)$$

When the fuel cell is recharged, the reaction occurs in the opposite direction.

The cathode and the anode, where reactions (22.4) and (22.5) take place, are inert (chemically stable), but catalytically active, so that the reactions would go fast.

The polymer anion-exchange membrane separating the two half-cells contains positive functional groups (most often the cations, of quaternary alkylammonium compounds). These positive groups electrostatically repel the positive iron and chromium ions, thus preventing their crossover. Therefore, it is not possible for these ions to get to the "wrong" electrode, and their direct chemical interaction is excluded as well. There is no hindrance, though, for the anions of iron and chromium salts to pass through the membrane. Therefore, each elementary act of the reaction, described by the Equation (22.5), is attended by the transfer of an electron through the external circuit from the negative to the positive electrode, and also by a transfer of one negative electric charge (in the form of an anion) through the membrane and also through the internal circuit, from the positive to the negative half-cell, in order to make possible reactions (22.4) and (22.5), and to close the electrical circuit.

The equilibrium potentials set up by the iron and chromium redox systems at the inert electrodes depend on the concentration ratios of the divalent $[M^{2+}]$ and trivalent $[M^{3+}]$ ions (M stands for iron and chromium, respectively), according to the Nernst equation:

$$E_M = E_M^0 + \left(\frac{RT}{nF}\right) \ln \frac{[M^{3+}]}{[M^{2+}]} \quad (22.6)$$

When divalent and trivalent ions are present in identical concentrations, then $E_M = E_M^0$, The electromotive force of the fuel cell $\varepsilon = E_{Fe} - E_{Cr}$ depends on the four ion concentrations, through the two Equations (22.6). When the fuel cell delivers electric charges, the iron(III) ion concentration decreases, while the chromium(III) ion concentration increases. Accordingly, the potential of the positive electrode becomes more negative, and the discharge voltage of the negative electrode becomes more positive. Therefore, the fuel cell discharge voltage gradually decreases. When the fuel cell is recharged, the iron(III) ion concentration increases, and the chromium(III) ion concentration decreases. Thus, the electrode potentials move in the opposite directions, and the cell discharge voltage U_d gradually increases.

Carbonaceous materials are usually used in iron–chromium redox flow fuel cells as (inert) electrodes (carbon fiber cloth, felt, etc.). The solution for the positive half-cell usually contains a certain concentration of hydrochloric acid in addition to iron(III) and iron(II) chlorides.

Many investigations into iron–chromium redox flow fuel cells were carried out by a research group in Alicante University, Spain. At a temperature of 44°C with a 2.3 M HCl + 1.25 M $FeCl_2$ + 1.25 M $CrCl_3$ solution they attained a maximum specific power of 73 mW/cm^2. Codina et al. (1994) reported building a 20-fuel cell stack delivering a power of 0.1 kW. National Aeronautics and Space Administration (NASA) considered using such batteries for spacecrafts.

Redox flow fuel cells have certain advantages over conventional electrical storage batteries, such as lead-acid batteries: (i) the electrode reactions are relatively simple; (ii) the energy losses because of polarization of the electrodes are low; (iii) aging phenomena at the electrodes do not exist, and hence the lifetime is long; (iv) the electrical capacity and energy content of the system can be increased simply using larger redox reactant solution volumes (and related storage tanks); (v) there is no gas evolved during overdischarge or overcharge. A certain disadvantage of the redox flow fuel cells is the necessity to spend part of the energy, both when delivering power, and when recharging, for the operation of pumps moving the solutions containing the redox systems.

22.2.2 All-Vanadium Redox Flow Fuel Cells

In vanadium redox flow fuel cells, a redox system of penta- and tetravalent vanadium ions is used in the positive half-cell, and a redox system of di- and trivalent vanadium ions in the negative half-cell. When the fuel cell delivers charge, the following reactions take place:

$$(+) \; VO_2{}^+ + 2H^+ + e^- \rightarrow VO^{2+} + H_2O \quad E^0 = 1.00 \, V \tag{22.7}$$

$$(-) \; V^{2+} \rightarrow V^{3+} + e^- \quad E^0 = -0.26 \, V \tag{22.8}$$

$$\text{Overall:} \; VO_2{}^+ + V^{2+} + 2H^+ \rightarrow VO^{2+} + V^{3+} + H_2O \quad \varepsilon^0 = 1.26 \, V \tag{22.9}$$

During recharging, the reactions occur in the opposite direction. As stated earlier, the values cited for the thermodynamic parameters of the reactions refer to the situation where the concentrations of the vanadium ions in the higher and lower oxidation states are equal.

A large program of investigations into such vanadium redox flow fuel cells was conducted between 1985 and 2003 by workers at the University of New South Wales in Kensington, Australia. Stack prototypes, having a power of 1 kW, were developed. In the two half-cells, solutions 1.5–2 M in vanadium sulfate and 2.6 M in sulfuric acid were used. Carbon felt was the electrode material. A stack consisting of 10 fuel cells in series could be cycled between voltages of 8 V (the terminal discharge voltage) and 17 V (the terminal charging voltage). At currents for discharge and charging of 20 A, the faradaic efficiency was 92.6%, and the voltage efficiency was 95%. The overall efficiency was 88%. At a discharge current of 120 A, and a charging current of 45 A, the voltage efficiency dropped to 73%, and the energy efficiency dropped to 72%. For electrochemical energy storage, such performance figures must be acknowledged as being quite high.

More specified data about the two systems described earlier can be found in a detailed review by Ponce de Léon et al. (2006).

22.3 BIOLOGICAL FUEL CELLS

Biological fuel cells are those in which at least one of the following two conditions is met:

(a) at least at one of the electrodes the electrochemical reactant is a substance found in biological fluids (for instance, in blood) or in other biological materials (for instance, in biomasses);

(b) at least at one of the electrodes, the catalyst for the electrochemical reaction are microorganisms (microbial fuel cells) or enzymes (enzymatic fuel cells).

At present, biological fuel cells are a subject of great interest for the following two reasons:

(1) The range of medical devices being implanted into the human body, such as pacemakers, defibrillators, insulin micropumps, and analyzers, periodically need electrical energy. Implanted batteries have a limited capacity and must be periodically replaced. This operation requires a repeated surgical intervention. A long uninterrupted function of these devices would be possible if implantable fuel cells could be built that use products of metabolism as the reducing agent and oxygen of hemoglobin as the oxidizing agent.

(2) While energy resources become really scarce in many countries, huge energy reserves in the form of biowaste (so-called biomasses) accumulate everywhere. Existing ways of utilizing them, for instance, by enzymatic transformation to ethanol and subsequent use of ethanol in heat engines are complicated and include many individual steps. A direct processing of biomasses in biological fuel cells would lead to a significantly more efficient utilization of their energy content. A definite solution to these final aims is still far away. So far, the research has been focused on individual, relatively narrow aspects of this problem.

22.3.1 Enzymatic Fuel Cells

Enzymes are natural, very active, and efficient catalysts for the chemical reactions occurring in the bodies of men and animals. The distinguishing feature of enzymes is their high selectivity. A given enzyme is active in the reaction of only one particular reactant (called *substrate*). The reactions occurring in the human body are homogeneous, proceeding in aqueous (physiological) solutions. A number of attempts have been made in recent decades to use enzymes for catalytic acceleration of electrochemical reactions (bioelectrocatalysis). To this end, the enzymes used are immobilized on the surface of the supporting electrodes (usually some carbonaceous supports). Special intermediate redox systems (mediators), serving as shuttles between the enzyme and the substrate, are used to secure electron transfer from the enzyme to the substrate (or vice versa). As a rule, the same ambient conditions that are found in biological systems should be maintained in the electrochemical systems, such as pH value of about 7, and a moderate temperature.

Topcagic and Minteer (2006) reported building a fuel cell that was working with simple reactants—ethanol and oxygen, but where the platinum catalysts that had usually been used were replaced by enzymes.

Alcohol dehydrogenase and aldehyde dehydrogenase were used as the enzymes for anodic ethanol oxidation. Together with the coenzyme NAD^+ (nicotinamide adenine dinucleotide), they formed the $NAD^+/NADH$ redox system. Aided by these enzymes, this reaction occurs in two steps. First, the former enzyme oxidizes ethanol to acetaldehyde, and then the latter enzyme oxidizes the aldehyde to acetic acid.

Bilirubin and bilirubin oxidase were used for cathodic oxygen reduction, while $Ru(bpy)_3^{3+}/Ru(bpy)_3^{2+}$ [$Ru(bpy)_3^{3+}$ and $Ru(bpy)_3^{2+}$ are complex tris (bipyridine)ruthenium(III) and tris(bipyridine)ruthenium(II) cations] was the mediator redox system. In the electrodes, these enzymes were immobilized with Nafion solution treated with quaternary ammonium salts, and put on a support of carbonized cloth, serving as the current collector. The treated Nafion solution helped to maintain enzyme activity for a long time.

The enzymes being highly selective, such electrodes were used in two types of fuel cells: (i) fuel cells with cathode and anode compartments separated by an ion-exchange membrane, and with an individual reactant supply; or (ii) fuel cells without separator, where the reactants are added as a mixture.

The membrane version had an open circuit voltage 0.68 V. The highest specific power achieved was $83 \, mW/cm^2$. In the undivided fuel cell, a buffer solution with a pH value 7.15 containing 1.0 mM ethanol and 10 mM NAD^+ was used. The solution was kept in contact with air, and after a number of days had an equilibrium concentration of the dissolved oxygen. The starting value of the open circuit voltage was 0.51 V. The highest energy density attained was $0.39 \, mW/cm^2$. The fuel cell worked for more than 30 days, at which point the specific power decreased by 20%. The reason was insufficient chemical stability of the used redox system. The enzyme systems themselves retained their activity for more than 90 days.

Results from investigations of electrochemical reactions occurring at individual electrodes with enzymatic catalysts have been reported in many older papers. Yaropolov et al. (1976) studied cathodic oxygen reduction at a pyrolytic graphite electrode with peroxidase as an enzyme. As the mediator of electron transfer, the quinone/hydroquinone redox system was selected. The electrode's polarization decreased by 30 mV, when the enzyme and the mediator were present in the solution. Betso et al. (1971) observed an increase in the oxygen reduction rate at a mercury electrode in the presence of cytochrome C. Wingard et al. (1971) saw an increase in the glucose oxidation current at a platinum electrode when glucose oxidase was present as an enzyme.

One of the properties of enzymatic systems that is of great interest for enzymatic fuel cells in general is their high selectivity. If universal but absolutely nonselective platinum catalysts could be replaced by other, highly selective and highly active catalysts, then fuel cells with a mixed reactant supply having a considerably higher technical and economic efficiency could be built.

22.3.2 Bacterial Fuel Cells

Bacterial fuel cells are less selective than enzymatic fuel cells and can be used for reactions with a great variety of reactants. A large group of microorganisms can be used for the oxidation of any carbohydrate compound in biomasses. Most often, a mixture of different kinds of microorganisms, rather than any particular single one were used. Zhang et al. (2006) wrote in their paper on bacterial fuel cells that samples of microorganisms were scooped up from ponds close to the laboratory.

However, in their mechanism and in their action nature bacterial and enzymatic fuel cells have much in common. In bacterial fuel cells intermediate redox systems are often used, as well, to facilitate electron transfer to (or from) the substrate. As the effect of microorganisms is much less specific than that of enzymes, a much wider selection of redox systems can be used, in particular, the simplest iron(III)/iron(II) system. The working conditions of these two kinds of biological fuel cells are similar as well: a solution with pH around 7.0 and a moderate temperature, close to room temperature.

Tests involving a bacterial fuel cell with anode and cathode compartment separated with a cation-exchange membrane were reported in the paper of Zhang et al. mentioned already. The compartments were filled with phosphate buffer solution of pH 7. The anode compartment was in addition made 20 mM in acetate or 10 mM in glucose. The cathode compartment was 50 mM in the salt $K_3Fe(CN)_6$. Carbonized cloth carefully freed of metallic contaminants was used as electrode material. Air was passed through the cathode compartment during the tests. A culture of pond microorganisms was grown in standard nutrient solutions and added to the compartments prior to the tests. The OCV in the fuel cell with glucose was about 0.55 V. At a load of 1000 ohms, the voltage was about 0.34 V (the maximum power was about 5 $\mu W/cm^2$). Figures of the same order of magnitude were obtained in the fuel cell with acetate. It is interesting to note that by electron microscopy, a morphological difference between the bacterial beds was noticed on the anodes after the tests with glucose and with acetate.

The effects of common baker's yeast were studied by Walker and Walker (2006). Here again, the fuel cell was divided into two halves by an ion-exchange membrane. The anodic part of the fuel cell contained a phosphate buffer solution with the yeast *Saccharomyces cerevisiae*, 0.1 M $K_3Fe(CN)_6$, and 0.2 M glucose. The cathodic part contained a mediator solution of the Fe^{3+}/Fe^{2+} type. When charge is drawn, the iron(III) ions are reduced to iron(II) ions, and the trivalent ions are, in turn, regenerated by air oxygen brought into this part of the fuel cell. At a temperature of 45°C, the maximum energy density was 130 $\mu W/cm^2$. At 10°C this value dropped to less than 20 $\mu W/cm^2$. Another version of a bacterial fuel cell, which worked with sugar industry effluents, was described by Prasada et al. (2006). A culture of *Clostridium sporogenes* was used in the anodic part of the fuel cell, and a culture of *Thiobacillus ferrooxidans* was used in the cathodic part. The open circuit voltage was 0.83 V, the maximum energy density was about 4 $\mu W/cm^2$.

22.4 DIRECT CARBON FUEL CELLS (DCFCs)

22.4.1 Purpose and First Efforts

When Wilhelm Ostwald, toward the end of the nineteenth century, formulated his idea of using an electrochemical mechanism for the direct conversion of natural fuels' chemical energy to electrical energy, coal was the chief kind of fuel in the hands of mankind. Even today, notwithstanding the widespread use of petroleum products and the development of nuclear power, coal remains a very important component of world energy supply. Its share of all known natural fuel reserves worldwide is about 60%. In China today, about 80% of the electrical energy is produced by coal-fired power stations, these being responsible for 70% of the carbon dioxide emissions and 90% of the sulfur dioxide emissions in this country (Cao et al., 2007).

An electrochemical utilization of the coal's energy would provide huge gains, not only and not so much because of higher conversion efficiencies, but also for other reasons. The thermal power stations emit into the atmosphere carbon dioxide mixed with air and other gases, as well as with uncombusted coal particles. In a fuel cell, coal would be oxidized anodically in separate compartments, closed off from the air. From the gases evolved in these compartments by-product hydrocarbons could readily be separated, and then utilized, while particulates could be filtered off. The remainder, almost pure carbon dioxide, could be sent to underground storages. This would make a huge contribution to solving the problem of global warming, caused by greenhouse gas emissions. Power plants with direct carbon fuel cells could, in principle, be set up directly in coal mines to eliminate numerous economic and ecological problems associated with long-distance coal transport.

There are two possibilities for an electrochemical utilization of the chemical energy of coal: (i) via prior coal gasification and the subsequent use of hydrogen and/or of carbon monoxide, thus produced, in various fuel cells; and (ii) by a direct electrochemical oxidation within the fuel cell. For the first of these possibilities no insurmountable technical or scientific problems arise. Gasification units and proton-conducting membrane fuel cells are well known and very reliable devices, so, their use is connected only with economic and lifetime problems.

Some attempts to realize the second approach that, basically, is a much simpler one-step process, are discussed in this section.

Antoine César Becquerel in France (1855) and Pavel Yablochkov in Russia (1877) have built electrochemical devices, using coal anodes in a molten potassium nitrate electrolyte. William Jacques (1896) obtained a US patent for his invention of a "coal stack" with a coal anode and an iron cathode immersed into molten alkali hydroxide. Despite the great doubts raised, as to the nature of the processes taking place in the stack, the electrical performance of this fuel cell stack, operating at temperatures from 400 to 500°C, had a rather impressive total power 1.5 kW, and current densities up to

$100\,mA/cm^2$. Much later, a carbon fuel cell prototype with a solid electrolyte, working at a temperature of 1000°C, was built by Baur and Preis in Switzerland (1937, 1938).

It was shown that an effective anodic oxidation of carbon cannot take place in low-temperature aqueous solutions, but only at high temperatures and in other electrolytes (particularly, in molten carbonates or alkali hydroxides). Thus, in molten carbonates at 600°C, the equation can be formulated as follows:

$$C + 2CO_3^{2-} \rightarrow 3CO_2 + 4e^- \quad E^0 = -0.02\,V \qquad (22.10)$$

At the same time, it must be remembered that at temperatures above 750°C, the so-called Boudouard equilibrium is established:

$$C + CO_2 \rightleftarrows 2CO \qquad (22.11)$$

which, during carbon oxidation, would lead to the formation of carbon monoxide, entailing some energy loss (only two electrons are liberated, instead of four, when a carbon atom is oxidized merely to carbon monoxide).

22.4.2 Work on Carbon Fuel Cells Since 1960

Despite the numerous problems encountered in earlier research, attempts to build versions of carbon fuel cells were continued when the "fuel cell boom" had started in the 1960s.

Two questions arise: when carbon (or coal) is used in fuel cells, what should be the nature and the physical state of the carbon material to be used. The notion of "carbon" is highly indefinite. It covers a large number of carbon materials, both of natural origin and manmade. It includes different kinds of graphite, coke, and carbon black, very strongly differing in their structure and in the content of additional components, both with volatile compounds (hydrogen, oxygen, nitrogen, sulfur, organic materials, etc.) and with involatile compounds (mineral salts). Carbon materials of natural origin could be used as such, or after having been subjected to pretreatments eliminating undesired components.

The physical state of carbon material in a fuel cell can be of two kinds: (i) relatively massive electrodes in the shape of plates or rods. Electrodes of this type may be cut out of graphite or pressed from powdered carbon materials (usually, with different binders), and may, at the same time, serve as current collector and as consumable electrode; (ii) highly disperse carbon powders, present as slurry in the liquid carbonate electrolyte, which constantly come in contact with a metal electrode, serving as current collector, where they take part in the electrochemical reactions with electron transfer.

In the development of direct carbon fuel cells, different directions were followed with respect to the electrolytes—high-temperature melts of carbonates or sodium hydroxide and high-temperature solid electrolytes.

Melts as Electrolytes

Studies of electrochemical carbon oxidation in carbonate melts at 700°C were performed by Weaver et al. (1981) at the Stanford Research Institute, Menlo Park, California. They used rods of different carbon materials as electrodes. The electrode potentials were measured relative to a gold reference electrode in an atmosphere of carbon dioxide mixed with oxygen at the same temperature. The electrodes proved to be more active the lower the degree of crystallinity of the initial powder (from which the rods were pressed). The electrodes had open circuit potentials around 1.1 V. At a current density of $100\,mA/cm^2$ the potential of the most active sample was 0.8 V (and 0.9 V when the temperature was raised to 900°C).

At the Lawrence Livermore National Laboratory, a fuel cell using a "semisolid" suspension of carbon powder in a molten carbonate was developed. Porous nickel was used as the cathode material. At a temperature of 800°C and a voltage of 0.8 V, current densities of 50 to $125\,mA/cm^2$ were obtained. In one test, a current density $27\,mA/cm^2$ was drawn for 30 h.

Hackett et al. (2007) reported experiments with direct carbon fuel cells, where rods prepared from different carbon materials were used as the anodes. The cathodes were made from iron–titanium alloy. Melts of sodium hydroxide at temperatures in the range of 600–700°C were used as electrolyte. The most stable performance with high-performance figures was obtained with graphite anodes. The open circuit voltage was 0.788 V. At the optimum temperature of 675°C, the voltage at a current density $140\,mA/cm^2$ was 0.45 V. Anodes made of other carbon materials had higher values of the open circuit voltage (up to 1.044 V), but exhibited an inferior and less stable performance when current was drawn. An undesirable side reaction is the carbonation of alkali hydroxide by the evolved carbon dioxide.

Solid Oxide Electrolytes

A difficulty of principle arises when using a solid electrolyte in direct carbon fuel cells. In fact, in fuel cells with a liquid electrolyte (solution or melt) the entire surface area of the carbon material is in contact with the electrolyte (is wetted by the electrolyte). In fuel cells with a solid electrolyte, to the contrary, the contact between the solid carbon material and the solid electrolyte is a mere point contact, and the working surface area is much smaller.

Two ways to overcome this difficulty have been suggested. The first (Pointon et al., 2006) makes combined use of a solid oxide electrolyte and a liquid (molten carbonate) electrolyte. The oxygen electrode was separated by the solid electrolyte from the carbonate melt that contained suspended carbon particles. The second way, described by Gür and Huggins (1992), used a fuel cell, consisting of two halves, separated by a solid electrolyte in the shape of a tube. Both the inside and outside surfaces of the tube were coated with a layer of platinum. The inside of the tube was in contact with ambient air. The outside was in contact with a closed compartment holding air and the carbon samples. The solid electrolyte was maintained at a temperature of 932°C, while the compartment holding carbon was maintained (in one experiment) at a level of 955°C. Because of the reaction between oxygen and carbon, a reduced oxygen equilibrium pressure was set up in the compartment. This led to a considerable oxygen pressure gradient between the two sides of the solid electrolyte, and hence, to a potential difference of about 1.05 V. When a current of $10\,\text{mA/cm}^2$ was drawn, this difference (the discharge voltage) dropped to a value of about 0.4 V.

REFERENCES

Codina G, Perez JR, Lopez-Atalaya M, Vasquez JL, Aldaz A. J Power Sources 1994;48:293.

Ponce de Léon C, Friss-Ferrer A, González-Garcia J, Szánto DA, Walsh FC. J Power Sources 2006;167:716.

Prasada D, Sivarama TK, Berchmans S, Yegnaramana V. J Power Sources 2006;160:991.

Topcagic S, Minteer SD. Electrochim Acta 2006;51:2168.

Walker AL, Walker CW. J Power Sources 2006;160:123.

Wingard LW, Liu CC, Nagda NL. Biotechnol Bioeng 1971;13:629.

Yaropolov AI, Varfolomeev SD, Berezin IV, Bogdanovskaya VA, Tarasevich MR. FEBS Lett 1976;71:306.

Zhang E, Xu W, Diao G, Shang C. J Power Sources 2006;161:820.

Baur E, Preis H. Z Elektrochem 1937; 44, 695 (1938);43:727.

Cao D, Sun Y, Wang G. J Power Sources 2007;167:250.

Gür NM, Huggins RA. J Electrochem Soc 1992;139:L95.

Hackett GA, Zondlo JW, Svensson R. J Power Sources 2007;168:111.

Howard HC. Direct Generation of Electricity from Coal and Gas (Fuel cells). New York: Wiley; 1945.

Jacques WW. US patent 555,511. 1896; Z Elektrochem 1910;4:286.

Pointon K, Irvine J, Bradley J, Jain S. J Power Sources 2006;162:750.

Weaver RD, Leach SC, Nanis L. *Proceedings of 16th Intersociety, Energy Conversion Engineering Conference.* New York: ASME; 1981. p 717.

MONOGRAPHS

Betso SR, Klapper MH, Andreasson LB. *Biological Aspects of Electrochemistry.* Basel and Stuttgart: Birkhäuser; 1971. p. 162.

Howard HC. Direct Generation of Electricity from Coal and Gas (Fuel cells). NewYork: Wiley; 1945.

23

ALKALINE FUEL CELLS (AFCs)

The low-temperature fuel cells described in Chapters 18 and 19 have (acidic) proton-conducting membranes as an electrolyte. On account of corrosion problems, only metals of the platinum group can be used as catalysts for the electrode reactions in such fuel cells.

In fuel cells having an alkaline electrolyte, or alkaline fuel cells ("AFCs" for short), catalysts advantageously may be selected from a much wider range of materials, some of them relatively inexpensive. Thus, highly dispersed nickel is a good catalyst for the electrochemical oxidation of hydrogen in alkaline solutions. Under certain conditions, this catalyst will even catalyze methanol oxidation. For the electrochemical reduction of oxygen in alkaline solutions, highly dispersed silver and gold are good catalysts, but again, considerably cheaper materials exist as well (various kinds of activated carbon, metal oxides of the spinel, and perovskite type).

In AFCs, like in other electrochemical devices, solutions of potassium hydroxide KOH prepared in different concentrations are usually used as the electrolytes.

Alkaline electrolytes offer further advantages over acidic electrolytes. Oxygen reduction is considerably faster in alkaline electrolytes, which implies that the working potential of the oxygen electrode is more positive, and a larger cell voltage can be realized. Also, apart from the advantages in catalyst selection, less severe corrosion conditions allow nickel and alloys of iron to be used as structural materials in AFCs.

Yet, alkaline electrolytes also have some severe disadvantages relative to acidic electrolytes. The most important disadvantage is their reactivity with carbon dioxide, which they readily absorb forming carbonates. As is always present in air, an

Electrochemical Power Sources: Batteries, Fuel Cells, and Supercapacitors, First Edition.
Vladimir S. Bagotsky, Alexander M. Skundin, and Yurij M. Volfkovich
© 2015 John Wiley & Sons, Inc. Published 2015 by John Wiley & Sons, Inc.

alkaline hydrogen–air fuel cell requires prior stripping of carbon dioxide from the air that reaches the oxygen electrode in contact with the electrolyte. Otherwise, KOH would be transformed to K_2CO_3, with two negative consequences: a decrease in the concentration of the free alkali needed for the electrochemical reactions and precipitation of crystals of potassium carbonate when its concentration has become high enough, which may mechanically upset the structure of the electrodes and catalyst. Another defect of alkaline solutions is their "creeping," or ability to permeate the smallest cracks and holes making tight sealing of AFC a tough problem.

23.1 HYDROGEN–OXYGEN AFCs

In hydrogen–oxygen fuel cells with an alkaline electrolyte, the reactions at the electrodes and the overall current-generating reaction can be formulated as follows[1]:

$$\text{Anode:} \quad 2H_2 + 4OH^- \rightarrow 4H_2O + 4e^- \quad E^0 = -0.828\,V \tag{23.1}$$

$$\text{Cathode:} \quad O_2 + 2H_2O + 4e^- \rightarrow 4OH^- \quad E^0 = 0.40\,V \tag{23.2}$$

$$\text{Overall:} \quad O_2 + 2H_2 \rightarrow 2H_2O \quad \epsilon^0 = 1.228\,V \tag{23.3}$$

23.1.1 Bacon's Battery

The modern era of fuel cell development started with the alkaline hydrogen–oxygen battery demonstrated in operation by the British engineer Francis Thomas Bacon (1960) in 1960. It had been under development since 1932. It was the first real fuel cell battery developing a power of more than 1 kW and able to operate for a relatively long period of time. The operating conditions were extreme in modern terms: the working temperature was above 200°C, the gases (hydrogen and oxygen) were compressed to over 20 bar, the pressure going even up to 45 bar. This necessitated a very bulky and heavy design. In Bacon's battery, potassium hydroxide solution of intermediate concentration (37–50%) served as the electrolyte. The electrodes were made of porous nickel. This worked without any problems for the hydrogen electrodes: hydrogen is readily oxidized at nickel as a catalytic anode. Problems had to be solved for the oxygen cathodes, as their surface is converted to poorly conducting nickel oxides in the presence of oxygen. Therefore, Bacon treated the positive electrodes with a lithium hydroxide solution and subjected them to a heat treatment. This produced lithiated nickel oxide as a corrosion-resistant electrically conducting oxide film on the nickel surface at which oxygen is readily electrochemically reduced.

---

[1] The potentials of the hydrogen and oxygen electrode have values that depend on the pH value of the solution, implying a change by -0.828 V between pH = 0 (acidic solutions) and pH = 14 (alkaline solutions). The electromotive force (EMF) value for reaction (27.3), which is the difference between the two electrode potentials, is independent of the solution pH.

Gas diffusion electrodes having a barrier layer for the gases were used for the first time in Bacon's battery. These electrodes were made of coarse nickel powder giving a structure with relatively large pores. A layer of finer nickel powder yielding finer pores was applied to the electrodes on the side facing the electrolyte. When in contact with the electrolyte solution, both layers soak up the solution by capillary forces. Gases supplied from outside under a certain pressure displace the solution from the coarse pores, but the pressure is not high enough to displace it from the fine pores where the capillary forces are much stronger. In this way, permeation of the gases through the electrodes, leading to unproductive losses, was prevented.

The electrodes in Bacon's battery measured $370 \, cm^2$, their thickness was 1.8 mm. At current densities from 200 to $400 \, mA/cm^2$, the voltage of an individual cell in the battery was 0.90–0.95 V, which is considerably higher than that in phosphoric acid fuel cell (PAFC) and in modern proton exchange membrane fuel cell (PEMFC). A variety of corrosion problems were responsible for the total lifetime of Bacon's battery not reaching more than a few hundred hours.

23.1.2   Batteries for the Apollo Spacecraft

In the early 1960s, the American company Pratt & Whitney, a part of United Technologies Corporation (UTC), acquired all rights to the patents of Bacon and started work on further development of alkaline hydrogen–oxygen fuel cells. The major goal was to give up the high gas pressures that had been used in Bacon's battery. High working temperatures had to be maintained in order to preserve performance. This was attained by using very highly concentrated (85%) potassium hydroxide solution, de facto, molten KOH. A battery working at 250°C was built. The electrodes were made by an improved and simplified technology.

A battery weighing about 114 kg was designed for delivering a power of 1.5 kW and sustaining short-term loads of up to 2.2 kW. The lifetime of this improved battery was extended to several thousands of hours, much more than that of Bacon's original battery. Batteries of this type were used in the Apollo space flights. The spacecraft included three batteries fully satisfying all of its electrical energy requirements. Successful return of the spacecraft would have been possible on merely one of the batteries, with the other two out of commission. The water generated as the reaction product by operation of the hydrogen–oxygen battery was recovered and used as drinking water supply for the crew. During the 18 flights of Apollo spacecraft, batteries of this type logged a total of more than 10.000 operating hours without any important failure.

23.1.3   First AFC with Matrix Electrolyte

In 1962, a research group in the American company Allis-Chalmers started developing a new type of hydrogen–oxygen fuel cell with an alkaline electrolyte. The distinguishing feature of this cell was to use, instead of a freely flowing liquid electrolyte (KOH solution or melt, as described above), a quasi-solid electrolyte in the form of potassium hydroxide solution immobilized in an asbestos matrix. Asbestos was found to be a material very well suited to this use. It has a very high chemical and thermal resistance. It is readily made into thin sheets (asbestos paper). It has a relatively large volume of fine pores strongly retaining the alkaline solution by capillary forces. Using this matrix electrolyte one need not have a special gas barrier within the electrodes: even at high pressures, the gases will not be able to overcome the capillary pressure in the matrix and push the alkaline solution out of the matrix pores. The Allis-Chalmers battery had filter-press design. It contained 5 M KOH solution in asbestos diaphragms 0.75 mm thick. The working temperature was 65°C, much lower than in the batteries described above. At current density of 200 mA/cm² and gas pressure of 1.35 bar the voltage per cell was 0.75 V, which is somewhat lower than in the battery types working at higher temperatures.

Allis-Chalmers being a manufacturer of agricultural machinery, they installed a battery in an experimental tractor from their product line. It actually was operated under field conditions for some time, and then was demonstrated in many exhibitions and fairs. This was the first use of fuel cells as the power source in an electrically powered vehicle.

23.1.4 Batteries for the Orbiter Space Shuttle

After the successful completion of all Apollo missions, the UTC began developing batteries for a new spacecraft, the Orbiter Space Shuttle. This spacecraft had much larger energy requirements and needed a much more powerful battery. It led to a radical redesign of the fuel battery. For a more compact design, 35% KOH solution immobilized in an asbestos diaphragm was selected, as in the Allis-Chalmers version. The electrodes contained large amounts of noble-metal catalyst: 10 mg/cm² of Pt + Pd for the anode, and 20 mg/cm² of Pt + Au for the cathode. This combination gave a very high specific power of more than 1 W/cm². The power plant installed in the spacecraft consisted of three independent blocks, each having 96 AFCs, a weight of about 120 kg, and a power output of 12 kW (with a short-term capability of 16 kW).

23.1.5 AFC with Skeleton-Type Catalysts

It was shown in the late 1950s (Justi and Winsel, 1959) that highly efficient hydrogen gas diffusion electrodes could be prepared using the so-called skeleton nickel (Raney nickel) as the catalyst. This is made from nickel–aluminum alloy subjected to leaching with hot alkali solution (Raney, 1927). Aluminum dissolves selectively, leaving behind a very highly disperse "skeleton" of nickel metal. The powder is highly pyrophoric and cannot be used as such for making strong electrodes. Justi and Winsel suggested pressing the electrodes from a powdered mixture of carbonyl nickel and nickel–aluminum alloy. Leaching the aluminum leaves a dual skeleton (double-skeleton electrodes). The sintered carbonyl nickel yields the supporting skeleton structure, while the Raney-type skeleton nickel produced by leaching provides the catalytic activity. Using the same principle, highly active oxygen electrodes were produced from skeleton silver. Laboratory models of such hydrogen–oxygen fuel cells worked for a long time in 6 M KOH solution at temperatures of 30–35°C, yielding current density of $30-50\,mA/cm^2$ and voltage of $0.75-0.85\,V$; the excess pressure of the gases was about 3 bar.

23.1.6   AFC with Carbon Electrodes

In 1955, the American company Union Carbide Corporation (UCC) started a large development program headed by Karl Kordesch for alkaline hydrogen−oxygen fuel cells with carbon electrodes. From the outset, a model with circulating alkaline solution was selected. In first experiments, the relatively thick pitch-bonded carbon electrodes (6.4 mm) developed previously for regular zinc−air batteries were used. Later, relatively thin multilayer carbon electrodes containing a metal support (a nickel screen or Exmet extended nickel sheet) were used. The electrodes included several hydrophobized diffusion layers (consisting of a mixture of carbon black, polytetrafluoroethylene (PTFE), and a variety of pore-forming and wet-proofing additives) and one or two catalytically active layers. In some versions, silver and platinum catalysts were added to the electrodes. The overall thickness of the electrodes was about 0.4 mm. Electrodes were prepared in the size of $17 \times 17\,cm^2$. They were made by rolling technology. At current density of $100\,mA/cm^2$, the voltage produced was about 0.6 V in air, and with additional catalysts it reached 0.67 V in oxygen, while the current density could be raised to more than $200\,mA/cm^2$.

Using fuel cells of this type, Dr. Kordesch built a power plant to power a converted Austin A40, a light passenger vehicle, which for a number of years he used for urban driving. The vehicle had hybrid power including a storage battery providing boost power for acceleration and hills. Otherwise, under normal road driving conditions, the full load was satisfied by the fuel cells that at the same time recharged the storage battery. The fuel cell power plant had an output of 6 kW at a voltage of 90 V (0.7 V per cell). There were 13 tanks with hydrogen providing fuel for producing 45 kWh of electrical energy. Air oxygen was the oxidizing agent. The total weight of the fuel cell battery was 250 kg; the additional storage battery weighed 150 kg. This was the first demonstration of a relatively long-term practical use of an electric car powered by fuel cells.

After the tragic 1984 Bhopal accident in India, UCC stopped all work on this project. Karl Kordesch then continued his research at the Technical University of Graz, Austria (Cifrain and Kordesch, 2004).

23.2   PROBLEMS IN THE AFC FIELD

23.2.1   Matrix Electrolyte or Circulating Liquid Electrolyte?

It can be seen from the historical review in the previous sections that, basically, two solutions for electrolyte management have emerged at the different stages of development of hydrogen−oxygen AFCs. One solution has the alkaline solution (or melt) immobilized in a porous matrix consisting of asbestos or similar material with fine pores. The other solution has the liquid electrolyte being pumped around. What are the advantages and disadvantages of these two versions?

The matrix version of AFC offers a more compact design. It lacks the electrolyte loop, needed in the other case, where pumps, valves, tanks, regulators, and so on must handle the caustic and, sometimes, hot liquid. Thus, many problems arising with the need for sealing not only the cells but also the loop do not exist in the matrix version. It is another advantage of the matrix version that, owing to the matrix tightly holding the electrolyte solution in its fine pores, the gas diffusion electrodes do not need an additional porous barrier layer for the gases, which greatly simplifies electrode manufacturing technology and makes it much cheaper.

The version with circulating liquid electrolyte has the following advantages: (i) it becomes much simpler to maintain the water balance in the cell, as there is no chance for the electrodes to dry out, and the electrolyte flow can be used to eliminate product water; (ii) the electrolyte flow can also be used to eliminate heat, so that special cooling plates or loops for a cooling fluid are not required; (iii) the electrolyte flow serves to level any gradients of electrolyte concentration that could arise within an individual cell or within the battery; (iv) the electrolyte flow may serve to eliminate "foreign" matter such as gas bubbles, corrosion products, and insoluble carbonates that turn up in the liquid.

In either version, cells of the bipolar type assembled to batteries of the filter-press type can be used. The version with circulating electrolyte is also feasible for building monopolar ("jar"-type) cells that are cheaper than bipolar types (as they do not require expensive bipolar plates) and offer a highly flexible series–parallel switching within the battery to accommodate specific user needs.

An important advantage of batteries with circulating electrolyte is the much simpler shutdown and restart procedure by simply stopping and restarting electrolyte circulation. For a cold start-up, in particular, only the tank holding the electrolyte solution has to be heated, rather than the entire battery.

The chief problems that arise in the design of fuel cells with circulating electrolyte are those of providing perfect sealing of the full loop, and of sufficient high corrosion resistance of all parts (pipe sections, pipe junctions, pumps, and valves) toward the hot alkaline solution.

23.2.2 The Problem of Electrolyte Carbonation

Any carbon dioxide getting into the KOH solution binds (neutralizes) alkali yielding potassium carbonate:

$$CO_2 + 2KOH \rightarrow K_2CO_3 + H_2O \tag{23.4}$$

This is a very dangerous event in AFCs with matrix electrolyte, as the total amount of alkali in the fine pores of the matrix is limited, and there is no way for removing carbonates from the pores. Loss of alkalinity because of a partial neutralization by carbon dioxide affects performance. The carbonate crystals disturbing the catalytic layer's integrity in the electrodes may be formed in the solution when the carbonate concentration becomes high enough. Therefore, in almost all practical applications of matrix-type fuel cells, pure oxygen was used as the oxidizing agent rather than air that always contains traces of carbon dioxide, particularly so in spacecrafts.

In fuel cells having free liquid electrolyte, and particularly a circulating electrolyte, it was found that initial fears were exaggerated. Individual AFC electrodes functioned without any important loss of performance for more than 3500 h at current density of 150 mA/cm^2, while being supplied with oxygen containing 5% of carbon dioxide, which is 150 times higher concentration than in the air. For the formation of crystalline precipitate of potassium carbonate at 70°C, the local KOH concentration should be, at least, 16 mol/l. When there is a real need, one may use relatively simple, cheap devices consisting of scrubbers filled with an alkali solution or other chemicals in order to remove CO_2 from the air.

23.2.3   The Lifetime of AFCs

The AFC-based power plants for Apollo and Orbiter spacecrafts described above have been used for numerous flights and attained a total operating time of tens of thousands of hours. Systematic long-term tests of simpler power plants for terrestrial applications have hardly been reported in the literature. It was said that, in a number of tests, the time to failure was 2000–3000 h.

23.3   THE PRESENT STATE AND FUTURE PROSPECTS OF AFC WORK

As with the other fuel cell types, periods of pronounced ups and downs can be distinguished in R&D for AFCs. Work in the United States, Europe, and a number of Asian countries evolved very vigorously after the 1960 demonstration of Bacon's battery and, more particularly, after the first flight of the Apollo spacecraft having the new power plant on board.

It has been entirely justified to build spacecraft power plants with AFCs. This has also to do with the fact that, in this application, cryogenic storage of the fuel and oxidant, hydrogen and oxygen, is feasible economically and technically. In addition, it was an important aspect that the water produced during operation could be used as drinking water for the crew. The same situation exists in a number of other manned mobile installations, such as bathyscaphs and submarines.

In terrestrial applications certain difficulties were encountered. It was a natural desire to use air oxygen rather than pure oxygen whenever this was feasible. This change is associated, however, with a marked drop in the electrical performance figures relative to those achieved in spacecraft power plants. Also, the complications associated with the presence of carbon dioxide in the air that were mentioned earlier come into play here. Only pure electrolytic hydrogen is used as the fuel in AFCs. It is not possible to use the much cheaper technical hydrogen produced by the reforming of hydrocarbons or other organic compounds, or by carbon gasification, as it contains not only traces of carbon monoxide CO (poisoning the platinum catalysts), but also appreciable amounts of CO_2. Eliminating the carbon dioxide from hydrogen would be complicated and expensive. Both gases, hydrogen and oxygen, pose problems in their storage and transport. A cryogenic storage is not feasible in many terrestrial applications, and storage of the compressed gases in tanks has a weight penalty that is admissible for stationary installations but undesirable or impossible for mobile and portable devices.

All these points, and also the uncertainties as to the potential lifetime, have led to a strongly reduced set of candidate applications for AFC-based power plants. The volume of research and engineering work has shrunk accordingly. Most organizations and companies stopped work in this area between 1990 and 1995, and some of them switched to other systems when work on fuel cells with acidic proton-conducting membranes started to take off in the mid-1990s. Still, the ideas concerning further improvements of hydrogen–oxygen AFCs and potential new practical applications have remained attractive.

23.4   ANION-EXCHANGE (HYDROXYL ION CONDUCTING) MEMBRANES

The development and industrial production of cation-exchange (proton-conducting) membranes of the Nafion type induced a "quantum jump" in fuel cell development. A number of important problems could be solved by using such membranes: a complete sealing of individual cells in the battery and of the battery as a whole, the avoidance of direct contact between the electrodes, the prevention of gas bubble forming in the interelectrode gap. With these membranes, very compact designs of fuel cell batteries would be possible. The cation-exchange membranes have the defect that the rates of a number of electrochemical reactions will decrease considerably when the electrode is in contact with them, rather than with an alkaline electrolyte. This is true, in particular, for cathodic oxygen reduction, anodic methanol oxidation, and the anodic oxidation of many other organic substances.

It is a natural question to ask then, why one could not make anion-exchange (hydroxyl-ion-conducting) membranes having properties similar to those of Nafion but, because of their alkaline nature, leading to a marked acceleration of the electrode reactions.

Despite great efforts, this has not been achieved until now. Existing types of anion-exchange membranes (used for electrodialysis) are far inferior to Nafion, both in their conductivity and in the chemical and thermal stability. Therefore, one cannot so far meaningfully discuss their potential as substitutes for the circulating or matrix electrolyte in alkaline hydrogen–oxygen fuel cells.

The situation is different, when looking at fuel cells using methanol (or other similar types of liquid organic fuel). Methanol fuel cells built with proton-conducting membranes of Nafion type are currently in an earlier stage of development than that of alkaline hydrogen–oxygen fuel cells. Many attempts to develop methanol fuel cells with proton-conducting membranes of the Nafion type are reported. They have the important defect of insufficient high performance because of slow anodic oxidation of methanol in fuel cells with a membrane electrolyte.

23.5 METHANOL FUEL CELL WITH AN INVARIANT ALKALINE ELECTROLYTE

In 1964, Cairns and Bartosik (1964) suggested an original way of solving the problem of carbonation of the alkaline electrolyte solution during the oxidation of organic substances to carbon dioxide. It involved using concentrated solutions of the alkali metal carbonates Cs_2CO_3 and Rb_2CO_3 instead of the pure alkali solutions of KOH or NaOH. Using such carbonate solutions, it is possible to work at high temperatures (up to 200°C). Bicarbonates are unstable at temperatures above 100°C. Therefore, the composition of these solutions will not change when carbon dioxide is produced in them. At current density of $126\,mA/cm^2$, a specific power of $40-45\,mW/cm^2$ was achieved (on an IR-free basis). In a test involving continuous operation for more than 560 h, no indications for a drop in electrical performance were seen. No information could be found in the literature about any further development of this idea.

REFERENCES

Bacon FT. Ind Eng Chem 1960;52:301.
Cairns EJ, Bartosik DC. J Electrochem Soc 1964;111:1205.
Cifrain M, Kordesch K. J Power Sources 2004;127:234.
Justi E, Winsel A. Abh Mainzer Akad, No. 8 1959, Steiner, Wiesbaden.
Raney M. US patent # 1628,190. 1927; Ind EngChem 1940;32:1190.

MONOGRAPH

Kordesch K, editor. *Brennstoffbatterien*. Berlin: Springer-Verlag; 1969.

24

APPLICATIONS OF FUEL CELLS

24.1 LARGE STATIONARY POWER PLANTS

According to a terminological distinction that is sometimes adopted, stationary fuel cell-based power plants may be divided into those of large power (more than 10 kW) and those of small power (less than 10 kW). Basically, the former are used primarily to generate grid power, that is, to provide power to settlements of various sizes, or to relatively large industrial sites. In their intended functions, they would not differ from thermal, hydraulic, or nuclear power plants. Fuel cell-based power plants of small power are built to supply electric power to individual residential or administrative quarters or to remote individual power customers. The dividing line is quite arbitrary, of course.

24.1.1 Basic Development Trends

As shown in previous chapters, in quite a few countries large power plants based on different fuel cell systems have been built. Using phosphoric acid fuel cells, the American company United Technology Corporation produced a large number of 200-kW power units. In Japan, multimegawatt power plants were built. Using molten carbonate fuel cells, the American company Fuel-Cell Energy built a power plant with an output of up to 2 MW in Santa Clara, California. Using solid oxide fuel cells, the company Siemens-Westinghouse built a 100-kW power plant in Netherlands.

Electrochemical Power Sources: Batteries, Fuel Cells, and Supercapacitors, First Edition.
Vladimir S. Bagotsky, Alexander M. Skundin, and Yurij M. Volfkovich
© 2015 John Wiley & Sons, Inc. Published 2015 by John Wiley & Sons, Inc.

This shows that all these systems have already left the experimental R&D stage and become an economic reality. They have already demonstrated lower consumption of natural fossil fuels per unit of power generated and lower emission of greenhouse gases and other harmful products. Other advantages over thermal power plants are associated with fuel cell-based power plants. They can be produced in modules, from which power plants of different size and output can be put together. This attribute, known as *scalability*, facilitates the design and construction of plants adapted to specific local requirements.

There has been a steady increase in the number of fuel cell-based power plants built and put in service. The online information newsletter Fuel Cell Today collects and analyzes this data in the preparation of its annual review of global fuel cell shipments.

In Table 24.1, the number of such plants constructed every year between 1996 and 2010 is shown, as well as the electric power added every year (megawatts per year). It can be seen that in individual years this figure was 20–25 MW. The average power output of these plants increased from 0.2 MW in 1997 to 0.57 MW in 2010. Table 24.2 shows percentages of the various fuel cell systems used to build these plants, year by year. It can be seen that, over a number of years, more than 40% of the new plants were molten carbonate fuel cells and solid oxide fuel cell-based high-temperature plants. It is notable that in many cases phosphoric acid fuel cells with concentrated phosphoric acid were used, a system that for a long time had been regarded as not promising, for which R&D efforts have ceased long ago.

A marked fraction of the plants was built using polymer electrolyte fuel cells that cost more than other fuel cell systems. A wider use of these fuel cells is probably

TABLE 24.1. Annual Number of Large Fuel Cell Units and MW Installed

Year	Number of Units	Installed Megawatt per Annum
1996	25	6.3
1997	50	9.6
1998	20	3.3
1999	41	7.6
2000	39	7.6
2001	36	5.7
2002	32	5.7
2003	68	15.7
2004	48	8.2
2005	63	16.1
2006	57	18.5
2007	50	27.4
2008	54	25.0
2009	39	25.4
2010	99	22.3

TABLE 24.2. Technology Type of Large Fuel Cell Units, by Percentage

Year	Percentage for			
	MCFC	PAFC	PEMFC	SOFC
2003	43	25	25	7
2004	21	42	23	14
2005	62	12	15	11
2006	46	24	13	17
2007	48	16	22	14
2008	24	44	13	19
2009	28	46	0	26
2010	7	21	4	68

justified in those cases where a cheap source of highly pure hydrogen is available, such as close to chlor-alkali industries where pure hydrogen is a free by-product.

Today, most large fuel-cell-based plants are produced and set up in the United States, following active government support. A similar tendency can be seen in South Korea. In Japan, where many large fuel cell-based power plants had been built and operated in earlier years, attention is focused, at present, on the development of small stationary power plants and on power units for electric vehicles.

In most cases, large fuel cell-based stationary power plants are used for power production and, at the same time, for heat supply to customers in nearby locations (combined heat and power (CHP) systems). This combined use of two different types of energy implies a very considerable increase in the total economic and energetic efficiencies of these plants.

24.1.2 Hybrid Stations

Recent years have seen large efforts to build hybrid plants combining high-temperature fuel cells and gas turbines. The basic idea is that of highly efficient coordination of gas and heat flows entering and leaving the two components of these combined power plants. This coordination serves to lower the energy losses and raises the overall plant efficiency dramatically.

The American company FuelCell Energy developed molten carbonate fuel cells with direct internal reforming of the hydrocarbon fuel (natural gas). The main source of electric power (over 80%) in these hybrid plants are the fuel cells. The gas turbine produces an additional amount of electrical energy from the heat evolved by the operating fuel cells. The turbine also compresses air fed into the fuel cells. Preliminary tests of such a hybrid power plant of 250 kW gave very promising results. The plant feeding power into the grid was operated more than 6000 h. Nitrogen oxide emissions into the atmosphere were minimal (less than 0.25 ppm) as the turbine was operated with heat from the fuel cells. These results are now used as a basis for building hybrid power plants of megawatt size. For instance, the Siemens-Westinghouse developed recently 3-MW SOFC-based hybrid power station.

24.2 SMALL STATIONARY POWER UNITS

Fuel cell-based power plants that have an output of up to 10 kW are under vigorous development as well, and they find ever wider practical uses. Table 24.3 shows the number of such plants produced every year from 2001 to 2010. Approximately half of the units produced in 2006 had a power of about 1 kW, and the other half had an output of 1.5 – 10 kW, the numbers being distributed evenly over this time interval. The overwhelming number (more than 50%) of these plants were produced and set up in Japan, with the United States taking the second place. Most of the low-power units were built with polymer electrolyte fuel cells. The fraction of solid oxide fuel cells has decreased gradually.

The range of applications of small power plants is very large. They are designed primarily for the needs of communities for a combined supply of electric power and heat (hot water and heating) to individual structures of different sizes: individual cottages, administrative and office buildings, hospitals. At present, the cost of these units, as a rule, is not competitive (it should be no higher than $1500/kW). Therefore, a part of the costs is supported by municipal or federal authorities.

Another important area of application is the use of small power units as backup power in situations of sudden loss of grid power because of natural or technical problems. Such backup units are extremely important to those consumers who cannot tolerate power interruptions. This is the case for various stationary telecommunications installations (for instance, receivers, transmitters, relay stations, signal amplifiers). Other users who can use units of this type are surgery wards in hospitals, computer units in traffic control, financial institutions, and emergency lighting systems in public

TABLE 24.3. Annual Number of Small Fuel Cell Units Installed

Year	Annual Number of Units
2001	271
2002	510
2003	714
2004	544
2005	815
2006	1600
2007	1900
2008	3500
2009	6530
2010	7260

spaces. Uninterruptible power is needed here to prevent severe accidents, loss of lives, or loss of valuable data that sometimes may not be recoverable.

Another important application of small power units is in remote-area power supply (RAPS). According to World Bank estimates, currently some two billion people around the world have no access to central power supply systems. For many of them, diesel electric generators are the only available power source. These generators are cheap, but the cost of diesel fuel, including its transport to remote places, may sometimes be very high. Also, diesel generators operate with large emissions of greenhouse gases and other harmful substances. A natural way out of this situation would be the use of natural energy sources, such as solar energy via photovoltaics, or wind energy via wind turbines. Using these natural resources, which are available only intermittently, one must have additional devices for temporary energy storage. Lead-acid batteries were used for some time for such energy storage in the Amazonian region in Peru.

An interesting project concerns plans to use fuel cells in Antarctic Region, on the island of Béchervaise. This project, supported by a grant of $600,000 from Australian government, involves production of hydrogen by electrolysis with wind energy and its use in low-power fuel cell units producing electric power and heat.

24.3   FUEL CELLS FOR TRANSPORT APPLICATIONS

For more than three decades, the possible use of fuel cells as a power source for automotive transport applications has been explored widely. This scenario arose in connection with fuel supply problems during the energy crisis of the 1970s. Since then, it has been raised repeatedly in connection with air pollution in cities with rapidly growing motor car populations. These problems have caused a number of countries (and, notably, in the US state of California) to make it mandatory for carmakers to produce and sell a certain number of low or even "zero-emission" vehicles. Carmakers were thus forced to see how they could replace the internal combustion engines (ICEs) by other power units.

24.3.1   Alternatives to Traditional Vehicles with ICEs

Early in the twentieth century, cars powered by storage batteries were quite common. About 35% of all cars registered at that time in the United States were battery powered. They were more convenient than vehicles with ICEs that had to be cranked to start the engine. The situation took a turn when an electric starter was patented in 1912 and was then widely introduced. This meant the practically complete displacement of electric cars by today's traditional ICE vehicles. Today, battery-powered vehicles are used only for transport within production plants and for certain home deliveries, such as milk products and mail.

When electric cars are used, one has practically no local emissions (other than hydrogen and oxygen while charging the batteries). The only ecological consequence of this transport mode is that the emission of greenhouse gas, carbon dioxide, is relocated from populated cities to a remote area during battery charging. Electric cars are user-friendly. They have two major advantages: (i) ease of changing the operating mode (e.g., the speed), and (ii) potential energy recovery while braking. These characteristics are, particularly, a possibility for their use in populated city (stop-and-go) traffic. Yet, battery-driven cars have two large defects that have prevented their wide introduction so far: (i) a limited driving range (of only 60–200 km between battery-charging stops, and (ii) the long time (several hours) needed to recharge the battery ("refuel" the vehicle).

The most important alternative to traditional cars with ICEs is hybrid cars, in which an ICE is combined with an electric drive and a storage battery. The electric drive is used primarily in city traffic, and the ICE is used mainly on a highway under constant driving conditions, while simultaneously recharging the batteries. This type of cars combines advantages of traditional cars with those of electric vehicles, while minimizing the defects of each.

24.3.2 Problems in the Use of Fuel Cell Vehicles

The advances made during the 1990s in the development of various types of fuel cells (mainly polymer electrolyte fuel cells) have, of course, come to the attention of people developing passenger vehicles and other transport means, and were seen as a promise that an electric car could be developed with all its user convenience, wide driving range, rapid start-up, and minimal emissions. Almost all the large carmakers embraced the problem and began design and experimental testing. Numerous joint projects were undertaken between car companies and companies developing and producing various types of fuel cells. Many demonstration vehicles having fuel cell power plants were developed in these projects. The additional financial resources of the carmakers stimulated further work on fuel cells and contributed to improve their electrical and operating characteristics.

It soon became evident that building an electric car with fuel cells is a task associated with numerous basic problems. These problems were first openly formulated in an article published in 2001 by McNicol et al. (2001) entitled: "Fuel cells for road transportation purposes—yes or no?" In the opinion of these authors, the major problem is associated with the selection of a suitable reactant (fuel) for road transport fuel cells. The only fuel that is acceptable ecologically (that does not have harmful emissions) and by its technical characteristics is hydrogen. However, use of hydrogen is associated with problems of onboard storage (a tank with highly compressed gas, a cryogenic vessel with liquefied hydrogen, or a metal hydride). Moreover, an infrastructure for hydrogen transport and distribution is almost nonexistent. There is, of course, the possibility of decentralized production of electrolytic hydrogen (for instance, in special service stations), but the cost of this hydrogen would be higher than the current cost of gasoline. Hydrogen could be produced from gasoline or other petroleum products (for which the distribution infrastructure already exists) by reforming directly onboard, but this approach was judged to be unrealistic by the authors because of the complexity of the reforming equipment needed to lower sulfur contaminants present in these fuels, and because of the ecologically harmful emissions that accompany the reforming. Thus, the answer to the question formulated in the article's title was "no" for the approaches discussed above, as well as for approaches involving other indirect uses of fuel (i.e., uses requiring fuel processing). In the opinion of the authors, the only promising approach is that of using fuel cell power plants working with straightforward reactant supply to the fuel cells. Such reactants could be hydrogen (when problems of its transport and storage would have been solved), or methanol and similar reducing agents (when the corresponding catalytic problems would have been solved).

Attention was also called to the fact that a definite changeover from traditional cars to electric cars with fuel cells would be associated with the need for a fundamental restructuring not only of the car industry and of the oil production and refining operations, but also of many other economic sectors related to them. It must be acknowledged that even today no solution is in sight for many questions raised in the mentioned article.

In a review entitled "Fuel cell vehicle: Status 2007," von Helmolt and Eberle (2007) discussed basically only two problems: onboard storage of hydrogen and possibilities for lowering the cost of the fuel cell plant. Concerning the first problem, the authors concluded that cryogenic hydrogen storage is not appropriate in transport applications, owing to large unavoidable losses while parking and even while driving. Practically, the only acceptable method of hydrogen storage would be in light cylindrical tanks under a pressure of 700 bar. Under the criterion of volume taken up, hydrogen storage in the form of metal hydride would be advantageous, but this method of storage is not practical for electric cars as refueling (metal hydride formation) and subsequent liberation of hydrogen are associated (in systems known at this time) with the uptake and evolution of large amounts of thermal energy, and that would need heat-exchange equipment of unrealistically high capacity in an electric car.

As to the cost of the power plant, the final goal would be $50/kW, or about $5000 for one electric car. The cost of the power plant depends not only on the cost of the fuel cell stack as such, but also on the cost of all auxiliary units and systems, including hydrogen storage and supply. The authors do not rule out the possibility that the goal could be achieved when changing over to mass production of such power plants (many millions of units).

In a review on this topic by Ahluwalia and Wang (2008) entitled "Fuel cell systems for transportation: status and trends" (2008) two questions are examined: possibilities for further cost reductions, and possibilities for a further performance increase of power plants with polymer electrolyte membrane fuel cells and a direct hydrogen supply. According to the American program DOE/FreedomCAR, the cost of said plants, which in 2005 was $125 per kilowatt of electricity, should have come down to $30 per kilowatt by 2016. By comparison, ICEs cost $25–35 per kilowatt of electricity equivalent. The specific power of fuel cell plants should increase from 500 to 650 W/kg during the same period. In the review, various technical solutions are discussed that might make it possible to approach these goals, if not to actually reach them. It is interesting to point out that, according to the data given in this review, in 2005 the cost of fuel cell stacks was 63% of the entire power plant costs. By 2010, this share has come down to 48%. In this review, as in many others, the question of hydrogen storage in vehicles was completely left out of consideration.

24.3.3   Perspectives for the Further Development of Fuel Cell Vehicles

Attention must be called to the fact that, in practically all work concerning the use of fuel cells for road transport vehicles, only fuel cells using hydrogen as a fuel were considered. There is no doubt that hydrogen–oxygen fuel cells (and, in particular, those of the polymer electrolyte membrane type), at present, have been developed to such a degree that in all their technical parameters they are fit for electric car's power plants. Tests of different types of electric cars with such power plants were already being performed for almost 10 years, and they will undoubtedly be continued and broadened (probably involving hundreds of cars). This is the only way to gather the needed data concerning all special technical and operational features of this kind of transport. Fundamental problems in the broad development of these electric vehicles have been, and continue to be, the hydrogen supply and the cost of their fuel cell power plants. Local electrolytic hydrogen generation would require a considerable power grid enhancement.

Probably, the first instances of wider practical commercial use of fuel cell power plants will take place in connection with heavier transport equipment used within cities and near cities—buses and trucks supplying merchandise to the local trade and materials to construction sites. This heavy transport equipment is responsible for large part of air pollution in big cities. They can be readily equipped with compressed hydrogen tanks. Creating an infrastructure for hydrogen distribution and refueling is easier within a city than all over the country.

The first demonstration-type buses having a fuel cell power plant were introduced toward the end of the twentieth century. In London, such buses started to operate in suburban commuter transit soon after 2000. During 2001–2006, a program of the European Commission called *Clean Urban Transport for Europe* was run involving regular use of 27 fuel cell-powered buses in nine European cities. The success achieved by this initiative led to the introduction of this type of bus transport in other cities of Europe, Asia, and North and South America.

As for the multimillion fleet of individual cars that are driven in and outside cities, a massive changeover to fuel cell-powered electric cars will become possible only when technical and economic figures of fuel cell vehicles become comparable to those of internal combustion and hybrid cars (which also undergo constant improvement). The changeover from petroleum-based automobile transport to electric car transport (requiring a radical restructuring of many sectors of economy) will, at any rate, be gradual and will stretch out for many years.

In the economically developed countries the first steps of such a changeover can already be seen. According to information from the German company Daimler AG (2011), three Mercedes-Benz fuel cell cars successfully accomplished a 125-day drive around the globe through Europe, North America, Australia, Asia and China, and Russia. Throughout the duration of the drive, the cars demonstrated versatility and reliability, with no problems related to the fuel cell-drivetrain.

In September 2009, the leading vehicle manufacturers in fuel cell technology from Germany, the United States, France, and South Korea signed a letter of understanding and gave a joint statement on the development and market introduction of electric vehicles with fuel cells. Shortly afterward, leading energy companies, equipment manufacturers, and vehicle manufacturers signed a memorandum of understanding for an initiative titled "H_2 mobility." This memorandum involves evaluation of the setup of hydrogen infrastructure in Germany to promote serial production of electric vehicles with fuel cells. This marks a major step toward commercialization of such locally emission-free vehicles. This memorandum envisages a significant expansion of hydrogen fueling station network by the end of 2011. Commercialization of fuel cell vehicles with several hundred thousand units is anticipated from 2015 onward.

24.3.4 Fuel Cell-Based Power Plants for Other Transport Means

As to the use of fuel cell-based power plants in other transport means, we have repeatedly spoken in the present book about practical applications of such plants in manned spacecraft. The first examples were 1-kW polymer electrolyte membrane fuel cell plants used in the 1960s in the now outdated Gemini spacecraft; then, in the 1970s, three 1.5-kW alkali fuel cell plants were used in Apollo spacecrafts, and, finally, three 12-kW alkali fuel cell plants were recently used in the Orbiter space shuttles. In these applications, where the power plant did not have to operate for more than a few days, the only possible alternative for powering the electrical equipment of the spacecraft would be storage batteries. Under these conditions, fuel cell-based power plants have much higher energy contents per unit volume and unit weight (mass) than batteries. Another great advantage is constant water production for the needs of the crew.

It is quite realistic to use fuel cell-based power plants for river, marine, and submarine vessels. In water transport, such difficulties as with placement of devices for storage or production of hydrogen, as in electric cars, are more readily overcome. The problems of heat exchange (removal of heat from an operating power plant) are easier to solve for such vessels. In the case of large vessels and ferries, diesel engines are provided not only for the propulsion of the vessel, but also for the generation of electrical energy needed for various operations (loading, unloading, lighting, communications, air conditioning, and other auxiliary needs). The energy for these needs must also be generated while in port, producing appreciable air pollution in the port area. Therefore, even here, considerable ecological effects would be gained with the use of fuel cells.

In a paper by Psoma and Sattler (2002), development work done in Germany on submarines with power generators based on polymer electrolyte membrane fuel cells was discussed in detail. Preliminary work in this direction began in the 1970s. By 2002, type 212A submarines were under construction, to be placed in service for German and Italian navies by 2003. Modules with power between 30 and 120 kW and with a total power of 300 kW are used in these submarines. Oxygen for the fuel cells comes from cryogenic vessels kept outside the submarine's pressure hull. These vessels resist potential impact loads, as well as the outside pressure changes occurring when the vessel dives to different depths. Hydrogen is stored in bound form, as metal hydrides. This storage method provides the largest amount of hydrogen per unit volume (even larger than cryogenic storage). The equipment holding the metal hydrides is also kept outside the submarine's pressure hull. The cooling needed during charging (introduction of hydrogen into the metal alloy) is handled by land-based equipment. During discharge (liberation of hydrogen from the hydride), thermal energy set free during fuel cell operation is introduced into the storage system. This reactant storage system secures longer underwater runs of the submarines than does traditional equipment based on storage batteries.

However, a further increase in the underwater range is not possible, owing to the large weight (and high cost) of the metal hydrides. For this reason, work on equipment for onboard steam conversion of methanol began in 1995. The carbon dioxide produced in the reforming process may be kept onboard the vessel in a liquid state, and would be ejected periodically into seawater. A plant of this type, capable of supplying 240 kW worth of fuel cells with hydrogen, has successfully passed all necessary tests. Production of type 214 submarines including such equipment began in 2001 for Greece, South Korea, and other countries. In the opinion of the authors, this indicates that power plants with fuel cells are the ultimate selection when building nonnuclear submarines.

For rail transport, the use of fuel cell-based power plants is technically possible, without any doubt. However, whether it will be appropriate to replace the diesel engines used today by new types of power plants will depend on various economic factors, including the situation in the oil markets.

24.4 PORTABLES

In all likelihood, what for the sake of brevity we shall designate as "portables" will become the most important field of large-scale use of fuel cells over the next decade. This field covers the most diverse portable electronic and electrotechnical devices for daily life, as well as for many portable manufacturing aids that need electric power to function. Historically, probably the first such devices were pocket flashlights, which needed power in the milliwatt range. The first walkie-talkie (portable transceiver or radio communication equipment) was developed in 1940, and such devices became important in World War II. The first electronic quartz wrist watches appeared toward the end of the 1960s, and they consumed power in the microwatt range. At about the same time, various hearing aids and portable radios appeared. It was typical for all these devices that almost immediately after they appeared they went into mass production, with millions of pieces made. The list of devices classified as "portables" has become considerably longer during the past two decades, and their versatility and functions have grown tremendously: mobile phones, notebook-type portable computers, digital still and movie cameras, camcorders, and function sensors for communications, out-of-office work, mobile Internet connections, personal navigation, sophisticated medical monitoring, and many others. Production volume rose together with the levels of sophistication and utility.

24.4.1 Power Sources for Portables

The only power sources feasible for all these portables are electrochemical batteries. Disposable batteries were the classical power source for flashlights, and still hold a very strong position (salt Leclanché; later, alkaline "dry cells" of zinc−manganese dioxide type). Rechargeable batteries became ever more important: first, nickel−cadmium, and more recently, nickel−hydride and lithium ion batteries. For convenient handling, a power source is usually placed somewhere inside the device, so it should respect certain limitations as to weight and volume. As a rule of thumb, a power source should not exceed 30–40% by mass and volume of the device powered by it. A similar upper limit 30–40% applies to the cost.

It became obvious, in view of newer high-performance electronic devices powered by batteries, that even the most perfect rechargeable lithium ion batteries hardly meet the desirable targets of operating time between recharges. Thus, a notebook may need recharging of its state-of-the-art lithium ion battery after 2–3 h. For a media player of the MP3 type less than 10 h is reached as a rule, while a cell phone will require recharging after 3–4 h of conversation. A major defect of rechargeable batteries is the time required for recharging, which may easily be more than 5–6 h, and people tend, for that reason, to keep a second battery pack, thus doubling the weight of the power source.

It is for these reasons that minifuel cells are looked on as a superior power source able to supply higher currents for much longer times. It should be pointed out at once that, from the outside, fuel cells (including their fuel and oxidizing agent) work by exactly the same principles of current generation as do conventional batteries. It is quite clear that we do not expect complete substitution of ordinary batteries by fuel cells in the portable field. Conventional batteries will continue to maintain their leading market position as power sources for a large number of devices. Thus, disposable batteries, for example, are expected to maintain their importance for pocket flashlights and for various medical devices. Simple electronic devices, such as portable radios, audio players, digital cameras, and so on, will continue to be powered by rechargeable nickel−cadmium and nickel−hydride batteries. Lithium ion rechargeable batteries are likely to continue to be used in simpler cell phones. Fuel cells will be attractive for more complex equipment, such as notebooks used for more than 2 h at a time, where even lithium ion batteries have insufficient energy density.

24.4.2 Development Work on Fuel Cells for Portables

Development work on fuel cells for portables evolves quite differently from that on fuel cells for other applications. Fuel cells intended for large and small stationary power plants and those intended for electric cars must be seen in the context of the global problems: lower consumption of hydrocarbon fuels (for which the resources are limited) and lower air pollution (greenhouse gases and other contaminants). For this reason, society at large and all governments take active interest in developing such systems. These works were initiated and/or financed by government programs.

More perfect power sources for portables are, to the contrary, strictly in the interest of individual manufacturers' and consumers. In the past, development activities in this area were mainly market-driven and did not benefit from any general programs involving government, with the important exception, however, of work toward power sources satisfying the needs of portable equipment for the military.

24.5 MILITARY APPLICATIONS

From the very beginning of modern fuel cell development, potential applications for military purposes were an important driving force and source of financing of this R&D work. In the early 1960s, for example, the work of General Electric on membrane-type fuel cells that led to the power plants for Gemini spacecraft was financed, in part, by the US Navy's Bureau of Ships (Electronic Division) and by the US Army Signal Corps.

Speaking of military fuel cell applications, we must first point out the following. Most versions of fuel cell-based power plants are of ambivalent utility: they are just as good for civil as for military purposes. Thus, the stationary power plants of different sizes used for uninterruptible emergency power supplies for military objects such as forts, command centers, radar stations, and the like do not differ in any way from similar power plants for civil use in hospitals, telecommunications installations, computer centers of banks, and so on. Power plants for automotive land and water-bound means of transport are equally good for civil and for military vehicles. This is true, even more directly, for power sources intended to supply portable equipment. A lower volume and weight of all equipment carried by soldiers in combat (e.g., as a means of communication, as a means of orientation, as night-vision devices) is a very important point for land forces.

Fuel cell power plants for military applications differ from their civil analogs, primarily, in higher demands on reliability and trouble-free operation. They should keep working under real conditions of military action, day and night, at any time of the year, and whatever the weather. They should be simple to attend to, and as insensitive as possible to faulty manipulations. They should admit all types of transport and shipping, including parachute dropping to destinations.

REFERENCES

Ahluwalia R, Wang X. J Power Sources 2008;177:167.
von Helmont R, Eberle U. J Power Sources 2007;165:833.
Psoma A, Sattler G. J Power Sources 2002;106:381.
McNicol BD, Rand AJ, Williams KR. J Power Sources 2001;100:47.

25

OUTLOOK FOR FUEL CELLS

Prior to 1960, only a few enthusiasts believed in Ostwald's idea of direct conversion of the natural fuels' chemical energy to electrical energy and tried to build fuel cells. For many others at that time, the idea was either altogether utopian or a matter of a very distant future. Decisive change came about after Bacon's demonstration of a working fuel cell and after the first practical use of fuel cells in the Gemini spacecraft. Many scientific groups, private companies, and, in the end, government bodies as well, became interested. This led to the broad development work described in this book. Numerous projects were started for testing the practical utility of fuel cells in various economic fields. Many fuel cell varieties attained a high degree of technical perfection. Industry began to produce certain fuel cell types in numbers attaining hundreds per year.

However, many problems remained without solution, hindering further progress. Primarily, these fundamental problems were either economic (the need to lower the production costs) or lifetime-related (the need to extend working life). A considerable part of these problems was associated not with the fuel cells' properties, but with auxiliary units, such as systems for reactant storage and feed, product removal, thermal management, humidity management, and the like. The auxiliary units of a fuel cell power plant account for about 40% of the total weight, for major part of the costs, and for a considerable amount of operating defects (King and O'Day, 2000).

Electrochemical Power Sources: Batteries, Fuel Cells, and Supercapacitors, First Edition.
Vladimir S. Bagotsky, Alexander M. Skundin, and Yurij M. Volfkovich
© 2015 John Wiley & Sons, Inc. Published 2015 by John Wiley & Sons, Inc.

25.1 ALTERNATING PERIODS OF HOPE
AND DISAPPOINTMENT—FOREVER?

At all past stages of fuel cell development, we have seen great enthusiasm about particular achievements, associated with optimistic predictions that fuel cells would be a cure for major ills of energy conversion and supply. When subsequent arduous work failed to fulfill high hopes, a period of calm ensued, and all efforts quickly waned.

In 1898, Wilhelm Ostwald proclaimed that energy supply in the future would be tied to electrochemistry and, thus, would be smoke-free. Subsequent work on fuel cells was hardly successful enough, largely—from what we know now—on account of material problems. The inevitable period of calm that followed did not end until 1959, when Francis Bacon demonstrated his fuel cell. His patents were basic for the Pratt and Whitney fuel cells and in the Apollo space flights. This situation led to the first real "fuel cell boom."

During 1960s and 1970s, work and progress were vigorous. Phosphoric-acid-based fuel cells were assembled and built, up to large power plants, first on the 100–200 kW scale, then even on the megawatt scale. However, predictions of continued strong growth and ever-wider use failed, and space flight and the energy crisis could not prevent another 20 years of slack. Few additional power plants were built.

This period ended only in the mid-1990s, when great success was achieved with membrane hydrogen fuel cells and with methanol fuel cells. Lamy et al. titled their 2000 review: "Direct methanol fuel cells: From a 20th century electrochemists' dream to a 21st century emerging technology."

25.2 DEVELOPMENT OF ELECTROCATALYSIS

Fuel cell development has always been related to progress in structural materials as well as progress in electrocatalysis, a field that had come into its own as a new field of theoretical and practical electrochemistry during 1970s. Like fuel cells, these fields have seen an alternation of great expectations and disillusionment.

Starting point for the emergence of electrocatalysis was the discovery that hydrocarbons could be oxidized at low temperatures (this fact had not been a part of the Ostwald scenario). Then it was discovered that synergistic effects were operative in the use of ruthenium–platinum catalysts for methanol oxidation, and that compounds such as platinum-free metalloporphyrins were useful catalysts for certain electrochemical reactions in fuel cells. Hopes were expressed that in the future expensive platinum catalyst could be replaced. Again, in the attempts of commercial realization of these discoveries considerable difficulties were encountered, which led to a period of disenchantment and pessimism in 1970s and 1980s. It had been demonstrated beyond doubt that, fundamentally, hydrocarbons could be oxidized at low temperatures, but practical rates that could be achieved were unrealistically low. It had also been demonstrated that fuel cells could be made to work without a platinum catalyst, but practical lifetimes that could be achieved were, again, unrealistically short. In real fuel cells platinum was still needed (even if in much lower quantities). To a certain degree, pessimism in the field of electrocatalysis may also derive from the fact that science is still far from having attained one of the final aims of theoretical work in this field, that is, of providing a full quantum-mechanical foundation for electrocatalysis.

At present, most researchers hold more realistic views of the promise and difficulties in the development of fuel cells, as well as in the development of electrocatalysis. We have concluded in Section 24.5 that an important near-term application is in portable power supply. This is a realistic expectation. It is based on need and on available solutions.

25.3 "IDEAL FUEL CELLS" DO EXIST

At this point we have reviewed the achievements and pointed out that, driven by the balance between requirements and performance, fuel cells are bound to find interesting uses in large numbers. We also have taken a more sober look at the ups and downs, the hopes and disillusionments. We now wish to reassure and position ourselves by a look at Mother Nature.

It is not unusual to compare manmade technical devices with their natural analogs. A well-known example of such a comparison includes certain parameters of sharks and of an electric torpedo used by navy. The two "objects" are of similar size and move under water with comparable speed. That implies that the power required for them to move is comparable as well. A shark gets energy by consuming "cheap" organic seafood, while complex storage batteries containing heavy metals, sometimes even large amounts of expensive silver, must be used to propel the torpedo. Successful torpedo designers will have to show how close the energy consumed by their torpedos comes to that employed by sharks.

In the same manner, fuel cell designers often noticed analogies between processes taking place in fuel cells and processes taking place in human body or certain other biological systems. Justi and Winsel included in their 1962 book, one of the first modern monographs on fuel cells, Chapter 10 "Similarities and differences between fuel cells and living beings." A fuel cell has the task of using (air) oxygen for a direct conversion of the chemical energy of fuels (including organic fuels) to electrical energy while bypassing the intermediate stage of producing thermal energy that, thermodynamically speaking, is not so useful. The heat produced in "cold combustion" in a fuel cell, as a result of various irreversible processes, is not an intermediate form of energy in the conversion process but a real, final product and can be used in addition to the electrical energy produced.

In the human body, for example, mechanical energy is obtained by an analogous direct conversion from the chemical energy of food products of organic nature, again by such "cold combustion," and in fact many food items are labeled with an indication of their caloric content. According to Srinivasan, the conversion efficiency has been estimated at about 35%. As an important point, note that mechanical energy, just like electrical energy, belongs to forms of energy that are thermodynamically useful, and, as such, admit interconversion with almost 100% efficiency.

On average, humans consume food products having energy content of 3000 kcal or 12.5 MJ per day. Spread out uniformly over 24 h, this provides an electrical equivalent of about 126 W. The mechanical energy derived by the human body is consumed in part for "internal needs": blood circulation, breathing, functioning of the brain, intestine, and other organs (certain data indicate that a brain runs on the equivalent of just 1 W). Tens of watts at a time may be available to perform "external work." While a modern man in some societies tends to delegate mechanical work to machines and to do the workout in a gym, we should remember here the useful physical work performed by slaves and criminals on galleys or that performed by the Volga boatmen.

Thus, from a purely energetic point of view, and abstracting from all other human activity (mental, etc.), one may look at a human as the equivalent of a 50-kg fuel cell working with an efficiency of 35% without interruption for an average period of 70 years, producing (after satisfying internal needs) tens of watts of useful power, and having a high degree of autoregulation. For fuel cell engineers, such parameters are a matter of envy and distant dreams.

It is remarkable that the analogy is not merely a superficial one. In fact, physiological processes taking place in the human body, including metabolism, are often of an electrochemical nature and reminiscent of the processes taking place in fuel cells. The first steps of the metabolic process are purely chemical breakdowns of nutrients into simpler low-molecular-weight compounds that occur in the digestive tract. The basic oxidation steps occur at the membranes of the cell mitochondria. Hydrogen atoms are split off from the low-molecular-weight compounds under the effect of the enzyme dehydrogenase and are transferred to molecules of the organic compound NAD (nicotinamide adenine dinucleotide), yielding $NADH_2$. The further oxidation (or dehydrogenation) of this compound occurs purely electrochemically. The electrons and protons are transmitted through a multienzyme system called *electron transfer chain*. The last step is electron and proton transfer from the enzyme cytochrome oxidase to an oxygen molecule existing in a hemoglobin complex. This oxygen reduction step is entirely analogous to the oxygen reduction reaction at the cathodes of fuel cells.

The high energy conversion efficiency found in the human body is the result of many detailed factors, the most significant of them being: (i) high activity and high selectivity of catalytic systems consisting of many different enzymes; (ii) the presence of exceptionally efficient organs for control and management, such as brain and nervous system, securing a high degree of coordination of all internal physiological processes and reacting rapidly and effectively to any perturbation.

It may be a long way to go for a fuel cell designer, but Nature does point the direction and destination (see Bagotsky's 2006 monograph, Chapter 30 "Bioelectrochemistry").

25.4   EXPECTED FUTURE SITUATION WITH FUEL CELLS

For further improvement of fuel cell performance, first of all obvious problems mentioned in the chapters of this book need to be solved, such as: raising the activity of catalysts for methanol oxidation, lowering the sensitivity of oxygen–electrode catalysts toward methanol, and building new improved membranes. In addition, a number of fundamental problems need to be solved in the field of electrocatalysis:

1. *Developing new catalysts for the oxygen electrodes.* It has been pointed out repeatedly that considerable energy losses occur in fuel cell operation on account of irreversibility of the electrochemical reduction of oxygen, a phenomenon also responsible for the considerable deviation of the oxygen electrode's open circuit potential from the thermodynamic value. (Analogous losses are seen in anodic oxygen evolution when electrolytically splitting water.) The metabolic reduction of oxygen in the human body that was mentioned earlier occurs under conditions close to reversibility and without energy losses, despite the fact that it, too, has strong electrochemical components. This is food for thought for researchers in the field of electrocatalysis, and reassurance that we are not facing a fundamental impasse here!

2. *Developing catalysts for the complete 12-electron oxidation of ethanol.* Such catalysts are needed for efficient utilization of biofuels in fuel cells producing electric power, which, in the final analysis, is derived from solar energy, involving consumption (rather than liberation) of carbon dioxide as a greenhouse gas.

3. *Developing highly selective catalysts for electrochemical reactions.* The platinum metal catalysts usually used in fuel cells are highly active but completely nonselective. They equally well catalyze reactions that are desired and reactions that are not needed or that are even detrimental. In contrast, enzymes as natural catalysts are highly active and at the same time highly selective. A given enzyme will catalyze only one particular reaction. If we had selective catalysts for the two electrodes in a fuel cell, then we could realize single-compartment fuel cells, that is, a type of fuel cells widely discussed at present where the fuel and the oxidizer are supplied together to the fuel cell. Such cells would have a design much simpler than current cells requiring separate reactant supply, and they would be easier and cheaper to manufacture than today's fuel cells. Such fuel cells would also be free of the numerous difficulties arising from the need for sealing, that is, for isolating the fuel compartment from the oxidizer compartment, which is a real headache for fuel cell makers. It may be possible, following again the leads provided by Nature, to achieve an adequate solution of these problems by developing catalysts with surface properties that have been tailored deliberately so as to provide a favorable catalytic action on all intermediate steps that need it. Such catalysts should be polyfunctional and exhibit a certain degree of chemical and structural microinhomogeneity. Nanoelectrochemistry and nanoelectrocatalysis may be a possible approach to

synthesizing such surfaces (see Bagotsky's monograph on fuel cells, Chapter Overlook).

REFERENCE

King JM, O'Day MJ. *J Power Sources*; 2000:86:16.

MONOGRAPHS

Bagotsky VS. *Fuel Cells: Problems and Solutions*. 2nd ed. Hoboken, NJ: Wiley, 1st ed., 2009; 2012.

Bagotsky VS. *Fundamentals of Electrochemistry*. 2nd ed. Hoboken, NJ: Wiley; 2006.

Justi EW, Winsel A. *Fuel Cells: Kalte Verbrennung*. Germany: Steiner, Wiesbaden; 1962.

Lamy C, Léger J-M, Srinivasan S. In: Bockris JO'M, Conway BE, editors, Vol. 34. *Modern Aspects of Electrochemistry*. New York: Plenum Press; 2000. p 53.

Srinivasan S. Fuel Cells: From Fundamentals to Applications. New York: Springer Verlag; 2006.

PART IV

SUPERCAPACITORS

26

GENERAL ASPECTS

Lately, quite a number of new capacitor types based on diverse electrochemical processes has been developed. According to Conway (1999), electrochemical capacitors are electrochemical devices, in which quasireversible electrochemical charging–discharge processes occur. The shape of galvanostatic charging and discharge curves corresponding to these is close to linear, that is, close to the shape of corresponding dependences for usual electrostatic capacitors. Electrochemical capacitors can be subdivided into film (dielectric), electrolytic capacitors, and supercapacitors (ECSCs) that, in their turn, are subdivided into electric double-layer capacitors (EDLCs), pseudocapacitors (PsCs), and hybrid capacitors (HCs). Capacitors are used as electric power batteries in pulse engineering devices and as reactive elements characterized by AC resistance practically without power losses in the sine-wave current technology. They are widely used in power and information portable devices (radio telephones, camera recorders, and players), and so on.

26.1 ELECTROLYTIC CAPACITORS

Electrolytic capacitors with electrodes based on aluminum foil and liquid electrolyte have been known already for several dozens of years and are characterized by high capacitance values (up to 150 F). The dielectric layer in these capacitors is a thin (about a micrometer) layer of aluminum oxide formed by means of electrochemical

Electrochemical Power Sources: Batteries, Fuel Cells, and Supercapacitors, First Edition.
Vladimir S. Bagotsky, Alexander M. Skundin, and Yurij M. Volfkovich.
© 2015 John Wiley & Sons, Inc. Published 2015 by John Wiley & Sons, Inc.

oxidation of aluminum foil. The faults of aluminum electrolytic capacitors are the dependence of their parameters on the temperature and considerable leakage currents. High capacitance values of electrolytic capacitors are obtained following the presence of a thin dielectric film on the anode and the developed surface area obtained using the technique of electrochemical or chemical etching. The surface area increases under electrochemical etching owing to appearance of a system of pores and cavities. Capacitance of aluminum foil after etching can increase almost by two orders of magnitude. The dielectric aluminum oxide film can be formed electrochemically on the aluminum surface in different solutions. The most frequently used solutions are those of inorganic phosphates and borates. The film thickness is usually $12-14$ Å/V of the voltage applied, which corresponds to the breakdown voltage of $\sim 10^7$ V/cm. Formation of a layer of porous hydrated oxide on the aluminum surface is possible before the oxide film is formed. The hydrated layer is partly transformed into oxide in the course of electrochemical treatment yielding a film with high dielectric properties and a higher capacitance. The formed film consists of a metal–oxide barrier layer with a residual hydrated layer at the oxide–electrolyte interface.

Solutions for electrolytic capacitors must have the following properties: high conductivity, a wide working range of temperatures, incapacity of the oxide film on Al for reduction. Quite diverse compounds are used as electrolytes for aluminum electrolytic capacitors: salts of adipinic, trifluoroacetic, maleic, salicylic, citric, tartaric, and formic acids. Organic electrolytes based on ethylene carbonate, γ-butyrolactone (GBL), and so on are also used in aluminum electrolytic capacitors.

The main fault of liquid electrolytes is their thermal instability. The best known solid electrolytes are salts of the 7,7,8,8-tetracyanoquinone dimethane (TCQD) complex. TCQD is an excellent acceptor of electrons. It can be bound to different electron donors forming a conducting salt complex. Thin films of the TCQD complex are applied onto plates of porous aluminum oxide using the ion sputtering technique. N-isopropyl-4,4′-bipyridine was also used as a donor of protons for the TCQD complex. Capacitance of the capacitor with this solid electrolyte was $600\,nF/cm^2$. However, salts of the TCQD complex are thermally unstable. N-n-butyl isoquinoline–$(TCQD)_2$ is characterized by high conductivity even after the melting and the following cooling. Its conductivity is higher than that of conventional electrolytes used in capacitors. Stable high conductivity is observed for this complex in a wide temperature range between $-55°C$ and $+105°C$. Capacitance of a capacitor with N-n-butyl isoquinoline–$(TCQD)_2$ was $2200\,\mu F$.

As compared to aluminum electrolytic capacitors, capacitors based on tantalum electrodes have a higher specific capacitance, lower leakage currents, longer storage time, better temperature–frequency characteristics, and high reliability. Tantalum oxide has a high oxide film thickness than that of aluminum oxide. A considerably higher mechanical strength of tantalum allows using it in the form of a thinner foil, which additionally enhances capacitance. However, tantalum capacitors are more expensive than aluminum ones. Chemical resistance of tantalum oxide allows using solvents that cannot be used in aluminum electrolytic capacitors. Application of MnO_2 as solid electrolyte allows using tantalum capacitors at low temperatures down to $-55°C$.

At present, the options for practical application of niobium and titanium oxides in electrolytic capacitors are studied as they are characterized by a high dielectric constant and lower cost than tantalum oxide. Improvement of properties of electrolytic capacitors is carried out along the following lines: a significant decrease in the size of capacitors, expansion of boundaries of working temperature ranges, improvement of electric characteristics, an increase in reliability and service life of capacitors under storage and when in operation.

REFERENCES

Conway BE. *Electrochemical Supercapacitors*. Kluwer Academic/Plenum Publishers; 1999.

Hand J, Bowling L. Improving Producibility IEEE Intercon. Two Day Manufacturing Works; New York; 1974; No 4.

Kovacs GY, Marai GY. Hiradastechnika 1984;35:454.

27

ELECTROCHEMICAL SUPERCAPACITORS WITH CARBON ELECTRODES

27.1 INTRODUCTION

General Electric Company first patented electric double-layer capacitors (EDLCs) in 1957. The capacitor consisted of porous carbon electrodes and used electric double-layer capacitance for the charging mechanism. Standard Oil Company in Cleveland, Ohio (SOHIO), patented a device storing energy at the interface of the electric double layer (EDL). Nippon Electric Company (NEC) in Japan obtained a license for the manufacturing technology from SOHIO and introduced first EDLCs to the market as memory power backup in computers in 1957. At the time, SOHIO acknowledged that the "double layer at the interface behaves as a capacitor with a relatively high specific capacitance." SOHIO patented a disk-shaped capacitor in 1970 using carbon paste impregnated by electrolyte. NEC had licenses to SOHIO technologies up to 1971 (Sharma P., Bhatti T (2010)). The first commercially successful EDLC with the name of "supercapacitor" was set off by NEC. A number of companies started producing electrochemical capacitors up to the 1980s.

At present, a great number of research works on development and improvement of electrochemical supercapacitors (ECSCs) are published. As pointed out above, ECSCs are subdivided into EDLCs, in which the energy of EDL recharging is used; pseudocapacitors (PsCs), in which pseudocapacitance of fast quasireversible faradaic

Electrochemical Power Sources: Batteries, Fuel Cells, and Supercapacitors, First Edition.
Vladimir S. Bagotsky, Alexander M. Skundin, and Yurij M. Volfkovich.
© 2015 John Wiley & Sons, Inc. Published 2015 by John Wiley & Sons, Inc.

reactions is used; and hybrid capacitors (HCs). In HCs, various fast processes occur on different electrodes (positive and negative); for example, EDL recharging occurs on a single electrode and a certain fast faradaic redox reaction occurs on the other electrode.

ECSCs are used in pulse engineering devices; as store of electric energy; in cars, railway engines, and ships for their starter firing and braking power regeneration; in electromobiles, electric cars, and wheelchairs; in sine-wave current technologies as reactive components with AC resistance with practically no power losses. They are widely used in energetic and information portable devices (radio telephones, camera recorders, players). It appears promising to use ECSCs for load leveling in electric mains.

27.2 MAIN PROPERTIES OF ELECTRIC DOUBLE-LAYER CAPACITORS (EDLC)

Figure 27.1 shows typical charging–discharge galvanostatic curves (i.e., dependences of DC potential on the charge value (or time of charging and discharge)) for one of the actual EDLC electrodes.

EDLC consists of two porous polarizable electrodes. The process of energy saving in EDLC is based on charge separation on two electrodes with a sufficiently high difference of potentials between them. The electric charge of EDLC is determined by

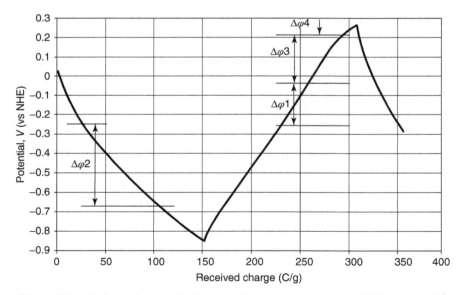

Figure 27.1. Typical galvanostatic charging–discharge curves of a real EDLC electrode.

the EDL capacitance. The electrochemical process in EDLC can be presented as

$$\text{positive electrode}: E_s + A^- \leftrightarrow E^+_s//A^- + e^- \tag{27.1}$$

$$\text{negative electrode}: E_s + C^+ + e^- \leftrightarrow E^-_s//C^+ \tag{27.2}$$

$$\text{overall reaction}: \quad E_s + E_s + C^+ + A^- \leftrightarrow E^-_s//C^+ + E^+_s//A^- \tag{27.3}$$

E_s represents the electrode surface; $//$ is EDL, where the charge is accumulated on both sides; C^+ and A^- are cations and anions of electrolyte, respectively.

In the course of charging, electrons move from the positive electrode to the negative one via the external power source. Ions from the electrolyte bulk move to the electrodes. Electrons move in the course of discharge from the negative electrode to the positive one through the load and ions return from the surface to the bulk of electrolyte. The charge density at the interface and concentration of electrolyte change during the discharge and charging.

Theoretical concepts of specific (per unit true electrode surface area) EDL capacitance are based on the known classical theories of EDL developed by Helmholtz, Stern, Gouy–Chapman, Grahame, and so on. One of the directions of modern studies of EDL is elucidation of ratios between different surface layer characteristics. These include specific capacitances in zero charge points, electronic work functions of metals, their liophilicity, zero charge potentials. Correlations are established between many of these characteristics in a number of metals and solvents. At the same time, there are significant deviations from main trends. Zero charge points were first determined for different carbon materials in the works of Frumkin et al.

For a more detailed electrochemical analysis of ECSC electrodes as compared to galvanostatic curves, same as those of other electrochemical systems, cyclic voltammetric curves (CVs), are measured, that is, dependences of current on potential at the given potential sweep rate (V/s). Figure 27.2 shows schematic CV dependences. The upper figure shows such dependences for the electrode of an ideal EDLC, in which only EDL charging occurs. To the first approximation, these dependences are direct lines parallel to the x-axis. The lower figure shows schematic CV dependences of a PsC electrode with pseudocapacitance and EDL capacitance.

For EDLCs, same as for all ideal capacitors, capacitance is inversely proportional to the thickness of a capacitor plate:

$$C = \frac{e}{4\pi d} \tag{27.4}$$

where e is dielectric permeability, d is the thickness of the capacitor plate. For example, the plate in classical paper capacitors is paper between the electrodes. Its thickness is several tens of micrometer and such capacitors have an accordingly low specific capacitance value. The plate thickness in an EDLC electrode is the EDL thickness being about several tens of nanometer. As a result, specific capacitance C_s per unit electrode/liquid electrolyte true interface surface area is many orders of magnitude higher than in the case of usual capacitors; it is about 10 µF/cm² for

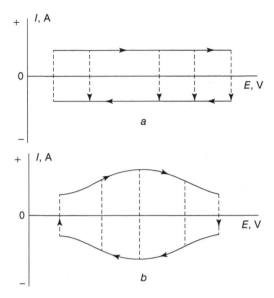

Figure 27.2. Schematic CV dependences for an electrode of ideal EDLC, in which (a) EDL charging occurs and (b) for an PsC electrode with pseudocapacitance.

aqueous electrolytes and several micofarad per square centimeter for nonaqueous electrolytes.

Specific capacitance per 1 g of the electrode:

$$C_g = C_s \times S \tag{27.5}$$

where S is the specific surface area (cm^2/g). To obtain high capacitance in EDLCs, electrodes with high specific surface area $S = 1000-3000 \, m^2/g$ are used; they are based on highly dispersed carbon materials (HDCMs): activated carbons (ACs), aerogels, carbon blacks, carbon nanotubes (CNTs), nanofibers, graphenes, and so on. Therefore, the maximum values of $C_g = 100-300$ F/g are obtained for EDLC electrodes in aqueous electrolyte using Equations (27.1) and (27.2). As the size of ions and molecules and therefore the EDL thickness are lower in aqueous electrolytes than in nonaqueous electrolytes, specific capacitance values for aqueous electrolytes is in most cases higher than in nonaqueous electrolytes.

The measured specific capacitance values of carbon materials used in supercapacitors are in most cases lower than these high values and are in the range of 75–175 F/g in aqueous electrolytes and 40–100 F/g using organic electrolytes, as a relatively large part of the surface in the majority of carbon materials is in pores that ions from electrolyte cannot penetrate. This is especially true for organic electrolytes. Porous carbons used in supercapacitors must have a large fraction of pores with the diameter of 1–5 nm. Materials with small pores (<1 nm) manifest a pronounced decrease in capacitance at the currents above $100\,\text{mA/cm}^2$ especially when organic electrolytes are used. Materials with large pore diameters can be discharged at the current densities above $500\,\text{mA/cm}^2$ with the minimum decrease in capacitance. Voltage of a supercapacitor cell depends on the electrolyte used. The cell voltage is about 1 V for aqueous solutions of electrolyte; the cell voltage is $3-3.5$ V for organic electrolytes.

27.3   EDLC ENERGY DENSITY AND POWER DENSITY

For ideal EDLCs with ideally polarizable electrodes, specific discharge energy is

$$A = \left(\frac{1}{2}\right) C[(V_{\text{max}})^2 - (V_{\text{min}})^2] \tag{27.6}$$

where C is the average capacitance of electrodes, U_{max} and V_{min} are the initial and final values of discharge voltage. If $V_{\text{min}} = 0$, then

$$A = A_{\text{max}} = \left(\frac{1}{2}\right) C(V_{\text{max}})^2 \tag{27.7}$$

Owing to the low solubility of nonaqueous electrolytes, energy density of EDLC with nonaqueous electrolyte depends not only on capacitance, but also on the concentration of electrolyte.

As seen from Equation (27.7), the maximum energy density of EDLC is proportional to capacitance to the first degree and squared maximum voltage. When comparing the values of energy density of EDLC with aqueous and nonaqueous electrolytes, one has to take into account that the values of specific capacitance and conductivity are higher in the case of aqueous electrolytes as compared to nonaqueous ones and the values of the maximum voltage are lower for aqueous electrolytes than for nonaqueous ones. Figure 27.3 shows for comparison the dependences of energy density on the discharge current density we obtained for symmetrical EDLC with similar electrodes based on activated carbon cloth (ACC) with the specific surface area of $600\,\text{m}^2/\text{g}$, but with either aqueous or nonaqueous electrolyte (Volfkovich Y, Serdyuk T, 2002). One can see that the energy density values at low current densities are higher for EDLC with nonaqueous electrolyte, while in the case of high current densities, the specific power values are higher for EDLC with aqueous electrolyte.

Supercapacitors were developed as an alternative to pulse batteries. To be an alternative, supercapacitors must have much higher power and cycling time. Characteristics of a number of supercapacitors and pulse batteries are shown in Table 27.1.

According to Burke (2000), there are two approaches toward the calculation of peak power density of ECSC and batteries pointed out in Table 27.1. The first and more standard approach consists of the determination of power at the so-called corresponding state of impedance, at which half the discharge energy is converted into electricity and half is converted into heat. The maximum power at the given moment

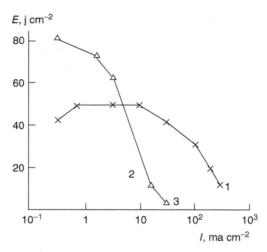

Figure 27.3. Dependences of energy density on the discharge current density for EDLC with the following electrolytes: (1) 35% H_2SO_4 and (2) 1M $LiAlF_4 + \gamma$-butyrolactone (GBL).

is determined by the following equation:

$$P_{mi} = \frac{V_{oc}^2}{4R_b} \tag{27.8}$$

where V_{oc} is the open-circuit voltage of a battery and maximum voltage of a supercapacitor, R_b, is the resistance of the corresponding device. The discharge efficiency in this state is 50%. For many applications, for which a considerable fraction of energy is stored in an energy storage before it is used in the system, efficiency of the charging/discharge cycles is of great importance for the performance of the system. In these cases, use of the energy storage must be limited by conditions resulting in high performance both of charging and discharge. Discharge/charge capacitance of a battery and EDLC as a function of performance is determined by the following equation:

$$P_{ef} = EF \cdot (1 - EF) \cdot \frac{V_{oc}^2}{R_b} \tag{27.9}$$

where EF is the efficiency of a high-power pulse. For $EF = 0.95$, $P_{ef}/P_{mi} = 0.19$. Thus, in applications, where efficiency is the main problem, useful power of a battery and ECSC is much lower than peak power P_{mi} most often stated by the manufacturer of the battery and ECSC. In the case of ECSC, the peak power for discharge between voltages V_0 and $V_0/2$, where V_0 is the nominal voltage of the device, corresponds to the following formula:

$$P_{pi} = \frac{9}{16} \cdot (1 - EF) \cdot \frac{V_{oc}^2}{R_{uc}} \tag{27.10}$$

TABLE 27.1. Main Characteristics of Some Supercapacitors and Pulse Batteries (Sharma P, Bhatti T (2010))

	Voltage (V)	Capacitance Ah	Weight (kg)	Resistance (mΩ)	Wh/kg	P_{ef} (95% Discharge) (W/kg)	P_{pi} (W/kg)
Supercapacitors							
Maxwell							
2700 F	3	2.25	0.85	0.5	4.0	593	5294
1000 F	3	0.83	0.39	1.5	3.1	430	3846
Panasonic							
800 F	3	0.67	0.32	2.0	3.1	392	3505
2000 F	3	1.67	0.57	3.5	4.4	127	1128
Superfard (250 F)	50	3.4	16	20	5.4	219	1953
SAFT							
Gen2 (144 F)	3	0.12	0.030	24	6.0	350	3125
Gen3 (132 F)	3	0.11	0.025	13	6.8	775	6923
Power Stor	3	0.083	0.015	10	8.33	1680	15000
Batteries							
Panasonic NiHD	7.2	6.5	1.1	18	42	124	655
	12.0	98	17.2	8.7	68	46	240
Ovonic NiHD	13.2	88	17.0	10.6	70	46	245
	12.0	60	12.2	8.5	65	80	420
	7.2	3.1	0.522	60	43	79	414
Sanyo Li ion	3.6	1.3	0.039	150	121	105	553
Hawker Pb acid	2.1	36	2.67	0.83	27.0	95	498
	12	13	4.89	15	29.0	93	490
Optima Pb acid	6	15	3.2	4.4	28	121	635
Horizon Pb acid	2.1	85	3.63	0.5	46	115	607

where R_{uc} is the ECSC resistance. This equation accounts for a decrease in voltage under discharge of the device. The peak power values in Table 27.1 are calculated for impedance and high-performance discharge of batteries and supercapacitors.

The power of supercapacitors is almost in all cases obviously higher than that of batteries. It is notable that it is not quite correct to compare high-performance specific power for supercapacitors with the corresponding impedance specific power of batteries, as it is usually done. Power options of both device types depend primarily on their resistance and knowing their resistance is the key for determining peak useful power. Thus, measurement of resistance of the device in the pulse mode of operation is critical for estimation of its ability to provide high power.

In addition to the ability to generate high power, there is another cause to consider supercapacitors for particular applications, namely, that they are characterized by high stability and high cycling times. This is especially true as regards supercapacitors based on carbon electrodes. Most secondary (rechargeable) batteries if left alone and unused for many months noticeably deteriorate and become in effect use-less after this period because of self-discharge and corrosion. Supercapacitors also self-discharge over a long period of time to low voltage, but they preserve their capac-itance and thus are capable of returning back into the initial state. Experience showed that supercapacitors can stay unused for several years and they still remain practi-cally in the initial state. Supercapacitors can be subjected to deep cycling at high sweep rates (discharge in seconds) up to 1,000,000 cycles with relatively small vari-ation of characteristics (10–20% of degradation in capacitance and resistance). This is impossible in the case of batteries, even for small discharge depth (10–20%).

Thus, supercapacitors as high-power pulse devices possess the following advan-tages as compared to batteries: high efficiency, long storage life, and good cyclability.

Difference ($U_{max} - U_{min}$) is denoted as the potential window. The wider this win-dow, the higher the values of energy density and power density of EDLC.

Using nonaqueous electrolytes in ECSCs with electrodes based on HDCMs allows obtaining high (up to 3–3.5 V) potential window values, which significantly enhances the energy but limits the power of capacitors because of low conductivity of these electrolytes. Aqueous solutions of H_2SO_4 and KOH with the concentrations of 30–40 wt% because of high conductivity allow reaching sufficiently high power values, but the small working voltage range (about 1 V) decreases the energy characteristics of ECSCs.

It was found that as opposed to batteries, ECSCs can operate in a very wide range of charging–discharge times from fractions of a second to hours. Accordingly, ECSCs are subdivided into two principal types: power ones with high power density and energy ones with high energy density. Power ECSCs include EDLCs. Power supercapacitors allow carrying out charging and discharge processes in very short time periods (from fractions of a second to minutes) and obtaining herewith high power characteristics usually from 1 to tens kW/kg in concentrated aqueous electrolytes. Measurements on highly dispersed carbon electrodes in the energy capacitor operation modes usually yield specific capacitance values in the range of 50–200 F/g. In the case of carbon materials, the limiting capacitance of 320 F/g was obtained because of a considerable contribution of pseudocapacitance of reversible redox reactions of carbon surface groups. Thus, this is not pure EDLC anymore.

A very wide range of characteristic charging–discharge times is illustrated in Figure 27.4 showing Ragone diagrams for different rechargeable electrochemical devices representing the regions of functioning of these devices in the coordinates of power density-energy density. This figure shows various types of batteries (lead, nickel–metal hydride, and lithium ion) and usual capacitors and ECSCs. One can see that the rage of ECSC operation reaches over seven orders of magnitude of char-acteristic times, which is much more than in the case of any battery type. ECSCs have their own "ecological niche." As compared to batteries, they are characterized by lower energy density values, but much higher power density values. Besides, they exceed batteries by many orders of magnitude as regards their cyclability that for some ECSC types (especially for EDLC) reaches hundreds of thousands and even millions of cycles, while in the case of batteries it is several hundreds to several thou-sands of cycles.

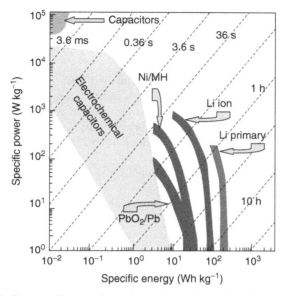

Figure 27.4. Ragone diagrams for various electrochemical rechargeable devices.

27.4 FUNDAMENTALS OF EDLC MACROKINETICS

According to Volfkovich and Serdyuk (2002), it appears important to consider the fundamentals of EDLC macrokinetics. Assuming that a porous EDLC electrode is ideally polarizable, the main processes occurring in the course of EDLC operation are generally: the nonsteady-state process of EDL charging in a developed electrode/electrolyte interface in pores, diffusion–migration ion transport in electrolyte within the pores, and internal ohmic energy losses. The following equation for the distribution of potential E over the electrode depth, that is, by the x coordinate, follows from the theory of porous electrodes:

$$\kappa \left(\frac{\partial^2 E}{\partial x^2} \right) = Si(E) \tag{27.11}$$

where κ is the effective conductivity of electrolyte in pores, S is the specific electrode surface area (cm^2/cm^3), $i(E)$ is the dependence of local current density i at the electrode electrolyte interface on the potential. Current density i is determined by the EDL charging:

$$i = C_s \left(\frac{\partial E}{\partial \tau} \right) \tag{27.12}$$

where C_s is the specific EDL capacitance per unit surface area and τ is the time. The boundary and initial conditions for the system of Equations (27.11) and (27.12) take the following form:

$$I = \kappa \left(\frac{\partial E}{\partial x}\right)_{x=L}; \left(\frac{\partial E}{\partial x}\right)_{x=0} = 0; \quad E_{(x,\tau=0)} = E_0 \qquad (27.13)$$

where I is the current density per electrode unit visible surface area. They correspond to the galvanostatic mode of operation of a flat porous electrode with the known initial ($\tau = 0$) potential value $E = E_0$. Here, L is the electrode thickness; $x = L$ is the coordinate at the front face of the electrode turned to the counterelectrode; $x = 0$ is the coordinate on its rear ("sealed") side. System of equations and boundary conditions (27.11)–(27.13) has an analytical solution.

We compared calculated and experimental discharge and charging curves for symmetrical EDLC with two similar electrodes of AC cloth AUT-600 with the specific surface area of $S = 600\,m^2/g$ and a separator of porous polypropylene. EDLCs with 10 M KOH aqueous electrolyte and 1 M LiAlF$_4$+γ-butyrolactone (GBL) nonaqueous electrolyte were studied. One can see that close coincidence of the calculated and experimental curves is observed, which points to the correctness of the chosen model for EDLC calculation and optimization. Therefore, this model can be used for the calculation of charging–discharge curves of various EDLCs.

Figure 27.5 visualizes the effect of ohmic losses in electrodes and the nature of the solvent on electric properties of EDLC. In the absence of ohmic losses in electrodes, charging and discharge curves must represent straight lines with the slope inversely proportional to EDLC capacitance. The EDLC capacitance in an organic solvent is much lower, which is manifested in the larger slope of the discharge curve. The higher or lower degree of deviation of the characteristics from a straight line corresponds to higher or lower ohmic losses and nonuniformities of the potential distribution over the porous electrode thickness. Hence one can also see that galvanostatic discharge and charging curves can differ from straight lines even in the absence of pseudocapacitance.

27.5 POROUS STRUCTURE AND HYDROPHILIC–HYDROPHOBIC PROPERTIES OF HIGHLY DISPERSED CARBON ELECTRODES

As already pointed out above, highly dispersed carbon electrode materials (HDCMs) with the specific surface area of $500–3000\,m^2/g$ are used as ECSC electrodes. It is known that carbon materials possess transient hydrophilic–hydrophobic properties. Figure 27.6 shows integral curves of pore size distribution for the CH900-20 AC cloth were measured using the method of standard contact porosimetry (MSCP) (Volfkovich et al., 2001) with octane and water as measurement liquids [8]. Porosimetric curves are measured using octane for all pores and using water only for hydrophilic pores. As follows from this figure, this cloth has a very wide pore spectrum: from micropores with radii $r \leq 1$ nm to macropores with $r > 100$ μm, that is, in the range of more than five orders of magnitude. This cloth contains micropores

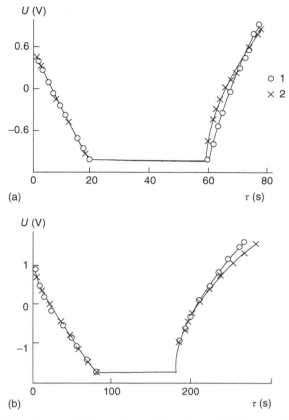

Figure 27.5. (1) Calculated and (2) experimental charging and discharge galvanostatic curves for EDLC: (a) electrolyte is 10 M KOH; $I = 160$ mA/cm^2 and (b) electrolyte is 1 M LiAlF$_4$ + GBL; $I = 16$ mA/cm^2.

and also macropores with $r > 1$ μm, but there are practically no mesopores with 1 nm $< r <$ 100 nm. Micropores provide high overall specific surface area with $S_p = 1500$ m^2/g and hydrophilic specific surface $S_{fi} = 850$ m^2/g. MSCP allows measuring overall and hydrophilic porosity and also overall and hydrophilic specific surface areas. For ECSC with nonaqueous electrolytes, the volume of electrolyte in pores is equal to overall porosity, while for ECSC with aqueous electrolytes, the volume of electrolyte in pores is equal to hydrophilic porosity. As follows from these curves, full porosity (by octane) was 86%, hydrophilic porosity was 78.5%, and hydrophobic porosity was 7.5%. Apart from such microporous carbons in ECSC, HDCMs containing besides micropores and macropores a significant fraction of mesopores are also used. Such carbons are called *mesoporous* and are mainly used for energy-type ECSCs. Microporous carbons are used in energy-type ECSCs.

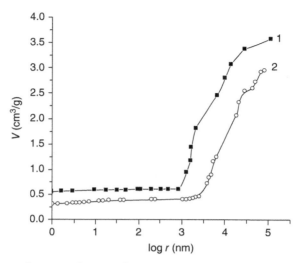

Figure 27.6. Integral curves of pore radius distribution measured in octane (1) and water (2) for a CH900-20 AC cloth.

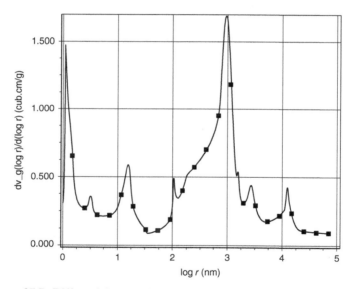

Figure 27.7. Differential curve of pore radius distribution for the Norit carbon.

Figure 27.7 showed the differential pore size distribution curve for the Norit AC. This carbon has the whole pore set: micro-, meso-, and macropores. The overall specific surface area of this carbon is $1700\,\mathrm{m^2/g}$.

As follows from formula (27.5), specific EDL capacitance C_g is proportional to specific surface area S. However, this regularity is rarely observed in practice, though specific capacitance is in most cases higher for HDCMs with a higher specific surface area than for corresponding HDCMs with a lower specific surface area. This issue is considered in more detail in the further parts of the chapter.

27.6 EFFECT OF RATIO OF ION AND MOLECULE SIZES AND PORE SIZES

In EDLCs, EDL formation on the carbon electrode surface is the main mechanism. Inagaki et al. (2010) have observed capacitance C_{obs}(F/g) and divided into two parts: capacitance because of the surface of micropores S_{micro} and capacitance determined by the surface of larger pores (mainly mesopores) that was denoted as external surface S_{ext}:

$$C_{obs} = C_{ext} \times S_{ext} + C_{micro} \times S_{micro} \qquad (27.14)$$

where C_{micro} and C_{ext} are capacitances per $1\,m^2$ for the region of micropores and larger pores, accordingly. Equation (27.14) can be rewritten as

$$\frac{C_{obs}}{S_{ext}} = C_{ext} + C_{micro}\left(\frac{S_{micro}}{S_{ext}}\right) \qquad (27.15)$$

This implies the linear dependence between the parameters of (C_{obs}/S_{ext}) and (S_{micro}/S_{ext}), both of which are determined experimentally. Capacitance was also measured on the same carbons in nonaqueous and aqueous electrolytes (1 mol/l TEMABF$_4$/PC and 1 mol/l H$_2$SO$_4$, accordingly) at different current densities between 100 and 1000 mA/g. It was shown that for aqueous electrolyte, $C = 0.12\,S_{micro} + 0.29S_{ext}$ at the current of 100 mA/g, $C = 0.10S_{micro} + 0.28S_{ext}$ at the current of 1000 mA/g, while in the case of nonaqueous electrolyte, $C = 0.05S_{micro} + 0.20S_{ext}$ at the current of 100 mA/g and $C = 0.03S_{micro} + 0.20S_{ext}$ at the current of 1000 mA/g.

The ratios between C_{obs}/S_{ext} and S_{micro}/S_{ext} are well approximated by a linear dependence both in aqueous and in nonaqueous electrolytes with different current densities, as shown for two current density values in Figure 27.8 [3]. The results show that the contribution of micropores and mesopores to the measured capacitance, that is, C_{micro} and C_{ext}, are different in aqueous and nonaqueous electrolytes. In nonaqueous electrolytes, the C_{ext} value does not change under current density variation and remains about 0.2 F/m^2, but C_{micro} is very low (0.04 F/m^2) and decreases at an increase in current density. On the other hand, in aqueous electrolyte, C_{ext} is almost constant at the level of 0.3 F/m^2 with a fast increase at the low current density below 100 mA/g; however, C_{micro} gradually decreases at an increase in current density. C_{micro} in nonaqueous electrolytes is much lower than in aqueous ones (about 0.04 and 0.1 F/m^2, accordingly), though the C_{ext} values do not differ so much (about 0.2 and 0.3 F/m^2). These results become understandable accounting for the fact that sizes of cations H$^+$ of the aqueous electrolyte are much lower than those of nonaqueous TEMA$^+$ cations.

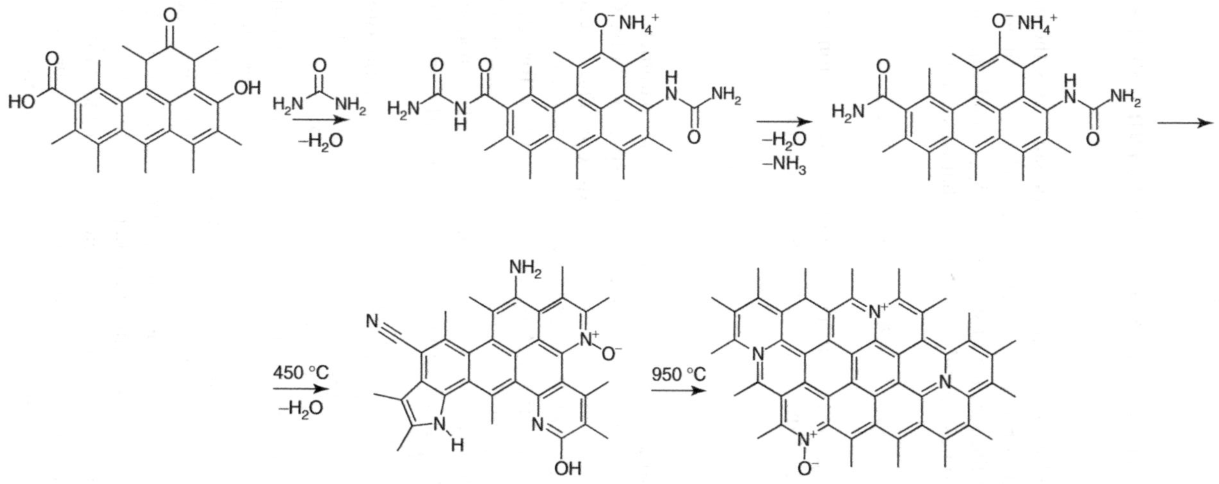

Figure 27.8. Possible chemical reactions with surface groups and thermal transformations.

27.7   EFFECT OF FUNCTIONAL GROUPS ON EDLC CHARACTERISTICS

For EDLCs, the presence of functional groups on the HDCM surface affects the electrochemistry of the carbon surface interface and EDL properties including: wettability, zero charge potential, electric contact resistance, adsorption of ions (capacitance), and self-discharge characteristics. Many surface groups of ACs have ion-exchange properties. Ion exchange capacitance of AC reaches approximately 0.5–2.0 mg-eq/g, which is comparable to exchange capacitance of polymer ionites and ion-exchange membranes. Thus, ACs are electron exchangers and ion exchangers. Carbons easily adsorb chemically molecular oxygen in air even at low ambient temperatures. Oxygen chemosorption (irreversible adsorption) grows at an increase in the temperature and results in the formation of different functional groups on the carbon surface, for example, $-COOH$, $=CO$, quinine–hydroquinone, and so on. Carbon–oxygen complexes are the most important surface groups on carbons. The types of functional groups strongly depend on the initial materials (precursors) and conditions of preparation of the carbon material. Acidic oxides are formed on the surface when carbons undergo oxygen exposure (activation) from 200 to 700°C or reactions with solutions of oxidants at the room temperature. Presence of oxygen in surface groups can result in the formation of melamine and urea and promotion of surface reactions, as shown, for example, in Figure 27.8, and can also lead to possible chemical reactions with surface groups and thermal transformations.

Thereof the curve of CV had the maximum similar as on a curve of Figure 27.2b.

These surface groups are considered less stable and include such groups as carboxyl, lactonic, and phenol ones. Basic and neutral surface oxides are considered to be more stable than acidic oxides and tend to contact oxygen at low temperatures. Functional groups that are electrochemically inert in the range of working potentials can cause an increase in the wettability of carbon electrodes and therefore enhance specific capacitance of carbon by improving access of aqueous electrolytes to pores that results in better use of the surface.

After electrochemical oxidation of graphite, the concentration of strong and weak acidic groups on its surface increases, which was confirmed using the titration technique. Electrochemical preoxidation results in an increase in the capacitance of graphite and ACC. Impedance measurements were used to calculate capacitance: 160 F/g for electrodes without oxidation and 220 F/g for electrodes after oxidation. An additional contribution into the overall capacitance is because of pseudocapacitance of redox reactions of surface groups.

The preservation degree of both physically adsorbed molecular oxygen and oxygen in the form of surface complexes strongly affects the rate and mechanism of capacitor self-discharge (leakage currents). In particular, carbons with a high concentration of acidic functional surface groups tend to manifest high self-discharge degrees. An increase in the leakage current implies that oxygen in functional groups can serve as active sites capable of catalyzing electrochemical oxidation or reduction of carbons or decomposition of the components of electrolyte. On the other hand, removal of surface oxygen functional groups under high-temperature treatment in an inert medium results in a decrease in leakage currents. The presence of oxygen functional groups can also promote instability of EDLC characteristics leading to an increase in resistance (ESR, equivalent serial resistance) and a change in capacitance. Removal of oxygen from carbons used in supercapacitor electrodes usually improves their stability.

Oxygen surface functional groups affect the rest potential (steady-state potential) of ACs. It turned out that the steady-state potential of AC was proportional to the logarithm of oxygen content or the concentration of acidic surface groups. Carbons with a high steady-state potential cause an undesirable high voltage under charging and this can result in gas generation.

At the same time, surface functional groups contribute to the main capacitance under charge accumulation on carbon electrodes; they can charge and discharge fast and result in a significant increase in pseudocapacitance.

One of the examples of reversible redox reactions of functional surface groups contributing to pseudocapacitance of carbon electrodes is reduction–oxidation of the cyclic diacyl peroxide group to two carboxyl groups according to the reaction in Figure 27.9.

Oxygen-containing functional groups are formed under usual activation and are obtained independently on the surface in the case of carbon oxidation by oxygen or HNO_3 and as a result of electrochemical oxidation. Intercalation of oxygen is detrimental for carbon materials used in nonaqueous solutions of electrolytes, as they produce a negative influence on the reliability of capacitors in respect to voltage stability, self-discharge, leakage current, and so on. Significant losses result from the removal of oxygen groups in production of ACs for commercial EDLCs. And vice versa, oxygen-containing functional groups enhance the overall capacitance in aqueous solutions of electrolytes, for example, in most cases, in H_2SO_4 solutions, as pseudocapacitance develops.

Electrochemical pretreatment of activated carbon fibers (ACFs) based on poly-acrylonitrile (PAN-based) in a $NaNO_3$ solution resulted in an increase in the content of oxygen-containing functional groups in the fiber surface and therefore led to an increase in specific capacitance C_g. ACFs oxidized for 6 h contained 1.3 mmol/g of O and Co was formed on 76% of them under thermal treatment. Various carbon materials, including anthracite and different carbon fibers, were activated by KOH, NaOH, CO_2, or water vapor at 650–750° C. The values of C_g measured for all carbon

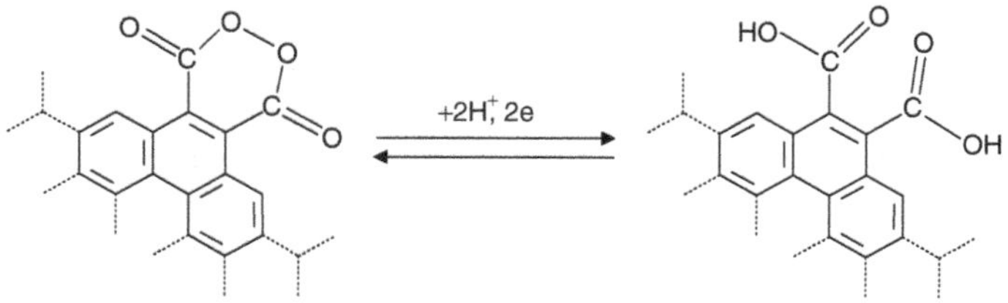

Figure 27.9. Variant of the mechanism of a surface redox reaction of peroxide groups on carbon.

materials activated under various conditions were closely related to the concentration of CO-desorbing complexes. It was observed that pseudocapacitance because of CO-desorbing complexes is more clearly traced in H_2SO_4 solutions as compared to KOH solutions.

Curves CV were measured for ACFs containing either COOH groups (the overall amount of oxygen-containing groups is 1.8 mmol/g) or phenol groups —OH (their overall is 0.8 mmol/g). ACFs rich in —COOH groups show hills in their part of the CV curves, which points to the appearance of faradaic reactions. However, similar ACFs rich in phenol hydroxyl groups demonstrate no pronounced hill in the CV curve. Linear dependences between specific capacitance and the amount of carboxyl groups were obtained for both carbon materials (AC and ACC).

Carbon materials pretreated by nitrogen, oxygen, or water vapors are frequently used for the development of EDLCs. Such pretreatment results in the modification of surface groups, changes the porous structure, removes impurities. At the temperature of $> 2300°C$, the degree of micrographitization increases and the specific surface area somewhat decreases. Oxygen-plasma treatment results in a significant increase in S_{BET}, which, in its turn, leads to an increase in C_g.

27.8 ELECTROLYTES USED IN EDLC

27.8.1 Aqueous and Nonaqueous Electrolytes

To obtain high energy density, electrolytes used in EDLCs must have the maximum high decomposition voltage and a wide stable potential range. Besides, they must be stable in the temperature range of -30 to $+70°C$. The main faults of aqueous solutions are low discharge voltage, narrow working temperature range, high corrosion activity. Nonaqueous liquid electrolytes have high decomposition voltage (> 2.3 V), wide working temperature ranges, high corrosion stability. Their faults include: low conductivity, low values of specific capacitance of carbon electrodes, necessity of thorough insulation from exposure, high cost. However, EDLC with aqueous electrolytes have a higher overall capacitance than those with nonaqueous ones because of their considerably higher specific EDL capacitance per unit true surface.

Capacitance of EDLCs with nonaqueous electrolytes increases at a decrease in the ion size: $BF_4^- > PF_6^- > Et_4N^- > B_4N^-$. The reduction potential of Et_4NBF_4/PC (propylene carbonate) is limited by decomposition of Et_4N^+ and PC. Many solvents have such a limiting potential of -3 B (SCE) related to decomposition of the Et_4N^+ cation. Properties of electrolytes based on cyclic aromatic imidazolium salts were studied. They have high conductivity ($>20 mS/cm$), a wide electrochemical stability range ($>3.5V$), high solubility. EDLC capacitance with carbon electrodes and electrolytes of imidazolium salts was about 125 F/g.

Shortage of electrolyte in dilute solutions can strongly affect the charging–discharge behavior of EDLC. The effect of resistance of electrolyte on the process of charge accumulation in EDLC carbon electrodes with solutions of tetraethyl ammonium tetrafluoroborate (TEABF$_4$) in PC was studied. The charge calculated on the basis of cyclic voltammograms decreases at an increase in the potential sweep rate. This effect becomes more noticeable at a decrease in the concentration of electrolyte, which is related to an increase in the inaccessibility of electrode matrix pores and at an increase in the resistance of electrolyte and also in the shortage of electrolyte during charging to high voltage values (3 V). A drastic voltage decrease was observed at a transition from charging to discharge current, which is explained by the effect of resistance of electrolyte and charge redistribution in pores.

27.8.2 Ionic Liquids

A special place among nonaqueous electrolytes is occupied by the so-called ionic liquids (ILs) representing organic salts liquid in the room temperature range. They consist of a large organic cation and a much smaller inorganic or organic anion. As compared to usual electrolytes, ILs contain no solvent. Focus on ILs is because of their featuring such properties as a wide liquid state range ($> 300°C$); some ILs are characterized by relatively high ionic conductivity ($> 10^{-4}$ S/cm) and a wide potential window, nonflammability, ability to function in high-temperature modes unavailable in other liquid electrolytes, nonvolatility (very low vapor pressure); and explosion safety; most ILs are nontoxic (Lazzari M et.al (2007)). Owing to the latter cause, they are also named "green liquids."

At present, hundreds of ILs have been obtained, but most of them are inapplicable in EDLC because of inappropriate physical properties: low conductivity, high viscosity, high melting point (Arbizzani et al. (2008)).

The materials most widely used in ECSC are imidazolium and pyrrolidinium salts that are characterized by the highest conductivity, while the electrochemical stability window (ESW) reaches even 5 V (these data, however, are obtained for smooth electrodes); then again, high cyclability has not been obtained as yet. Table 27.2 presents the values of conductivity, ESW, melting points with ILs in the limit, formula weight and density of N-butyl-N-methyl pyrrolidinium

TABLE 27.2. Values of Conductivity (σ), ESW, Melting Points with ILs in the Limit, Formula Weight and Density of N-Butyl-N-Methyl Pyrrolidinium (PYR$_{14}$+)$^-$ and N-Methoxyethyl-N-Methylpyrrolidinium (PYR$_{1(201)}$+)-Based ILs with Bis (trifluoromethanesulfonyl)imide (TFSI−) and Trifluoromethane Sulfonate (Tf−) Anions

IL	σ(mS/cm) 25°C	σ(mS/cm) 60°C	T (°C)	ESW at 60°C (V)	Formula Weight (g/mol)	Density (g/ml)
(PYR$_{14}$)$_{TFSI}$	2.6	6.0	−3	5.5	422	1.41
(PYR$_{14}$)$_{Tf}$	2.0	5.5	+3	6.0	291	1.28
(PYR$_{1(201)}$)$_{TFSI}$	3.8	8.4	−90	5.0	424	1.43

$(PYR_{14}+)^-$ and *N*-methoxyethyl-*N*-methylpyrrolidinium $(PYR_{1(201)} +)$-based ILs with bis(trifluoromethanesulfonyl)imide (TFSI−) and trifluoromethane sulfonate (Tf−) anions. The $PYR_{14}+$ -based ILs have the freezing point of about 0°C and addition of a methoxyethyl group at the nitrogen atom of the pyrrolidinium results in the case of $PYR_{1(201)}$ in a decrease in the freezing point to −95°C and an increase in conductivity rendering these ILs most promising for development of high voltage supercapacitors operating in a wide temperature range.

To obtain high capacitance of EDLC, carbon electrodes must have high specific surface area easily accessible for electrolyte with large-size cations; carbon porosity and surface chemistry are also of great importance. A carbon electrode must have a large pore size, at least wider than the size of counterions. Thus, mesoporous aero/cryo and xerogel carbons and ACs with pore sizes regulated by chemical conditions of synthesis must be especially promising electrode materials for EDLCs. Functional groups are also important as they affect the very electronic properties of carbon and surface polarizability and can favor good carbon wettability. Surface chemistry is also responsible for faradaic processes and cyclability of EDLC.

Studies of capacitance of negatively charged electrodes based on AC and aero/cryo/xerogel carbon in two ILs with the same anion and other cations with an almost similar size, that is, 1-ethyl-3-methylimidazolium bis (trifluoromethanesulfonyl) imide (EMITFSI) and *N*-butyl-*N*-methylpyrrolidinium bis(trifluoromethanesulfonyl)imide (PYR14TFSI) ILs were made (Fig. 27.10).

Main results of structural and capacitive measurements at 60°C for six carbons and two ILs were obtained for next carbon materials: C1 and C2 (cryogels), XI (xerogel), A1 (aerogel carbon), AC, and ACT (ACs and cloths), the initial one (AC) and the one treated at 1050°C in air for 2 h. This treatment removed surface functional groups containing oxygen and nitrogen that are often present in initial ACs and impart repellent (similar to hydrophobicity) properties toward EMITFSI and PYR14TFSI with a negative capacitive effect.

The main result of these measurements is that each carbon manifests double specific capacitance at 60°C in EMITFSI as compared to $PYR_{14}TFSI$. This cannot be

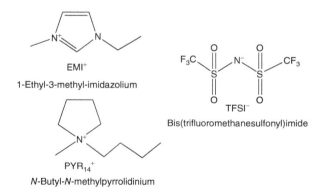

Figure 27.10. Structural formulas of the anion and cations of two ionic liquids.

explained by the difference in the accessible surface of these carbons in EDL formation for these two ILs. Indeed, the maximum sizes of counterions EMI+ and $PYR_{14}TFSI$ within EDL are close: they are less than 1 nm and all measured pore size distribution curves show that the principal pore size is larger than 2 nm. The cause for different capacitance in two ILs thus consists of different dielectric constants and EDL thicknesses that are because of different properties of EMITFSI and $PYR_{14}TFS$ at the interface because of a stronger interaction between the cation and anion in the latter IL, which, in its turn, depends on different polarizabilities of the cations and affects polarizability of the IL bulk. The main conclusion of this work is the considerable effect of the chemistry of ions on the capacitive properties of EDLC with ILs.

In most works, coulombic efficiency < 95% for EDLC with symmetrical IL electrolyte and with two similar electrodes at V_{max} (ESW) > 3.5V. In such EDLC, the mean voltage value is twice lower than V_{max}. Therefore, only a single PsC electrode material is used for further increase in energy density of ECSC with IL, that is, a hybrid ECSC is manufactured. PsC electrodes have a much higher specific capacitance as compared to carbon electrodes and can undergo several thousand charging–discharge cycles in aprotic ELs at the potential reaching the ESW limits. This can lead to an increase in ECSC energetic characteristics without compromising V_{max} and cycle life. In this case, poly(3-methylthiophene) (pMeT) can be used that can be reversibly p-doped/undoped in pyrrolidinium-based ILs at the correspondingly high electrode potentials with the specific capacitance of 200–250F/g when prepared in IL.

It is shown in a number of works that specific EDLC capacitance in EL electrolytes is higher on the same carbons in IL electrolytes than in usual solvent-containing non-aqueous electrolytes. In our opinion, this is because of the fact that the EDL thickness decreases in the absence of the solvent.

27.8.3 Polymer Electrolytes

Increased value of energy density and power density were obtained for ECSCs with polymer electrolyte functioning as a separator and ionic electrolyte, but also capable of creating a light and flexible multicell assembly of individual electrochemical capacitors. Polyvinyl alcohol (PVA)–KOH, (PVA)–H_3PO_4 gel electrolyte, polyacrylate (PAK)–KCl, poly(ethylene oxide)–tetraethyl ammonium tetrafluoroborate (PEO–Et_4NBF_4), Nafion, or ILs with PEO have been used in ECSC systems. Such systems manifest promising results for devices with polymer electrolytes only for relatively low potential sweep rates below 100 mV/s. Proton-conducting polymer electrolytes with heteropoly acids (HPAs) for symmetric and asymmetric ECSC devices at the rates of up to 1 V/s, as well as a PVA–H_3PO_4 device operating at 10 V/s on pseudocapacitive electrodes were also used.

Many attempts to replace liquid electrolyte by polymer electrolyte in ECSC were published. Polymer electrolytes are aqueous proton-conducting systems or they are based on polymers plastified with propylene carbonate and ethylene carbonate. Thus, alkaline polymer electrolytes were prepared for fully solid-state ECSC batteries with electrodes based on AC powders. Conductivity of alkaline electrolytes is based on polymers (PVA, PEO) or the copolymer of epichlorohydrin and ethylene oxide and was up to approximately 10^{-1} S/cm. A quinine-based supramolecular oligomer of 1,5-diamino-anthraquinone (DAAQ) has been lately developed as a promising electrode material for redox-type ECSCs manifesting high specific capacitance (50–60 Ah/kg) and high cyclability of 20,000 cycles at a small capacitance loss of about 5% with respect to the initial capacitance.

Much attention has been recently paid to the manufacturing of solid-state capacitors using solid polymer/gel electrolytes because of their high ionic conductivity and predominant mechanic properties including flexibility with a good contact between the electrode and electrolyte and ability to form thin films with the given surface (Gao H, Lian K (2011)).

Solid-state ECSCs were reported in several recent studies including various polymer/gel electrolytes, for example, $PMMA-EC-PC-LiClO_4$, polyethylene glycol $(PEG)-PC-TEABF_4$, $PEO-PEG-LiCF_3$ SO_3, $PVA-H_3PO_4$, $PMMA-EC-PC-NaClO_4$, and so on.

Fully solid-state ECSCs based on an supramolecular conducting oligomeric 1,5-diaminoanthraquinone electrode with the $PVA-H_3PO_4$ solid polymer electrolyte and $PMMA-EC-PC-$tetraethyl ammonium perchlorate polymeric gel electrolyte were manufactured and compared. The optimized composition of polymer/gel electrolytes showed at the room temperature conductivity of about 10^{-4} to 10^{-3} S/cm and good mechanical strength for its application in redox capacitors. These capacitors based on gel electrolyte manifested very high capacitance of approximately 3.7–5.4 mF/cm, which is equivalent to single-electron capacitance of $125 - 184$ F/g of a DAAQ electrode. An estimate of energy density of about $92 - 35$ Wh/kg was made for the maximum potential window of 2.3 V for a DAAQ oligomeric electrode.

The relatively lower capacitance of 1.1–4.0 mF/cm equivalent to a single-electrode capacitance of $36 - 136$ F/g for DAAQ was obtained by Hashmi et al. (2004) for a proton-conducting $PVA - H_3PO_4$ polymer ECSCs. As an illustration, this capacitor manifested the highest capacitance at high potential sweep rates up to 400 mV/s, which corresponds to very good kinetics.

Proton-conducting polymer electrolytes are promising electrolyte systems for both double-layer and PsCs. The HPA–PVA systems with various additives were optimized. Results obtained on a fully solid-state EDLC with silicotungstic acid H4SiW12O40 (SiWA)-based $HPA-H_3PO_4-PVA$ polymer electrolyte were reported. Herewith, very high charging–discharge rates at the sweep rate of 20 V/s were obtained. A significant advantage of ECSCs with polymer electrolyte is their very low self-discharge.

27.9 IMPEDANCE OF HIGHLY DISPERSED CARBON ELECTRODES

It is known that the impedance technique is one of the most informative electrochemical methods for studying electrodes and electrochemical systems in general. However, as ECSCs are used in a very wide range of characteristic discharge and charging

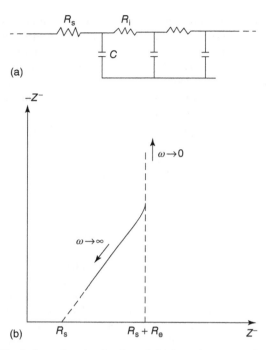

Figure 27.11. (a) Equivalent electric circuit and (b) impedance complex plane plot for an ideally polarizable porous electrode.

times in the range of five to seven orders of magnitude, the impedance technique is also used for studies of operation of ECSCs with different characteristic times.

According to the theory of de Levie, the impedance complex plane plots of an ideally polarizable porous electrode take the form shown in Figure 27.11a.

An equivalent electric circuit in such a pore was modeled by a transmission RC circuit containing similar resistances R_i of electrolyte within a pore with branching electric capacitances C of EDL of similar values within a pore (Fig. 27.11b). At high frequencies, the phase angle is 45° and the intercept at the Z' axis is equal to solution resistance outside the pore. At the frequency of $\omega \to 0$, extrapolation of the dependence to the Z' axis yields the sum of ionic resistance inside the pore structure and outer serial resistance.

However, the impedance complex plane plots measured in practice for real electrodes based on HDCMs differ from an ideal complex plane plot suggested by de Levie (1967). Firstly, the slope of a quasidirect line differs considerably from 45° and, secondly, a complex plane plot often contains an approximate semicircle (see Fig. 27.12).

Figure 27.12 shows a typical impedance complex plane plot for AC and Figure 27.13 shows the corresponding typical ladder-type equivalent circuit.

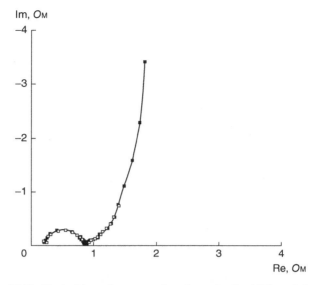

Figure 27.12. Typical impedance complex plane plot for AC-based electrodes.

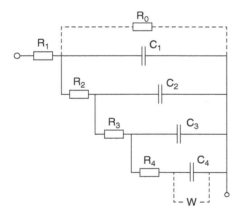

Figure 27.13. Typical equivalent circuit corresponding to the impedance complex plane plot shown in Figure 27.12.

The so-called ladder equivalent circuit shown in Figure 27.13 is characteristic of many ACs. It represents a set of several R–C parallel circuits and also Warburg diffusion impedance. Herewith, apart from the proper distributed line related to a porous structure of the studied object, one or several circuits in the ladder characterize parallel faradaic redox reactions of surface groups on the electrode. It was shown theoretically that phase angle $\phi = 45°$ independent of frequency ω is observed only for EDL charging in direct pores. Accounting for more complex shapes of pore cross-section, including its "corrugation," results in a change in the value of ϕ and in various dependences of ϕ on ω in general (i.e., in nonlinearity of the shape of the complex plane plot in the $Z''-Z'$ coordinates) for each particular pore shape. The mechanism of processes on an ECSC electrode resulting in an equivalent circuit of the type shown in Figure 27.13 is very complicated. This scheme probably reflects the pore size distribution, especially for micropores. It is known, in particular, that no normal EDL appears in micropores with especially small sizes (several angstroms). Therefore, the EDL capacitance value generally depends on the pore size.

27.10   NANOPOROUS CARBONS OBTAINED USING VARIOUS TECHNIQUES

27.10.1   Activated Carbons (ACs)

ACs are one of the most frequently used electrode materials for ECSCs. ACs are usually obtained in the course of carbonization and further activation of a whole series of natural and synthetic carbon-containing materials. Materials for AC preparation can be vegetable and animal raw materials (wood, sugar, coconut, nutshell, fruit kernels, coffee, bones, etc.), mineral raw materials (turf, coals, pitch, resin, coke), synthetic resins, and polymers. Decomposition of the initial materials (precursors) and removal of noncarbon elements occurs in the course of carbonization. Oxidation, burning of a part of nonorganized carbon and elementary crystallites takes place in the course of gas activation by water vapors, carbon dioxide, and oxygen at the temperatures of 500–900°C and a developed porous structure of AC particles is formed. As dependent on the technology of synthesis, the specific surface area of ACs measured using the BET method is in a very wide range: approximately from 500 to 3000 m^2/g. ACs contain three types of pores: micropores, mesopores, and macropores; herewith, the largest contribution into the value of the specific surface area is introduced by micropores and then mesopores. Simultaneously with a change in the porous structure of carbon materials, a chemical surface (2D) structure is formed in the course of their treatment and electrophysical parameters are changed. The chemical surface composition of ACs, their electrophysical characteristics, and porous structure produce the strongest effect on electrochemical and capacitive properties of ECSCs. The values of specific capacitance because of EDL charging are usually in the ranges of 50–200 F/g.

According to one of the variants of the technique for preparation of microporous ACs, a mixture of coke with a KOH solution as an activation agent at the ratio of 1 : 3 was heated in a microwave at 700–900°C. It was shown that the volume of micropores and mesopores grows at an increase in the activation time and the volume of micropores and mesopores practically stops changing after 30 min.

Figure 27.14a and b shows dependences' specific capacitance and ESR on current density for three types of carbons.

We can see that the values of specific capacitance and resistance decrease at an increase in current and the potential sweep rate. At $w = 20$ mV/s, the capacitance

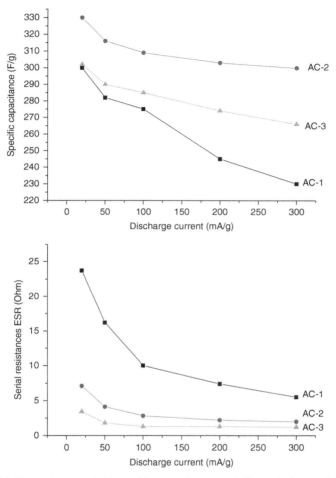

Figure 27.14. Dependences of (a) specific capacitance and (b) equivalent serial resistance (ESR) on current density for three types of carbons.

values reach 330 F/g. Capacitance grows at an increase in specific surface of these carbons AC-1, AC-2, and AC-3 (accordingly: 869, 1590, and 1760 m^2/g).

Activated Carbons with High Pseudocapacitance

Carbon electrodes are usually cycled in aqueous solutions in the potential range from 0 to 1 V (RHE, reversible hydrogen electrode). Herewith, as pointed out above, average values of specific capacitance of not more than 200 F/g are usually obtained. The lower limit of potentials is because of hydrogen evolution on metallic current taps and the upper limit is related to the oxidative corrosion of carbon. ACs were used in our work of Volfkovich et al. (2012) in studies of electrodes based on AC ADG and ACCs CH900 and TSA. This allowed cycling these electrodes in the maximum range of potentials from -1 to 1 V (RHE) because of high overpotential of hydrogen evolution on graphite. It is convenient to present the data obtained using the CV method for more graphic representation of capacitance properties, especially those measured at different potential sweep rates (w), in the form of capacitance–voltage curves constructed in the coordinates of differential capacitance (C)–potential (E), where $C = dQ/d\tau = I\, d\tau/dE = I/w$, I is the current, Q is the amount of electricity, $w = dE/d\tau$, τ is the time. Figure 27.15 compares cyclic capacitance–voltage curves measured in 48.5 H_2SO_4 for ACC CH900 at different w values in two ranges of potentials: in the reversibility range (from 0.25 to 0.8 V) and in the deep charging range (from -0.5 to 0.8 V). The w values used here are rather small: 0.5, 1, and 2 mV/s.

As follows from curve 4 measured in the reversibility range, only EDL charging occurs here, while pseudocapacitance of redox reactions of surface groups is very low in this case. As follows from this curve, the value of EDL capacitance is approximately 170 F/g.

Faradaic processes with a very high pseudocapacitance are observed in the range of negative potentials (< -0.1 V) (curves 1, 2, and 3).

The method of galvanostatic curves was used to measure the dependence of electricity amount Q on charge time in 40.3% H_2SO_4 after prolonged charging at the potential of $E = -250$ mV. The Q value grew very fast at very low charging times of seconds and minutes and then continued increasing further for dozens of hours. Such very slow growth can be explained by very slow diffusion. It is known that the lowest diffusion coefficients are observed in the solid phase; they are lower by orders of magnitude than in the liquid phase. In this connection, it was assumed that the process of hydrogen intercalation into AC carbon is controlled by slow solid-phase hydrogen diffusion. This is also indicated by the proportionality of the limiting current to the square root on the potential sweep rate and also a number of other experimental data. It was shown that the compound of C_xH was formed at deep cathodic charging

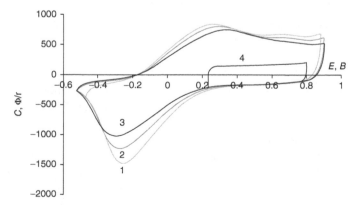

Figure 27.15. Capacitance–voltage cyclic curves for CH900-20 ACC. The potential sweep rates were (1) 0.5, (2) 1.0, and (3, 4) 2.0 mV/s.

of ACs in the course of hydrogen intercalation into carbon and C_6H was formed in the limit similar to the formation of the C_6Li compound under deep cathodic charging of carbon negative electrodes of a lithium ion battery. It was found that resistance of an AC electrode increased with the deep cathodic charging time. This dependence is explained by a change in the chemical composition of the electrode solid phase in the course of charging: from C to C_xH and, in the limit, to C_6H. The maximum specific charge of 1560 C/g was obtained after charging for 22 h at the potential of $E = -0.31$ V (RHE) in 56.4% H_2SO_4, which corresponds to specific capacitance of 1110 F/g. This value of capacitance exceeds much the corresponding values obtained in the literature for carbon electrodes. As shown above, the specific charge because of EDL charging is 240 C/g for CH900. The difference is 1560 C/g – 240 C/g = 1320 C/g. As it is just 1320 C/g that are to be spent according to the Faraday law to obtain the C_6H compound, this proves generation in the cathodic charging of ACs of the very compound of the earlier unknown C_6H compound, that is, carbon hydride or hydrogen carbide. The assumed structural formula of this compound is shown in Figure 27.16.

It was found that the maximum specific charge grows at an increase in the sulfuric acid concentration (in the range of 34–56%) and reaches the maximum of 1560 C/g at its concentration of 56%. High specific charge values in concentrated H_2SO_4 solutions were also obtained for a number of other ACs: ADG and TSA. The following facts are known from the literature: (i) Its intercalation into carbon materials growing at an increase in this concentration occurs in concentrated sulfuric acid solutions. (ii) Their swelling to a certain degree occurs under adsorption of different adsorptives on ACs. It was shown on the basis of these data that *double intercalation* occurs in this case. Sulfuric acid is intercalated into ACs forcing apart carbon atoms and thus facilitating the further intercalation of hydrogen. For CH900 in 100% H_3PO_4, the maximum specific charge of about 1200 C/g, that is, the average specific capacitance of 860 F/g, was obtained.

A mathematical model describing the processes of AC charging–discharge with account for EDL charging, intercalation of hydrogen into carbon, and its nonsteady-state solid-phase diffusion, electrode kinetics, ion transport over the electrode thickness, and characteristics of its porous structure was developed and confirmed experimentally. It is shown that the maximum path of diffusion of hydrogen atoms all other conditions being equal is inversely proportional to the hydrophilic specific surface area of the electrode.

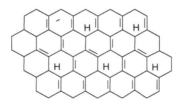

Figure 27.16. Structural formula of compound C_6H.

27.10.2   Templated Porous Carbons

The template method is used to prepare both microporous and mesoporous carbons as dependent on the very templates and raw materials. The method consists of the impregnation of finely porous precursors by an organic solvent (e.g., furfuryl alcohol), its carbonization in fine pores, and in dissolution of a precursor.

Carbonization of precursors in nanochannels of various types of zeolites served to prepare microporous carbons (zeolite-templated carbon ZTC), including carbon with the highest S_{BET} value and the overall pore volume of 4000 m^2/g and 1.8 cm^3/g, accordingly. As pores formed in ZTC appeared from zeolite channels, they had micropores of uniform size and morphology and also a small fraction of mesopores. CV curves of ZTC in a nonaqueous solution showed nonrectangular curves, which points to presence of pseudocapacitance mainly because of oxygen functional groups on the surface of carbon; the C_g, however, does not correlate with the content of oxygen. In the work by Inagaki et al. (2010) dependences of capacitance on current density for different ZTCs together with a similar dependence for commercial ACFs were measured. For most ZTCs, the results showed a very good discharge rate, that is, an almost constant capacitance (155–170 F/g) even at high current density, in 2 A/g in addition to high C_g values. Such a discharge rate is related to the 3D-ordered system of micropores, which yields low resistance in micropores. Other microporous ZTCs were manufactured using a two-stage cast process with zeolite 13X as a template with further activation of KOH by zeolite dissolution. High capacitances were preserved up to 94–100% of the initial capacitance at the current density of up to 2 A/g for nonactivated and activated ZTCs.

Mesoporous carbons were manufactured using mesoporous silica gel: 2D hexagonal mesoporous silica gel (e.g., SBA-15) yields carbons with 2D-ordered mesopores and 3D cubic mesoporous silica gel (e.g., MCM-48) yields carbons with 3D-ordered mesopores.

The first impregnation of mesoporous silica gel by furfuryl alcohol resulted in bimodal mesoporous carbon with the maximum pore size of 2.9 and 16 nm and further impregnation yielded unimodal carbons with the pore size of 2.8 nm. S_{BET} was 1540–1810 m^2/g and more than half the overall pore volume was attributed to mesopores and capacitance C_g was 200 F/g at the current density of 1 mA/cm^2, but decreased noticeably at an increase in current density. A 3D network for mesopores in carbons provided better performance than a 2D one. The combination of sugar and MCM-48 demonstrated the best performance among a variety of combinations of carbon initial materials (precursors) with templates. A study carried out with 25 silica gel template carbons with various pore structures and the surface area of 200–1560 m^2/g showed that these carbons show no obvious advantages before ACs in aqueous and nonaqueous solutions.

Mesoporous carbons were manufactured using MgO particles as a template. This process has a number of advantages because MgO particles can be dissolved in a weak acid, such as citric acid, and can be used again as the initial template and therefore carbons have a narrow pore size distribution in the mesopore regions. MgO– template mesoporous carbon is characterized by high percentage of capacitance retention at high charging/discharge rates in aqueous and nonaqueous solutions. High degree of capacitance preservation is because of high values of S_{ext} of carbons that are determined by mesopores.

27.10.3 Carbide Derivatives of Carbon

Various metal carbides (TiC, B_4C, SiC, etc.) can be used to manufacture highly porous carbons by thermal treatment at the temperature of 400–1200°C in a Cl_2 flow dissolving metals and silicon. S_{BET} is 1000–2000 m^2/g. The porous structure of these carbons strongly depends on carbide precursors and the thermal treatment temperature (TTT). It is mainly micropores that are formed in these carbide derivative carbons up to 800°C, but mesopores start prevailing under heating above 800°C and therefore S_{BET} has a maximum at 800°C in most carbons. EDLCs were developed using these carbide microporous carbons and their behavior in various nonaqueous and aqueous electrolytes was studied and the effects of the size of ions of electrolytes and organic solvent molecules were discussed (Rychagov et al., 2012). TiC-derivative carbons were manufactured at 500–1000°C because of which the average micropore sizes and S_{BET} changed from 0.7 to 1.1 nm and from 1000 to 1600 m^2/g. B_4C carbon derivatives yielded good results regarding the operation rates in a KOH solution and 86% retention of capacitance at variation of the potential sweep rate of 2–50 mV/s.

We carried out comparative studies of the effect of the porous structure of carbon materials on electrochemical electrode characteristics using various carbide carbons (CCs). Main structural characteristics for CCs based on silicon carbide are presented in Table 27.3 and those for titanium carbide are in Table 27.4. Specific surface areas were calculated on the basis of the nitrogen adsorption data with calculation using the DFT technique. This method is used to measure micropores and mesopores, but not macropores.

TABLE 27.3. Main Structural Characteristics for CCs Based on Silicon Carbide

TTO	S_{DFT}, (m^2/g)	V_Σ for $p/p_{0\sim0.99}$ (cm^3/g)	$V_{\Sigma DFT}$ (cm^3/g)	$V_{pores\ <2\ nm}$ (cm^3/g)	$V_{pores\ from\ 2\ to\ 50\ nm}$ (cm^3/g)
600°C	1183	0.52	0.42	0.38	0.04
900°C	1440	0.58	0.47	0.47	0
1500°C	621	0.47	0.37	0.23	0.14

TABLE 27.4. Main Structural Characteristics for CCs Based on Titanium Carbide

TTO	S_{DFT}, (m^2/g)	V_Σ for $p/p_{0\sim0.99}$ (cm^3/g)	$V_{\Sigma DFT}$ (cm^3/g)	$V_{pores\ <2\ nm}$ (cm^3/g)	$V_{pores\ from\ 2\ to\ 50\ nm}$ (cm^3/g)
600°C	1492	0.62	0.50	0.49	0.01
1000°C	777	0.62	0.52	0.28	0.24

Analysis of capacitance of CCs based on SiC (electrolyte is IL) in a wide range of potentials showed approximate proportionality of average integral specific capacitance and specific surface area for CC-600 C and CC-1500C (Tables 27.3, 27.4 and 27.5).

It was shown that the charging of CC SiC-900C characterized by the maximum specific surface area occurs with high irreversibility observed even at very low potential sweep rates. Comparison of cyclic capacitance–voltage curves measured in different electrolytes showed the general decrease in capacitance by 20–40% at a transition from aqueous electrolytes to the 1 Me3BuImBF$_4$ ILs. Herewith, the minimum differences are registered on mesoporous CC SiC-1500C and reversibility loss together with a decrease in capacitance manifested in the deformation of cyclic curves at an increase in the potential sweep rate is observed for microporous CC SiC-600C (Table 27.5). The limiting potential sweep rate corresponding to the symmetrical cyclic curve shape for nanostructured porous carbon (NPC) SiC-600C is 10 mV/s (Fig. 27.17).

As the implied cause of loss of capacitance and charging reversibility in microporous materials is the large sizes of organic ions in ILs, the cation of 1-methyl-3-butyl imidazolium was replaced by the lithium cation. Change of electrolyte resulted in a significant increase in reversibility that allowed carrying out cycling tests at the potential sweep rates of 50 and 100 mV/s.

The charging of mesoporous CC SiC-1500C in IL occurs without noticeable kinetic hindrances allowing increasing the charging rates of CC SiC-1500C as compared to microporous CC SiC-600C by more than an order of magnitude. This is explained by the fact that more than 50% of the surface for CC SiC-1500C are formed by pores with sizes exceeding 1 nm.

Thus, it can be stated that pore sizes primarily affect the kinetics of CC charging. The contribution of micropores completely inaccessible for IL is estimated as negligible (less than 10% of the overall CC surface). In most cases, accessibility of micropores grows at an increase in the temperature of thermal treatment and at a decrease in the charging currents.

TABLE 27.5. Comparison of Capacity Characteristics for Different Carbide Carbons in Different Electrolytes at Potential Rate 1 mV/s

HΠY	50% H$_2$SO$_4$		75% 1Me3BuImBF$_4$/PC		1N LiBF$_4$/PC	
	C_g (F/g)	C_s (mcF/cm^2)	C_g (F/g)	C_s (mcF/cm^2)	C_g (F/g)	C_s (mcF/cm^2)
SiC-600C	140	11.8	100	8.5	70	5.9
SiC-900C	120	8.3	70	4.8		
SiC-1500C	45	7.2	35	5.6	25	4.0
TiC-600C	190	12.7			100	6.7
TiC-1000C	80	10.3	65	8.4		

C_s, capacity of EDL per unit of true surface; C_g, integral specific capacity per unit of mass; PC, propylene carbonate.

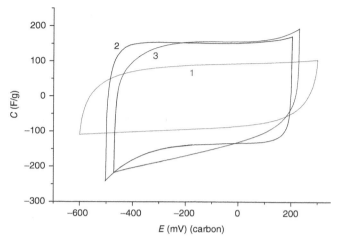

Figure 27.17. Capacitance–voltage curves of CC SiC-600C at the potential sweep rate of 10 mV/s: (1) IL; (2) 50% H_2SO_4; (3) 30% KOH.

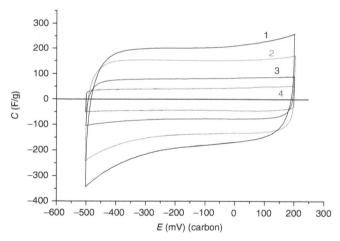

Figure 27.18. Capacitance–voltage curves for CCs in 50% H_2SO_4 at 20 mV/s: (1) TiC-600C; (2) SiC-600C; (3) TiC-1000C; (4) SiC-1500C.

A distinctive feature of CCs based on titanium carbide as compared to CCs based on silicon carbide is a higher specific surface area under similar synthesis conditions. Figure 27.18 shows comparison of capacitance–voltage dependences (potentials are measured vs. a porous carbon electrode with high capacitance) for four CCs in a sulfuric acid solution for the range of maximum reversibility, that is, the range of EDL charging.

As can be seen in Figure 27.18 and Tables 27.3 and 27.4, an increase in the CC-specific surface area results in deviation of curves from the shape characteristic of a classical double-layer capacitor. In the case of microporous CCs, the sections are absent in which capacitance is independent of the potential, as the potential of initiation of reduction and oxidation processes is shifted at an increase in the specific surface areas to the positive and negative, accordingly.

An increase in the cathodic polarization depth results in a significant increase in discharge capacitance (Fig. 27.19) similarly to the one described in Section 27.9.1.1 for ACs. As we see, the shape of CV curves differs significantly from the double-layer one. Herewith, the approximate proportionality of the maximum anodic current and specific surface area is observed for microporous CCs, while the values of capacitances of deep cathodic charging prove to be much lower for mesoporous CCs.

The value of capacitance of deep cathodic charging of microporous CCs grows significantly at a decrease in the potential sweep rate (Fig. 27.20).

The measured curve family obtained by galvanostatic discharge of CCs SiC-600C in 50% H_2SO_4 at different times of potentiostatic charging was used to construct a series of curves of the dependence of the electricity amount on the logarithm of the charging time for different CCs (Fig. 27.21) (Rychagov et al., 2012).

Extrapolation according to the sigmoid model allowed estimating the limiting values of the deep charging capacitance being 1500 C/g for SiC-600C and 1250 C/g for TiC-600C. They agree with the limiting capacitance value of 1560 C/g (or 1100 F/g) obtained earlier for ACC CH900 (see Section 27.9.1.1). This value is because of the formation of the C_6H compound.

Analysis of the general character of the capacitive behavior of CCs allowed drawing the following conclusions: (i) The maximum decrease in specific surface

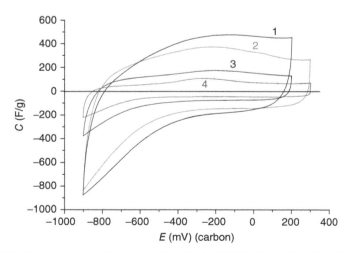

Figure 27.19. Capacitance–voltage curves for CCs in 50% H_2SO_4 at 20 mV/s: (1) TiC-600C; (2) SiC-600C; (3) TiC-1000C; (4) SiC-1500C measured at 20 mV/s.

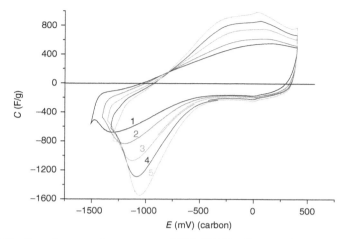

Figure 27.20. Capacitance–voltage curves of SiC-600C in 50% H_2SO_4: (1) 200 mV/s; (2) 100 mV/s; (3) 50 mV/s; (4) 20 mV/s; (5) 10 mV/s.

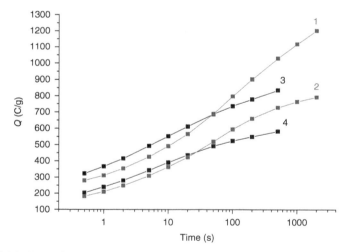

Figure 27.21. Dependence of discharge amount of electricity on the logarithm of the discharge time: (1) SiC-600C at discharge to 900 mV; (2) SiC-600C at discharge to 600 mV; (3) TiC-600C at discharge to 900 mV; (4) TiC-600C at discharge to 600 mV (NHE).

capacitance ($\mu F/cm^2$) is observed for microporous carbons at a transition from sulfuric acid to an IL. (ii) Despite the relatively low specific capacitance values, mesoporous CCs can be applied in pulse systems requiring the maximum charging–discharge rates.

27.10.4 Carbon Aerogels and Xerogels

Aerogels

Aerogels are a class of materials representing a gel, in which the liquid phase is completely replaced by the gas phase. Such materials have a record-breaking low density and are characterized by a number of unique properties: hardness, transparence, resistance to heat, extremely low thermal conductivity, and so on. The first carbon-based aerogel samples were obtained early in the 1990s. Aerogels belong to the class of mesoporous materials, in which pores occupy more than 50% of the volume. As a rule, this percentage reaches 90–99% and the density is 1–150 kg/m^3. As regards the structure, aerogels represent a tree network of nanoparticles with the size of 2–5 nm united in clusters and pores with the size of up to 100 nm. Carbon aerogels with a large amount of mesopores were manufactured mainly by pyrolysis of aerogels of resorcin and formaldehyde. Primary particles of carbon aerogels are connected to each other forming a network of interparticle mesopores. A change in the conditions of synthesis of carbon aerogels resulted in a strong effect on the EDLC capacitance. Activation of primary particles is required for the application of carbon aerogels in EDLC electrodes. Activation by CO_2 results in the transformation of a large amount of micropores to mesopores and leads to an approximately double increase of capacitance in the nonaqueous solution. Modification of the surface of this carbon aerogel can yield higher capacitance at a higher current density that can be related to the improvement of wettability of the carbon surface in organic electrolytes.

Xerogels

Xerogels represent gels from which the liquid medium is removed, so that the structure becomes compacted and the porosity value is to a certain degree decreased following surface tension forces in effect in the course of the removal of the liquid. Xerogels constitute an assembly of contacting spherical particles, with sizes and packing density depending on the manufacturing method. The value of C_g for carbon xerogel increased from 110 to 170 F/g under activation by CO_2. Such a change in C_g is related to a significant increase in S_{micro} from 530 to 1290 m^2/g together with an increase in S_{meso} from 170 to 530 m^2/g.

27.10.5 Carbon Nanotubes

CNTs are promising electrode materials for ECSCs. Their distinctive features are not only a large open surface area and different spaces for storage of ions of electrolyte, but also their high conductivity. Cavities in CNTs are shown in Figure 27.22.

The external surface of the CNT walls (1 in Fig. 27.22) is mainly composed of the basal plane of graphite. High EDL capacitance is observed at a wide potential window on such an ideal surface as in the case of CNTs because of high polarization. Most CNTs are connected together because of Van der Waals forces where only outer tubes in the bunch are exposed to electrolyte and the space between the tubes is layered (3 in Fig. 27.22) being only with difficulty used for EDL formation. The internal surface of nanotubes (2 in Fig. 27.22) is also basically suitable for the penetration of

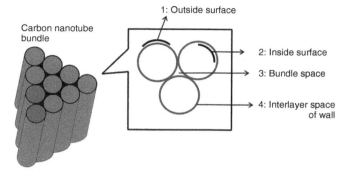

Figure 27.22. Schematic drawing of spaces for storage of electrolyte ions in CNTs.

ions of electrolyte; however, there are certain limitations: firstly, because of a very low internal diameter ($1.3-1.6$ nm), most ions cannot penetrate nanotubes; secondly, aqueous electrolytes do not wet the internal space because of hydrophobicity of the internal surface (see below). The interlayer space within the walls of multiwall nanotubes (4 in Fig. 27.22) can be accessible for intercalation of ions of electrolyte, such as Li^+. Thus, both intercalation and a faradaic reaction are possible corresponding herewith to a pseudocapacitance. Nevertheless, it will be noted that excess expansion of interlayer spaces under intercalation causes certain electrode degradation, which results in strong reduction of cyclability of a capacitor.

Single-Wall Carbon Nanotubes (SWCNTs)

The most widespread method of obtaining SWCNTs is electric-arc synthesis using different catalysts, for example, an Y/Ni catalyst. Nanotubes often have herewith a narrow diameter distribution of $1.4-1.6$ nm and the average length of about 0.7 μm. This is followed by the stages of the opening of nanotube ends and their cleaning by oxidation in air and washing, for example, in hydrochloric acid. Following high surface energy, nanotubes in a SWCNT powder are aggregated into strands (bunches) of a simplified structure with the length of up to $5-10$ μm and a wide transverse size distribution of strands from 6 nm to approximately 100 nm. Another method of obtaining SWCNTs is pyrolysis of ethylene.

Capacitive properties of SWCNTs were described in a lot of literature using various electrolytes, both organic and aqueous ones. In most papers, a conclusion is generally made that SWCNTs have the best properties as electrode materials for EDLCs at high charging and discharge rates as compared to ACs.

High power characteristics, more than 20 kW/kg at very high current densities to several hundreds of ampere per gram, were obtained in a number of works for EDLCs based on SWCNTs in sulfuric acid electrolyte. Such high power characteristics are explained by a regular structure of pores located between separate nanotubes and their strands (see Fig. 27.23). The porous structure regularity means the practical absence of tortuous and corrugated pores and therefore the maximum conductivity of electrolyte in pores.

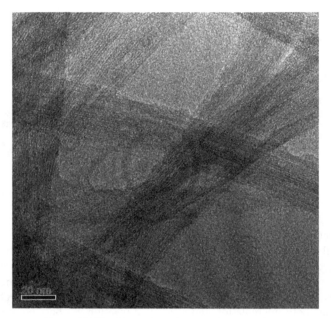

Figure 27.23. SEM microphotograph of SWCNTs.

Figure 27.24 shows capacitance–voltage CV curves measured for activated mesoporous carbon FAS and SWCNTs at a very high potential sweep rate of 1 V/s (Rychagov et al., 2012).

As seen from Figure 27.24, the curve for SWCNTs has a practically ideal shape characteristic of the EDL charging (cf. with the top image in Fig. 27.2) and the curve for CC FAS (it is commercial brand) has a distorted shape with a large hysteresis because of the fact that equilibrium charging–discharge processes do not have enough time at a high potential sweep rate.

For some SWCNTs thoroughly cleaned from admixtures, appearance of a characteristic region of potentials with a drastic decrease in capacitance was observed in the CV curves, which distinguishes SWCNTs from other carbon materials. This can be seen in Figure 27.24, in which such curves are shown for alkaline and acidic electrolytes.

Comparison of average capacitance and capacitance in the pit region shows the ratio of about 3 : 1 and the width of the pit region is about 500 mV. Analysis of the possible explanations of such voltammetric behavior of SWCNTs manifests direct correlation between the observed effect and semiconductor properties of nanotubes. Besides, Figure 27.25 shows the difference in the location of the pit region potential for alkaline and acidic electrolytes. This potential region is more negative for acidic electrolytes as compared to alkaline ones.

At an increase in the sulfuric acid concentration, the potential of the pit localization is shifted into the negative range, which results in a certain increase in the hydrogen

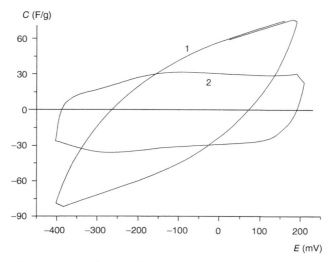

Figure 27.24. Capacitance–voltage cyclic curves at the potential sweep rate of 1 V/s measured in the range between −400 mV and 200 mV: (1) FAS, (2) SWCNTs. Electrolyte is 35 wt% H_2SO_4.

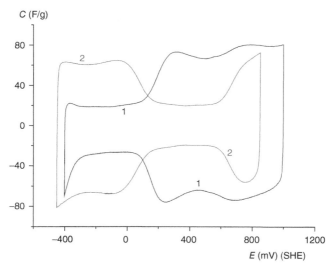

Figure 27.25. Comparison of capacitance–voltage dependences for SWCNTs in aqueous electrolytes: (1) 0.5 M H_2SO_4; (2) 1 M KOH.

evolution overpotential. Owing to such an effect, the maximum voltage for supercapacitors with SWCNT electrodes can reach 1.4 V in sulfuric acid. This shape of CV curves was also observed for electrolytes based on ILs.

The opening of holes at the tube ends was carried out by means of oxidation to use the SWCNT inner surface. If the inner surface of open SWCNTs can be fully used for EDL formation, then the limiting capacitance value must be doubled. Nevertheless, optimization of the oxidation process is necessary because excessively strong SWCNT oxidation can cause deterioration of the tube quality because of the introduction of various defects and/or tube shortening and disintegration, even if the specific surface area herewith were increased. At the optimum conditions of supergrown SWCNT oxidation, S_{BET} of above 2000 m^2/g was obtained and a capacitor using TEABF$_4$/PC yielded very high values of energy density (24.7 Wh/kg) and power density (98.9 kW/kg). Nevertheless, an increase in capacitance is not proportional to an increase in S_{BET}. The main fault of SWCNTs is their great expensiveness that does not allow applying them in practice in the near future.

Double-Wall and Multiwall Carbon Nanotubes

As compared to SWCNTs, the properties of capacitors based on double-wall carbon nanotubes (DWCNTs) were not published so frequently, as pure DWCNTs are very hard to obtain. Basically, DWCNTs and multiwall carbon nanotubes (MWCNTs) have a lower surface area for EDL formation as compared to SWCNTs. On the other hand, a variety of works on capacitive performance of MWCNTs has been published, as they are relatively easily synthesized and are much cheaper than SWCNTs. As dependent on the methods of synthesis and modifications, different types of MWCNTs with different specific surface area values were obtained. Their specific capacitance values obtained in aqueous and nonaqueous electrolytes are from 10 to 100 F/g. However, they are not so high as in the case of ACs. On the other hand, one must point out that bulk capacitance C_v is relatively high because of high bulk density of MWCNTs. Synthesis of MWCNTs is often carried out by pyrolysis of ethylene using catalysts, for example, Fe–Co.

Figure 27.26 shows a TEM microphotograph for MWCNTs with the diameter of 5 nm and specific surface area of 340 m^2/g.

27.10.6 Graphenes and Their Derivatives

Graphenes and their derivatives started being used lately as promising electrode materials for ECSCs. Graphenes were discovered only several years ago, for which their discoverers received the Nobel prize. And remarkable properties of graphenes were recognized practically immediately opening wide possibilities of their applications in various branches of national economy, including, in particular, chemical power sources and electrochemical capacitors. A single graphene sheet layer provides the specific outer surface area of up to 2675 m^2/g accessible for liquid electrolyte, as compared to the outer specific surface of approximately 1300 m^2/g for a single SWCNT. The inner surface of SWCNTs was practically inaccessible for electrolyte. In practice, graphene layers form plates of several

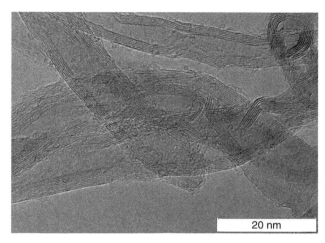

Figure 27.26. TEM microphotograph for MWCNTs with the diameter of 5 nm.

individual graphene layers, as a result of which the surface accessible for electrolyte decreases. Nevertheless, promising results have been obtained lately for ECSCs with graphene electrodes. Graphene electrodes often have, apart from EDL capacitance, pseudocapacitance of faradaic redox reactions.

Reduced graphene oxide (RGO) was obtained using the modified Hammers method. According to this method, the initial graphite powder was added under mixing and cooling into a mixture of concentrated sulfuric and nitric acids, and a triple amount of potassium permanganate in respect to graphite was added into this mixture after a while. Then a hydrogen peroxide solution was added to it and a triple amount of distilled water was used to dilute the mixture. After settling for several hours, the top transparent layer was decanted. The suspended deposit was filtered, washed by distilled water, and dried at the room temperature until the weight was constant. As a result, dry powdered graphite oxide was obtained. The obtained graphite oxide was subjected to reduction using the fast heating technique to the temperature of 1000°C. The layering of the material occurred in the course of reduction with a multiple increase of its volume (exfoliation). The final product represented powder of thin nanolayers containing 1–10 graphene monolayers per plate with the size scatter of 1–10 μm in the lateral direction (see Fig. 27.27). Figure 27.28 showed a schematic drawing of graphene layers of graphene oxide.

These monolayers, in their turn, are grouped into agglomerates, between which pores with a wide spectrum of sizes (four to five orders of magnitude) are formed: micropores, mesopores, and macropores. Here is one of the variants for the preparation of electrodes based on graphene oxide. Each electrode contained 85% of graphene, 5% Super-P (conducting additive), and 10% PTFE as a binder. This electrode mass was applied onto aluminum foil. The electrode thickness was usually 80–150 μm. The electrode was dried in a vacuum furnace at 120°C for 12 h. In another variant of the technique, a graphene electrode is made by filtering graphene

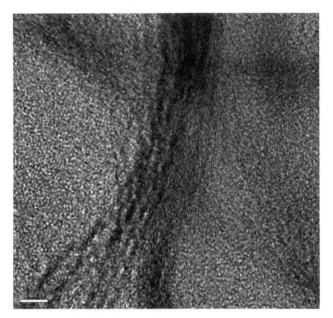

Figure 27.27. TEM microphotograph of graphene oxide.

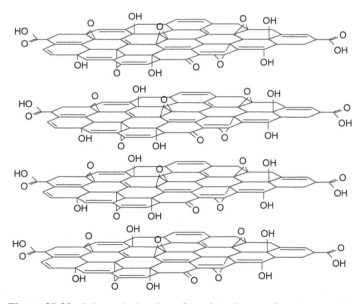

Figure 27.28. Schematic drawing of graphene layers of graphene oxide.

oxide into carbon paper. In some works, the technique of graphene modification using KOH was used. In a typical experiment, 150 ml of an aqueous KOH solution (10 M) was added to 1.5 g of graphene powder under ultrasonic mixing for 3 h. Then, the mixture was washed by distilled water and filtered until pH of the filtrate water becomes neutral. The filtered product was dried at 100°C for 12 h. Electrodes were manufactured as follows. The modified graphene powder was mixed with carboxymethyl cellulose as a binder at the weight ratio of 90 : 10 and then the mixture was introduced into the nickel foam. The thus obtained nickel–graphene electrodes were dried at 80°C under vacuum and then were compacted at the pressure of 5 MPa for 1 min. For comparison, the same method was used to manufacture the electrode of nonmodified commercial graphene powder. The initial graphenes manifested the specific capacitance of 101 F/g, which corresponds to the energy density of 14.0 Wh/kg, while KOH-modified graphenes manifested the specific capacitance of 136 F/g (which corresponds to 18.9 Wh/kg).

Graphene electrodes are characterized by high reversibility of charging–discharge processes. In the work by Du et al. (2010) were measured dependences of specific capacitance on current density and the number of galvanostatic charging/discharge cycles for two different electrodes with the specific surface area of 450–520 m²/g made on the basis of grapheme oxide applied on nickel felt. Electrochemical measurements were carried out in 30 wt% KOH. It was shown that capacitance remains practically unchanged when current density is changed by an order of magnitude. It was also shown that capacitance remains practically constant for 500 cycles.

Analysis of the electrochemical behavior of graphenes in IL 1Me3BuImBF$_4$ showed that development of the surface accessible for electrolyte and an increase in capacitance in the course of deep cathodic polarization is observed. An increase in the grapheme surface accessible for electrolyte can be explained by the fact that their superstructure, that is, their agglomeration structure, is not rigid and unchangeable. It consists of graphene layers without any strong bonds between them. Therefore, a change in the initial agglomeration structure is possible in the course of various adsorption electrochemical processes.

In the conclusion of this section, it should be noted that graphenes are characterized by fast kinetics of charging–discharge processes as opposed to most ACs, ACCs, and carbon blacks and therefore they can be used in power (pulse-type) ECSCs.

Graphenes are close to SWCNTs by their capacitive properties, but are much cheaper.

27.11 HIGH-FREQUENCY CARBON SUPERCAPACITORS

EDLCs can be used in devices requiring very small charging and discharge times. Such devices are, for example, pagers, notebooks, and so on. Besides, EDLCs are used for starter firing and regeneration of braking power of a combustion engine. Herewith, such EDLCs must have very low resistance and high capacitance. They must also be designed so as to have sufficiently high power at rather high frequencies. Though the specific surface area of carbon plays a great role in EDL formation, the pore size distribution and sizes of ions of electrolyte are also critically important. The larger fraction of interface in ACs is formed by micropores that owing to their size are inaccessible for ions for the formation of normal EDL at high frequencies. Therefore, optimization of carbon pore sizes in high-frequency EDLCs must provide high interface surface area at high frequencies, using also mesopores. Typical precursors for obtaining ACs are coal, anthracites, petroleum cokes, lignocellulosic materials or synthetic carbon precursors, and natural polymers. One of the cheap precursors used to obtain ACs with the required pore size distributions for high-frequency EDLCs is furfurylic alcohol (FA) synthesized on the basis of agricultural waste. Using synthetic precursors for obtaining carbons has some advantages before the conventional natural precursors: homogeneity of the structure and absence of inorganic admixtures (ashes). Synthesis of polyfurfurylic alcohol (PFA) was carried out by FA polycondensation under the action of an acidic catalyst. Thus, a highly cross-linked viscous product was obtained. The obtained polymer was chemically activated by KOH as an activating agent. PFA was mixed with KOH at the ratios of $1:1, 2:1, 3:1, 3.5:1$. The mixture was heated at 700°C for 1 h under a nitrogen flow (90 ml/ min). This product was cooled to the room temperature, neutralized by HCl, and washed by distilled water up to pH = 7. The thus obtained AC was labeled as PFA-KX, where X means the amount of the chemical agent used.

The porous electrode structure was studied using the method of low-temperature nitrogen adsorption. Table 27.6 shows the obtained characteristics of the porous electrode structure.

As seen from the table, the average size L_o of micropores with 0.6 nm for K1 to 1.6 nm for K3 grows at an increase in the activation degree. Herewith, the mesopore fraction increases. Despite this, the S_{BET} value for K3 is very high: 2600 m²/g. It is important to note that all studied carbon samples have a narrow small pore distribution, that is, a rather regular structure.

Electrochemical impedance measurements were carried out in the frequency range of 10 mHz–100 kHz to study the resistance of carbon electrodes.

Electric resistance decreased at an increase in the electrode compaction pressure. A K1 sample had low resistance of 0.041 Ω at 36 MPa corresponding to 130 S/cm,

TABLE 27.6. Characteristics of the Porous Electrode Structure on the Base of Polyfurfurylic Alcohol

Sample	S_{BET} (m²/g)	S_{mic} (m²/g)	L_o (nm)	V_{meso} (cm³/g)	V_{mi} (cm³/g)	Micropore Content (%)	O (wt%)
K1	1070	1448	0.6	0.03	0.43	95	3.1
K2	1600	1696	0.8	0.13	0.67	84	4.6
K2.5	2180	2423	1.0	0.21	0.82	80	4.8
K3	2600	1145	1.6	0.73	0.93	56	7.6

S_{mic}, specific surface area of micropores; L_o, average of micropores; V_{meso}, specific volume of mesopores; V_{mi}, specific volume of micropores.

but resistance increased as a result of further activation. An increase in the content of oxygen at an increase in the activation degree led to an increase in resistance.

The solution of 1 M TEABF$_4$ in acetonitrile (Et$_4$NBF$_4$/CAN) was used as electrolyte in electrochemical studies. The galvanostatic cycling of a supercapacitor was carried out between 0 and 2.5 V at current densities in the range of 0.3–225 A/g (per weight of the active material of a single electrode). Table 27.7 presents the specific capacitance values for various current values.

As seen from Table 27.7, capacitance decreases at an increase in current, but to a different degree for different electrodes. For some samples, for example, for K1, this decrease is too fast because of the very small pore size. For the K2.5 and K3 samples, the effect of current on capacitance is much lower because of the larger size of micropores. Table 27.8 presents the values of specific capacitance measured using the impedance method for the studied electrodes at three frequencies: 1 Hz, 10 Hz, and 100 Hz.

As seen from Table 27.8, capacitance decreases at an increase in AC frequency, but to a different degree for different electrodes. For some samples, for example, for K2, this decrease is too fast because of the small pore size. For the K3 sample, the effect of frequency on capacitance is much lower because of the larger size of micropores.

Thus, owing to the porous structure optimization, the studied carbons of the PFA type as compared to commercial carbons have capacitance at 1 Hz that is by 25% higher and capacitance at 10 Hz that is by 125% higher. At 100 Hz, the K3 carbon manifests capacitance above >30 F/g and thus it is promising for high-frequency supercapacitors. Long-term galvanostatic cycling was carried out on the corresponding ECSCs at the current densities of several hundreds of ampere per gram.

TABLE 27.7. The Values of Specific Capacitance Values (F/g) for Various Current Values from 0.3 to 225 A/g

Sample	$C_{0.3}$	C_{15}	C_{90}	C_{180}	C_{225}
K1	65	15			
K2	100	80	70	55	
K2.5	130	110	100	91	88
K3	150	135	120	110	104

TABLE 27.8. The Values of Specific Capacitance Measured Using the Impedance Method for the Studied Electrodes at Three Frequencies: 1, 10, and 100 Hz

Sample	1 Hz	10 Hz	100 Hz
K2	30	3	0
K2.5	110	70	20
K3	130	105	35

TABLE 27.9. The Specific Capacitance Values Obtained Using the Impedance Method for Different Frequencies

Frequencies, Hz	1	10	100	1000
Specific capacitance, mF/cm^2	0.9	0.6	0.3	0.05

Another direction in development of high-frequency ECSCs was using thin proton-conducting polymer electrolytes. Polymer electrolyte was manufactured by mixing a PVA (MW = 85,000–124,000) solution with $H_44SiW_{12}O_{40}$ (SiWA) and orthophosphoric acid (H_3PO_4). This electrolyte had the conductivity of 8 mS/cm. Stainless steel foil was used as electrodes, on which water-based graphite ink was applied. The foil thickness was 50 μm, the coating thickness was 25 μm. The thickness of polymer electrolyte was 45–50 μm. Therefore, the thickness of this completely solid-state individual cell was 200 μm. This cell was cycled at the sweep rates of up to 20 V/s and had the time constant of 10 ms. Table 27.9 presents the specific capacitance values obtained using the impedance method for different frequencies.

Thus, solid polymer electrolyte of this type can be used in high-frequency supercapacitors.

27.12 SELF-DISCHARGE OF CARBON ELECTRODES AND SUPERCAPACITORS

Self-discharge is an important common phenomenon in the behavior of electric rechargeable devices, such as batteries and ECSCs, and is determined as the rate of voltage drop with time t under open-circuit conditions (Kowal J, et al, 2011). As self-discharge means potential drop (V) with time for an individual electrode or voltage drop ΔV between electrodes of a charged device, it must result in a drop in the available electric energy according to Equation (2.2) for EDLCs:

$$A_t = \left(\frac{1}{2}\right) C(\Delta V)t^2 = \left(\frac{1}{2}\right) q \ (\Delta V_t) \tag{27.16}$$

Therefore, the rate of the available energy drop is determined by the rate of voltage drop:

$$\left(\frac{dA_t}{dt}\right) = \frac{1}{2} C \, d\frac{[(\Delta V)_t^2]}{dt} = C \, \Delta V_t \ d\left(\frac{\Delta V_t}{dt}\right) \tag{27.17}$$

Thus, self-discharge behavior is obviously a practically important characteristic, especially in stand-alone or stand-by applications. On the whole, capacitance with no leakage theoretically does not undergo self-discharge that can occur if there are certain faradaic processes with electron transfer, for example, "shuttle reactions" because of the presence of admixtures or reactive particles/groups formed from the material within the device or because of some ohmic leakage. Thus, the rate of self-discharge can considerably depend on the system chemistry and electrochemistry, origin of the HDCM, purity of reagents and electrolyte, presence of oxygen, cleanliness in the process of device manufacturing, cell design, and temperature.

As an example, one can mention the presence of iron admixtures in ACs of natural materials. These admixtures are often found in ashes. In this case, the following "shuttle reaction" occurs with the transport of Fe^{3+} and Fe^{2+} ions from the anode to the cathode and back under exposure to the difference of potentials:

$$\text{(anode) } Fe^{3+} + e^- \leftrightarrow Fe^{2+}\text{(cathode)} \tag{27.18}$$

The actual self-discharge rates are determined by the mechanisms of processes occurring in self-discharge. Conway (1999) formulated three kinetic cases for which he obtained diagnostic criteria for identification of such processes. They can be stated as follows:

(a) When self-discharge is not because of diffusion-controlled by to kinetically controlled faradaic processes caused by reactions of oxidation or reduction of recharging products or redox reactions of admixtures, such as Fe^{2+}/Fe^{3+} (see Eq. (27.18)) or other ions of elements with variable valency, oxygen solution in electrolyte, and so on, then the dependence of voltage drop V on $\log t$ is linear:

$$V_t = V_i - A \ \log(t + \theta) \tag{27.19}$$

where θ is the integration constant.

(b) If self-discharge is controlled by semi-infinite diffusion to a flat electrode of reagents of faradaic process reagents with participation of admixtures, potential drop in time is determined by a square root dependence:

$$V_t = V_i - \frac{[2zFSD^{1/2}\pi^{1/2}C_0 t^{1/2}]}{C} \tag{27.20}$$

(c) If self-discharge was caused by the "short-circuit" leakage current between the two electrodes, for example, as a faulty bipolar capacitor assembly, the dependence of $\ln V_t$ on t is linear:

$$\ln V_t = \ln V_i - \left(\frac{t}{RC}\right) \tag{27.21}$$

where R is the additional ohmic load resistance, for example, because of leakage in a bipolar circuit.

The electrochemical kinetic behavior described in Equations (27.19)–(27.21) are easily discernible and therefore provide a basis for diagnostics for the establishment of self-discharge mechanisms after recording the dependence of potential or voltage of an electrode or cell in an open-circuit voltage on time. The presence of admixtures including water traces in organic electrolytes was related to the causes of self-discharge and leakage current in ECSCs. It was shown that self-discharge was also related to the presence of carbon functional groups and accessibility of a microporous surface. Researcher considered the effect of variation of the composition of electrolyte, nature of conducting additives into the electrode, same as the composition of the separation material to control and limit self-discharge. A significant cause of ECSC self-discharge is the mechanical electrophoretic transport of small colloid particles of the carbon material from one electrode to another.

At present, two procedures are used to characterize the self-discharge behavior of ECSCs and batteries. One of them consists of the measurement of the time dependence of voltage drop under open circuit to distinguish between the self-discharge mechanisms, as described earlier. Another procedure is "float current" measurement of the dependence of current on time at a constant potential for an individual electrode or at constant voltage V on the cell.

Self-discharge of an ACC electrode was studied in 5 and 0.5 M H_2SO_4 with DC charging to various potentials and then opening the circuit recording potential drop in time (Niu et al., 2006). The resulting data were plotted in the form of dependences of $\ln V_t$ versus t or V_t versus $t^{1/2}$. It turned out that both dependences are nonlinear, that is, do not correspond to Equations (27.20) and (27.21). This shows that the causes of self-discharge are neither diffusion-controlled faradaic processes, nor "short-circuit" leakage currents between the two electrodes. However, dependences of V_t on $\log t$ are not strictly linear after $t \gg 0$ (see Eq. (27.21)) as shown in Figure 27.29.

This corresponds to the mechanism for case (a) (Eq. (27.21)), that is, the mechanism of a nondiffusion-controlled faradaic process. As seen in Figure 27.29, slopes of the straight lines depend on initial potential V_i, as opposed to the behavior predicted in Equation (27.21). These slopes decrease at a decrease in the initial potential. Ultimately, at $V_i \leq +0.9$ V, the slope ceases to depend on V_i. It was shown that the linear dependence of V_t on $\log t$ (Eq. (27.21)) at $V_i \geq +0.9$ V corresponds to a Tafel dependence and is explained by the reaction of the formation of oxygen that is a part of functional groups. Thus, the self-discharge rate grows at an increase in the concentration of surface groups. Besides, ohmic potential drop iR-drop also grows.

It was also shown that the charging–discharge history of an ACC electrode, that is, the maximum potential to which the electrode was earlier discharged, also affects the rate of potential restoration ("memory effect"). The higher the potential to which the ACC electrode was discharged, the higher the rate of the further potential restoration after circuit opening. This effect is explained by the process of potential redistribution over the porous electrode depth in the direction of its leveling after polarization, as a result of which a nonuniform potential distribution was established. This regularity was confirmed on the basis of the theory of porous electrodes.

Self-discharge was also studied for three types of commercial ECSCs. It was found that self-discharge greatly depends on the history, that is, on the rum of charging and discharge to the circuit opening. It also depended on the temperature, although not much. It was found that dependences of voltage drop in time can be divided into three phases, two of which can be described by exponential functions and the third one is

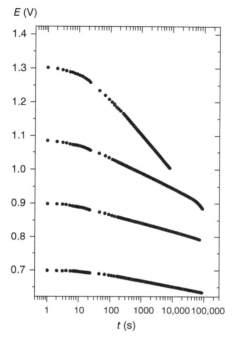

Figure 27.29. Potential decay relationships for self-discharge at an individual C-cloth electrode versus RHE started from the following potentials: +0.7, +0.9, +1.1, +1.3, and +1.35 V; 5.0 M H_2SO_4, after positive charging at 10 mA.

linear. The first exponential function has the time constant of about several hours and stops after sufficiently long charging or after short discharge. By contrast to this, the exponential function has the time constant of about several days. After a very long time (several months), voltage drop becomes linear with the slope of several volts a week.

A number of works suggests an explanation for the self-discharge mechanism that is based on redistribution of ions within pores. This explanation agrees well with the results of measurements, but requires further studies. A mathematical model of these processes has been recently developed.

The term of self-discharge requires clarification. Self-discharge is always loss of ampere-hour capacitance. As applied to batteries, self-discharge means that a charged active material turns spontaneously in the course of self-discharge into a discharged active material. In the case of supercapacitors, such a process plays only a small role. An almost linear voltage drop observed after prolonged rest periods can be responsible for such a process. However, studies showed that the larger fraction of voltage drop in ECSCs occurs because of charging voltage redistribution or, in other words, because of overpotential relaxation caused by the concentration gradients of charge carriers. Charge carriers are not lost, but are deep within pores, mainly within micropores. Charge carriers can again be extracted from ECSCs during very slow discharge. Therefore, it is not such a self-discharge process that occurs in batteries.

Thus, one can state that the mechanisms of self-discharge processes of ECSC carbon electrodes are quite diverse and complicated.

27.12.1 Float-Current Measurement Behavior

In batteries, current at the given polarization potential remains practically constant because of the fact that herewith an electrochemical reaction with consumption of reagents occurs and current is unchanged while the reagents are not exhausted. However, the situation for ECSCs is fundamentally different. Experimental potentiostatic measurements of current in time were carried out for ACCs at different concentrations of H_2SO_4 for a set of potentials V from $+0.6$ to $+1.1$ V (Niu et al., 2006). As a result, interesting information was obtained that sheds fresh light on the nature of self-discharge in porous highly dispersed ACC or in other carbon material. Potentiostatic float-current curves were measured for a C-cloth electrode in electrolyte of 0.05 M, 0.5 M, and 5 M H_2SO_4 from seven polarization potentials ($+0.6$, $+0.7$, $+0.8$, $+0.9$, $+1.06$ $+1.0$, and $+1.1$ V (RHE)). There are two important parameters in float-current measurements. This float current i_f and accumulated charge q_f at different potentials in the range corresponding to self-discharge. Current i_f is directed to compensation of the self-discharge rate and q_f represents total charge that can be transferred (or accepted) to the surface of a porous C-cloth electrode to establish steady-state charge compensating EDL self-discharge at the given potential. Therefore, q_f reflects complete self-discharge for a porous material with a high surface area. Figure 27.30 shows the curves of $\log q_f$ on V obtained by integration of potentiostatic

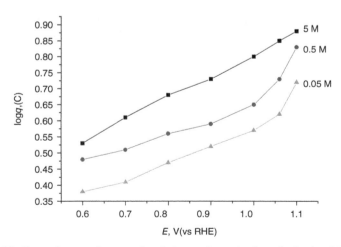

Figure 27.30. Dependences of accumulated charge determined on the basis of float-current experiments from the polarization potentials in 0.05 M (triangles), 0.5 M (circles), and 5 M (squares) H_2SO_4.

curves of i_f versus t of the type for different potentials and for three concentrations of sulfuric acid electrolyte.

As seen in Figure 27.30, q_f grows at an increase of polarization potentials and a decrease in the concentration of electrolyte pointing to an increase in the self-discharge rate at an increase in polarization potentials and a decrease in the concentration of electrolyte. The degree of electrolyte penetration into pores at its high concentrations and the electrolyte/carbon contact surface is higher than at low concentrations, so that charging compensating self-discharge in float-current measurements is lower for high concentrations as compared to low concentrations at similar polarization potentials, as seen in Figure 27.30.

Thus, float-current measurements can provide a direct and simple procedure for the choice of optimum conditions of operation of porous carbon materials with a high surface area used in ECSCs.

27.13   PROCESSES OF EDLC DEGRADATION (AGING)

Though EDLC cyclability is extremely high (hundreds of thousands and millions of cycles), they are, however, characterized by certain degradation and corrosion processes, especially if too high charging voltage is used: more than 1 V for aqueous electrolytes and more than 2.5 V for nonaqueous electrolytes. The aging processes result in a gradual decrease in capacitance of the electrodes and an increase in EDLC resistance (Zhu M et al, 2008). There are rather few works in the literature on studies of the EDLC aging processes because of the extremely slow rates of these processes. Therefore, various artificial conditions for acceleration of the aging processes are used in such studies available in the literature.

A particular feature of operation of EDLC with AC electrodes is the formation of a certain amount of CO_2 in the course of operation at the potentials above 1 V (NHE). Analysis of the gas collected in the case of EDLC with electrolyte of 38% H_2SO_4 showed that it consisted predominantly of N_2 and O_2 with traces of CO_2 and H_2O at potentials below 1 V. For charged EDLC, the gas consists mostly of CO_2 with small amounts of N_2, O_2, and H_2O vapors. Presence of water vapors shows that the performance of the capacitor can decrease with time because of loss of electrolyte, which results in a decrease in capacitance of the capacitor. This, however, can occur only in unsealed EDLCs. Carbon dioxide is a product of oxidation of carbon electrodes. Its amount strongly depends on the voltage applied. To prevent CO_2 accumulation in the course of EDLC operation, a release valve is used in some schemes. In most schemes, however, no valves are used.

The aging of carbon materials in acetonitrile-based electrolyte is very slow and at least several weeks are required to register the measurable changes even when conditions for accelerated aging (increased potential) are provided.

In the performed study, samples A and B made of natural precursors (coconut shell) and carbon C of synthetic resin were used. Let us point out that accelerated "chemical aging" was achieved in the experiments by the immersion of the electrode into acetonitrile heated to 50°C and conditioning for 1 month and usual electrochemical aging was carried out by EDLC conditioning at the voltage of 2.9 V for 45 days. The obtained results were compared to electric characteristics of ECSCs produced by EPCOS. Porosimetric and Raman spectroscopic data obtained after "chemical aging" did not reveal any significant changes for AC powders under conditioning at a relatively high temperature of 50°C for a prolonged time. As opposed to this, electrochemical aging affected the porous structure noticeably. Table 27.10 shows porosimetric characteristics for fresh and electrochemically aged samples of three carbons A, B, and C. Here, S_{BET} and S_{DR} are the specific surface area values obtained using the BET technique and the Dubinin–Radushkevich method; V_{mic} is the volume of micropores, V_{total} is the total volume of the adsorbate (N_2) in pores.

As seen from this table, the result of electrochemical aging for all the studied electrodes was a decrease in the specific surface area and the micropore volume; herewith, these changes in the anode are much greater than in the cathode. Additional studies showed that a decrease in the volume of micropores and mesopores is the consequence of their degradation and also the result of their clogging up by the products of decomposition of electrolyte. AC made of a synthetic precursor (sample C) shows the high structural stability as compared to ACs made of natural precursors (samples A and B).

According to chemical analysis carried out using a number of methods, especially X-ray photoelectron spectroscopy (XPS) and infrared spectroscopy, the chemical composition of carbons for different electrodes changes variously. All changes depend on the electrode polarity; negative electrodes (cathodes) age much less than positive electrodes (anodes). This must be taken into account in new supercapacitor schemes, for example, anodes must be made thicker than cathodes in an asymmetric scheme to compensate losses of capacitance or, in the case of presence of temperature gradients, to position an anode from the colder side. The most important measurements showed formation as a result of the process of the aging of C–H bonds on the cathode and bonds C–N and C–F on the anode and also polymerization of acetonitrile on the cathode and decomposition of tetrafluoroborate. The practically unavoidable

TABLE 27.10. Porosimetric Characteristics for Fresh and Electrochemically Aged Samples of Three Carbons A, B, and C

Sample		S_{BET} (m^2/g)	V_{micro} (m^2/g)	V_{total}(m^2/g)
A	Fresh	2000	0.70	1.00
A	Aged anode	530	0.20	0.35
A	Aged cathode	1700	0.70	0.80
B	Fresh	1700	0.64	0.75
B	Aged anode	370	0.13	0.25
B	Aged cathode	1100	0.40	0.55
C	Fresh	2100	0.80	0.90
C	Aged anode	900	0.35	0.40
C	Aged cathode	1600	0.60	0.70

presence of water traces affects all processes of "chemical aging" by supplying oxygen, which results in the formation of various chemical groups, for example, CO, OX, COOH, CONH, and F. This, in its turn, leads to enhancement of resistance.

As the aging processes ultimately result in the evolution of gases, mainly CO_2 and CO, considerable aging, for example, at extremely high charging voltages, release valves are required in the ECSC battery schemes, while in the case of very low aging rates such valves are not used.

REFERENCES

Arbizzani C, Biso M, Cericola D, Lazzari M, Soavi F, Mastragostino M. J Power Sources 2008;185:1575.

Du X, Guo P, Song H, Chen X. Electrochim Acta 2010;55:4812.

Gao H, Lian K. J Power Sources 2011;196:8855.

Hashmi SA, Suematsu S, Naoi K. J Power Sources 2004;137:145.

Kowal J, Avaroglu E, Chamekh F, Senfelds A, Thien T, Wijaya D, Sauer DU. J Power Sources 2011;196:573.

Lazzari M, Mastragostino M, Soavi F. Electrochem Commun 2007;9:1567.

Rychagov AY, Volfkovich YM, Vorotyntsev MA, Kvacheva LD, Konev DV, Krestinin NV, Kryazhev YG, Kuznetsov VL, Kukushkina YA, Mukhin VM, Sokolov VV, Chervonobrodov SP. Electrochem Energ (Russian) 2012;12:167.

Volfkovich YM, Mikhailin AA, Bograchev DA, Sosenkin VE, Bagotsky VS. Chapter 7. *Recent Trend in Electrochemical Science and Technology*, INTECH open access publisher. 2012. www.intechopen.com. p 159.

Zhu M, Weber CJ, Yang Y, Konuma M, Starke U, Kern K, Bittner AM. Carbon 2008;46:1829.

MONOGRAPH AND REVIEWS

Burke A. J Power Sources 2000;91:37.

Conway BE. *Electrochemical Supercapacitors*. Kluwer Academic/Plenum Publishers; 1999.

Inagaki M, Konno H, Tanaike O. J Power Sources 2010;195:7880.

Levie R. Adv Electrochem Electrochem Eng 1967;6:329.

Niu J, Pell WG, Conway BE. J Power Sources 2006;156:725.

Sharma P, Bhatti T. Energy Convers Manag 2010;51:2901.

Volfkovich Y, Bagotzky VS, Sosenkin VE, Blinov IA. Colloid Surface A 2001;187–188:349.

Volfkovich Y, Serdyuk T. J Electrochem 2002;38:935.

28

PSEUDOCAPACITOR ELECTRODES AND SUPERCAPACITORS

28.1 ELECTRODES BASED ON INORGANIC SALTS OF TRANSITION METALS

As pointed out earlier, according to Conway (1999), electrochemical supercapacitors (ECSCs) are electrochemical devices, in which quasireversible electrochemical charging–discharge processes occur. The shape of galvanostatic charging and discharge curves corresponding to these is close to linear, that is, close to the shape of the corresponding dependences for usual electrostatic capacitors. ECSCs include, apart from double-layer ECSCs, the so-called pseudocapacitors with the electrodes, on which fast quasireversible faradaic processes, that is, redox charging–discharge reactions occur. Capacitance of pseudocapacitor electrodes, as a rule, is higher than the capacitance of purely double-layer electrodes. As PsC electrodes are manufactured with high specific surface area, both pseudocapacitance of faradaic reactions and EDL capacitance contribute to their capacitance. Pseudocapacitor electrodes are subdivided into three main types: electrodes based on inorganic compounds of transition metals (oxides, sulfides, nitrides, etc.), electrodes based on electron-conducting polymers, and electrodes based on monomer redox systems. Oxides of variable valency metals are often used in pseudocapacitor electrodes. Historically, oxides of platinum metal groups were first studied. Cyclic voltammograms (CVs) for metallic Ru with an oxide-free surface are similar to CVs of Pt. However, as opposed to Pt, oxide film growth on Ru occurs under further cycling. Electric charge on the electrode

Electrochemical Power Sources: Batteries, Fuel Cells, and Supercapacitors, First Edition.
Vladimir S. Bagotsky, Alexander M. Skundin, and Yurij M. Volfkovich.
© 2015 John Wiley & Sons, Inc. Published 2015 by John Wiley & Sons, Inc.

of Ru oxide is accumulated because of EDL capacitance and faradaic pseudocapacitance related to a change in the oxidation degree of Ru in the oxide film. A reversible oxidation–reduction process occurs in the range of potentials of 0.05–1.5 V (E_i) with formation of Ru(II), Ru(III), and Ru(IV). Formation of Ru(VI) is also possible. It is important that no reduction to metallic Ru occurs. The current is practically independent of the potential in the above range of potentials. Similar behavior is also demonstrated to IrO_2, but its capacitance depends on a greater degree on the potential. Processes of proton intercalation and diffusion in a layer of Ru oxide occur in the course of oxidation–reduction resulting in a change in the oxidation degree of Ru ions in an oxide film. It is assumed that there are three regions in the oxide film: the bulk, the surface-adjacent region, and the electrode surface. Ru in these regions has different oxidation degrees. Redistribution of the oxidized Ru state between the above regions occurs after discharge. Ru oxide and Ru oxide hydrate were studied as electrodes for PsCs. Electrochemical reactions occurring in such PsCs have the following form:

$$\text{Positive electrode}: \ HRuO_2 \leftrightarrow H_{1-\delta}RuO_2 + \delta H^+ + \delta e^- \tag{28.1}$$

$$\text{Negative electrode}: \ HRuO_2 + \delta H^+ + \delta e^- \leftrightarrow H_{1+\delta}RuO_2 \tag{28.2}$$

$$\text{Overall reaction}: \ 2HRuO_2 \leftrightarrow H_{1-\delta} \ RuO_2 + H_{1+\delta} \ RuO_2 \tag{28.3}$$

where $0 < \delta < 1$ and RuO_2 and H_2RuO_2 represent the positive and negative electrodes in the fully charged state. Protons move from one electron to another through a separator in the course of charging and discharge. At the same time, movement of electrons occurs through the power source or external load.

The main difference between PsC and electric double-layer capacitor (EDLC) is that there is no overall ion exchange between the electrode and electrolyte in PsCs. That is, the concentration of electrolyte remains constant during the charging and discharge. As pointed out above, EDL capacitance in such systems always coexists with pseudocapacitance. However, the amount of ions participating in the process of EDL formation is much lower than the amount of protons exchanged between the two electrodes. Thus, pseudocapacitance is much higher than EDL capacitance in the same system.

A mathematical model of an electrochemical capacitor with double-layer and faradaic processes has been developed. It was assumed that a capacitor consists of two $RuO_2 \cdot xH_2O$ electrodes separated by an ion-conducting separator. It was also assumed that the concentration of electrolyte was constant and the side reactions and thermal effects were not taken into account. It was shown by means of mathematical analysis that the discharge time even for the faradaic process alone is much higher than the discharge time for a double-layer process (Fig. 28.1). This points to the importance of the faradaic process for an increase in the energy density of the capacitor. The capacitor, on the electrodes of which the double-layer and faradaic processes occur, usually has a higher energy density as compared to a capacitor, in which either the double-layer or faradaic processes occur. The suggested model

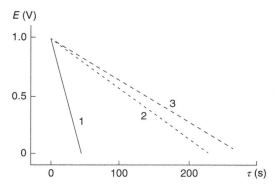

Figure 28.1. Discharge curves for (1) the double-layer, (2) faradaic, and (3) double-layer + faradaic processes for the $RuO_2 \cdot xH_2O$ particles with the diameter of 5 nm; $i = 0.001$ mA/cm^2.

correctly predicts that the faradaic process causes a considerable increase in the energy density of PsC as compared to EDLC.

Specific capacitance of the electrode consisting of a RuO_2 film grows at an increase in its specific surface area. In a crystalline RuO_2 electrode, energy storage takes place only in its surface layers. Redox reactions in the $RuO_2 \cdot xH_2O$ amorphous phase occur not only on the surface, but also in the bulk. Therefore, specific capacitance of the electrode of a $RuO_2 \cdot xH_2O$ amorphous phase (770 F/g) is much higher than capacitance obtained for crystalline RuO_2 films (380 F/g). However, high specific capacitance of the electrode of amorphous RuO_2 may result in depletion or supersaturation by protons from the internal region of the electrode, which can limit the maximum capacitor charging–discharge rate and power. To increase the conductivity of the electrode, amorphous RuO_2 powder was mixed with highly porous carbon black. The specific capacitance of such a composite electrode consisting of 80 wt% of RuO_2 and 20 wt% of carbon black was 570 F/g. Film electrodes of amorphous RuO_2 have a lower contact resistance than electrodes of powdered amorphous RuO_2. Film electrodes of amorphous RuO_2 deposited on a Ta support were manufactured. The maximum capacitance of 590 F/g was obtained for a film electrode deposited at 2000°C. PsCs with such oxide electrodes are highly reversible. Their capacitance varied negligibly even after prolonged cycling for 800 thousands of cycles.

Properties of RuO_2 composite electrodes with a Nafion ion-exchange resin were also studied. Capacitance of an electrode of this type with the RuO_2 content of up to 24 mg/cm^2 was 0.6 F/cm^2. Although the absolute capacitance values are low, there is a dependence between the fraction of RuO_2 in the mixture and the electrode capacitance value, which allows developing electrodes of this type with higher capacitance. In our opinion, the relatively low value of capacitance of composite electrodes with solid polymer electrolyte (SPE) as compared to liquid electrolyte is explained by low values of internal electrode/electrolyte interface specific surface area, as SPE particles cannot penetrate micropores because of their rather large sizes. However, the advantages of capacitors with SPE are the possibility of developing thin capacitors with good mechanical properties and their very low self-discharge.

C–Ru xerogels obtained under carbonization of resorcin–formaldehyde resins and RuO_2 were also studied. In the course of carbonization, the $RuO_2 \cdot xH_2O$ oxide is reduced to metallic Ru. Ru is oxidized in the course of the first charging–discharge cycle forming the active form of Ru oxide. Electrochemical capacitance of these materials grows at an increase in the Ru content in xerogel. Specific capacitance of the electrode containing 14 wt% Ru was 256 F/g.

Deposition of Ru particles onto highly dispersed carbon supports considerably improves their energy-saving characteristics. It is shown in one of the works that the presence of Ru particles (35 wt%) on the surface of carbon aerogel caused an increase in the specific capacitance of the electrode to 200 F/g because of pseudocapacitance. In a number of works in which RuO_2/activated carbon composites electrodes were studied, good capacitive characteristics for positive RuO_2/activated carbon composites electrodes in alkaline electrochemical capacitors were obtained. An AC powder with the specific surface area of $1500 m^2/g$ was mixed with a $RuCl_3$ solution (0.05 mol/l). After calcination of the mixture of AC and ruthenium oxide at 150°C for several hours, a carbon–ruthenium composite was obtained. Then, this composite was mixed with carbon black (as a conducting additive) and with polytetrafluoroethylene (PTFE, as a binder); this mixture was rolled into films of different thickness. The film was pressed into nickel foam using the pressure of 0.6 MPa and then it was cut into round electrodes. Electrodes were dried under vacuum for 12 h at the temperature of 100°C. Two electrodes were separated by a polypropylene membrane that was first impregnated by a KOH solution (6 mol/l). Assembled ECSC was conditioned for approximately 24 h before electrochemical measurements were started (at 25°C).

In the work of Yue-feng et al. (2007) were measured CV curves for two such alkaline electrochemical capacitors. One of the electrodes was an AC electrode and the other was a RuO_2/AC electrode containing 15% Ru. The curve for RuO_2/AC had an oxidation peak at approximately 0.95 V and a reduction peak at about 0.8 V; the comparative curve for AC contained neither oxidation nor reduction peaks. This confirmed that electrochemical oxidation and reduction occurred at the RuO_2/AC electrode when ECSC was charged and discharged. The AC electrode showed the ideal characteristics of an EDLC. The effective capacitance of the electrode can be estimated from the area under the curve. It can be seen from the CV curves that capacitance of RuO_2/AC as a positive electrode in alkaline AC was significantly larger than that of the AC-positive electrode.

In the work of Yue-feng et al. (2007) were measured three galvanostatic curves for three positive electrodes containing various amounts of Ru. Capacitance of PsCs grows at an increase in the content of ruthenium. Besides, these curves are close to straight lines and this confirms that electrodes with ruthenium oxides feature capacitive behavior described by Conway (1999) (see above).

At an increase in discharge current, capacitance of AC and RuO_2/AC electrodes gradually decreased, as shown in Figure 28.2.

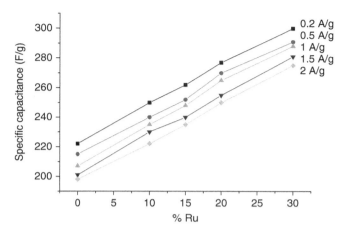

Figure 28.2. Specific capacitances of AC and RuO_2/AC electrodes at different current densities.

As seen in Figure 28.2, a carbon–ruthenium composite electrode had the characteristics similar to those of an AC electrode, but higher capacitance. It was also concluded that the rate of RuO_2 oxidation and reduction are as fast as the charging–discharge rate of the electric double layer on an AC electrode in an alkaline solution. Nevertheless, capacitance of RuO_2/AC electrodes was larger than that of an AC electrode. When the content of Ru in the composite was increased to 30%, specific capacitance of the electrode reached 300 F/g (at the discharge current decrease of 0.2 A/g) corresponding to the specific capacitance of 333 F/g of active materials. The capacitance of the AC electrode was 220 F/g at the same current density corresponding to specific capacitance of 241 F/g of active materials.

Though capacitance of the electrodes and the active materials obviously increased at an increase in the Ru content in electrodes, specific capacitance contributed by $RuO_2 \cdot H_2O$ herewith decreases. For example, this value is 701.7 F/g for 2% Ru and 434.5 F/g for 30% Ru. Close results were obtained for the system of RuO_2/AC in H_2SO_4 in electrolyte.

The high cost of oxides of noble metals limits their wide practical application. Therefore, other materials with similar properties have been searched for lately. These can be mixed with metallic oxides of noble and other metals deposited on metallic supports and supports of porous carbon. The suggested materials for electrodes in PsC capacitors are, for example, $RuO_2–VO_x$ electrodes with highly active surface. The specific charge of a RuO_2 (33%)$–VO_x$(67%)/Ti electrode with an oxide coating of $0.7 \, mg/cm^2$ was $162 \, mC/cm^2$, which corresponds to 0.76 of a proton required for the adsorption of each Ru ion. It is assumed herewith that charging capacitance is determined only by proton desorption. The charging capacitance value for a RuO_2 (33%)$–VO_x$(67%)/Ti electrode causes a considerable increase in the charging capacitance value obtained for a RuO_2/Ti electrode. High pseudocapacitance values were also obtained for $RuO_2–ZrO_2/Ti$ electrodes with the RuO_2 content of 50 mol%

(200 mC/mg). Electric charge in the electrode corresponds to 0.53 protons adsorbed on each Ru ion.

Of late, manganese oxides have become intensively studied as a material for a pseudocapacitive electrode as oxides of yet another metal with variable valency that is much cheaper than ruthenium. MnO_2 is studied most often. It is usually applied onto highly dispersed carbon electrodes. In particular, a technique was developed for the preparation of hydrothermally reduced graphene/MnO_2 (HRG/MnO_2) composites by immersion of HRG in a mixed solution of 0.1 M $KMnO_4$ and 0.1 M K_2SO_4 for various periods of time. The deposition mechanism of MnO_2 in HRG is suggested corresponding to the following reaction:

$$4MnO_4^- + 3C + H_2O = 4\,MnO_2 + CO_3^{2-} + 2HCO_3^- \qquad (28.4)$$

Morphology and microstructure of the manufactured composites were studied using field-emission scanning electron microscopy, X-ray diffraction, Raman microscopy, and X-ray photoelectron spectroscopy. The results of these studies showed that well-deposited MnO_2 is uniformly distributed over the HRG surface. Capacitive properties of synthesized composite electrodes were studied using the CV method and the impedance technique in an aqueous solution of 1 M Na_2SO_4. The main result of electrochemical studies was as follows: specific capacitance of an HRG/MnO_2−200 electrode (HRG was conditioned in a mixed solution of 0.1 M $KMnO_4$ and 0.1 M K_2SO_4 for 200 min) reached 211.5 F/g at the potential scan rate of 2 mV/s. Besides, the electrodes demonstrated good cyclability and stability.

Integration of CV curves for HRG, HRG/ MnO_2−10, and HRG/ MnO_2−200 electrodes in an aqueous solution of 1 M Na_2SO_4 at the potential sweep rate of 40 mV/s yielded the capacitance values of 74.5, 102, and 135.5 F/g. It is important to point out that according to the rectangular shape of the CV curve for HRG close to that in Figure 28.2a, capacitance of this electrode is clearly of the double-layer character. Meanwhile, capacitance of MnO_2 is clearly of the pseudocapacitance character.

Li et al. (2011) studied impedance measurements for HRG, HRG/MnO_2−10, and HRG/MnO_2−200 electrodes in an aqueous solution of 1 M Na_2SO_4. Complex plane plots for all three electrodes contain a semicircle in the range of high frequencies and a direct line in the range of low frequencies. Intersection of a semicircle with the real resistance axis in the high frequency range represents equivalent serial resistance (R_s) of the electrode, and the semicircle diameter corresponds to the charge transfer resistance (R_{ct}) through the electrode/electrolyte interface. Comparison of complex plane plots of these electrodes obviously shows that R_{ct} gradually grows with the increase in the load of MnO_2. This result is partially because of a decrease in the conductivity of the HRG/MnO_2 material because of deposition of MnO_2. A direct line in the range of low frequencies corresponds to diffusion resistance of electrolyte within the electrode and to ionic diffusion into the electrode. The almost vertical shape of this direct line representing fast ion diffusion in electrolyte and adsorption on the electrode surface implies an ideal capacitive behavior of electrodes. Thus, the impedance complex plane plot for HRG/MnO_2 electrodes characterizes both the presence of pseudocapacitance of the faradaic reaction because of the presence of a semicircle and the EDL capacitance.

It was suggested that synthesized HRG/MnO_2 composites because of their sufficiently high capacitive characteristics, high cyclability, and stability and cheapness would find wide application in supercapacitors.

Among the great amount of manganese oxides discussed in the literature, the systems of Li–Mn–O and Na–Mn–O attracted much attention because of their tunnel or multilevel crystalline structures that facilitated lithium/sodium intercalation–deintercalation. Nevertheless, the previous studies were largely focused on the system of lithium/sodium–manganese oxides as a positive electrode for lithium batteries. A number of communications are related to their application as electrode materials for supercapacitors. In this connection, such a compound as $Na_2Mn_5O_{10}$ is of interest for supercapacitors. $Na_2Mn_5O_{10}$ is synthesized by hydrolysis of $[Mn_{12}O_{12}(CH_3COO)_{16} \cdot (H_2O)_4]$ with the following thermal calcination. Amorphous $Na_2Mn_5O_{10}$ is obtained at a relatively low temperature (200°C). An increase in the calcination temperature results in the formation of crystals in the form of nanorods. Specific capacitances of 178, 173, and 175 F/g were obtained in an aqueous $0.5 M Na_2SO_4$ solution for $Na_2Mn_5O_{10}$ at different temperatures (200, 400, and 600°C), accordingly, at the charging and discharge current of 0.1 A/g. Capacitance loss after 1000 cycles was less than 3%.

CV curves measured for amorphous samples at the sweep rate of 2 mV/s in the range of 0–0.8 V have an ideal rectangular shape. However, a highly reversible couple of redox peaks was observed for crystalline samples calcinated at 600°C. Two mechanisms of charging–discharge processes for oxides in aqueous electrolytes were suggested; both of them explain the following reaction equation:

$$MnO_2 + M^+ + e^- \leftrightarrow MnOOM \tag{28.5}$$

where M represents hydrated protons or cations, for example, Li^+, Na^+, and K^+.

The first mechanism is the process of cation adsorption/desorption on the surface of the material characteristic of amorphous manganese oxides. The second mechanism is intercalation/deintercalation of cations into the bulk of the material characteristic of crystalline manganese oxides.

Electrochemical studies led to the conclusion that the $Na_2Mn_5O_{10}$ compound is a good candidate for a positive electrode material in ECSCs.

Vanadium oxides, same as oxides of yet another metal with variable valence and also with intercalation ability in respect to lithium cations, have lately been widely studied for their application as an active material for positive electrodes in ECSCs (Chen and Wen, 2003). V_2O_5 is applied on highly dispersed carbon materials as supports, as specific electron conductivity of V_2O_5 is rather low.

High-rate electrode characteristics of V_2O_5 sol/carbon composites were studied. The V_2O_5 sol was prepared by a reaction between metallic vanadium and a hydrogen peroxide solution. The acetylene carbon black powder was added to the sol together with acetone in the form of a homogeneous suspension. A composite of amorphous V_2O_5 and carbon black was introduced into a macroporous nickel current collector and heated at 120°C to manufacture the electrode. Electrochemical measurements were carried out in organic electrolytes of the type of $LiClO_4/PC$ or $LiPF_6/\gamma$-butyrolactone (γ-BL) at the room temperature. It was shown that a composite V_2O_5/carbon electrode at the weight ratio of 0.7 yielded a 54% of ideal capacitance in a 1 M $LiClO_4$/propylene carbonate electrolyte in 360 mAh/g (4.2 − 2.0 V) as per V_2O_5 even at a very high discharge rate at 150°C, that is, at 54 A/g of V_2O_5 (Kudo T, 2002). Reversibility was also quite satisfactory and practically no losses of capacitance were observed after several thousands of cycles between 4.2 and 3.0 V at the rate of 20 C. An electrode layout was manufactured by the application of a thin composite layer on an Al foil current tap. It had the electrode capacitance of 40 mAh/g at a sufficiently high current density of 30 mA/cm^2.

Electrodes consisting of triple oxides of Ti−V−W−O/Ti and Ti−V−W−O/C were also suggested. The oxide layer of the Ti−V−W−O/Ti electrodes consisted of WO_3 and $Ti_xV_yO_2$ oxides of the rutile type. The maximum capacitance of 125 F/g was also found for electrodes of this type.

Thin films of oxides and sulfides of transition metals, TiS_2, $LiMn_2O_4$, $LiNiO_2$, $LiCoO_2$ with the thickness of 5−15 μm deposited onto supports of Al or stainless steel were studied as electrodes for PsCs. The electrolyte was the solution of $LiAsF_6$ in PC. The faradaic process determining pseudocapacitance was the reaction of Li$^+$ intercalation into the matrix of a transition metal oxide. The capacitance of the studied PsCs was 400 μF/cm^2 for $LiNiO_2$ and $LiCoO_2$ and 1500 μF/cm^2 for TiS_2.

Electrodes of Ti nitride were electrochemically stable in alkaline electrolyte in a wide temperature range (−550°C to 100°C). Their specific capacitance was 125 mF/cm^2.

Mo nitrides were suggested as substitutes for RuO_2 : Mo_2N and MoN obtained by deposition on a Ti support. Electrochemical stability of films depends on the composition of the reaction mixture and deposition temperature. It was found using the cyclic voltammetry technique that films of Mo nitrides manifest a capacitive behavior similar to that of RuO_2. Herewith, the main contribution into the measured capacitance is introduced by the redox process occurring in the surface region of nitride films:

$$Mo^{3+} + N_3^- + 2H^+ + e^- \leftrightarrow Mo^{2+} + NH_2^- \tag{28.6}$$

A double-layer capacitance exists in parallel to pseudocapacitance. An important difference between nitride films and RuO_2 consists of the fact that the range of potentials of reversible capacitance for nitride films is only ∼0.7 V, while for RuO_2, this region is ∼1.4 V. Dissolution of nitride films occurs outside this working region. This leads to a certain limitation of its application as substitutes for RuO_2.

28.2 ELECTRODES BASED ON ELECTRON-CONDUCTING POLYMERS (ECPs)

28.2.1 Main Properties of ECPs as Regards Their Application in PsCs

One of the achievements of electrochemistry in the past 20 years has been the development of electron-conducting polymers (ECPs). Electron conductivity of

ECPs appears in the course of its doping by counterions because of formation of delocalized π-electrons or holes and their transport under the action of electric field through the system of polyconjugated double bonds characteristic of any ECPs. ECPs include polyacetylene (Pac), polyaniline (PAni), poly(p-phenylene) (PPh), polythiophene (PT), polypyrrole (PPy), polyporphyrin (PP), and their derivatives. Figure 28.3 shows structural formulas for some ECPs used in ECSCs.

One can see that all of these have polyconjugated double bonds providing electron conductivity as a result of doping by counterions. ECPs of the PP type have been lately developed by Vorotyntsev et al. (2011).

Table 28.1 shows the values of specific conductivity for a number of ECPs used in ECSCs.

The values of specific capacitance of all ECPs in Table 28.2, despite significant differences between them, are much higher than specific ion capacitance for most conventional electrolytes, especially nonaqueous ones. Table 28.2 shows the following

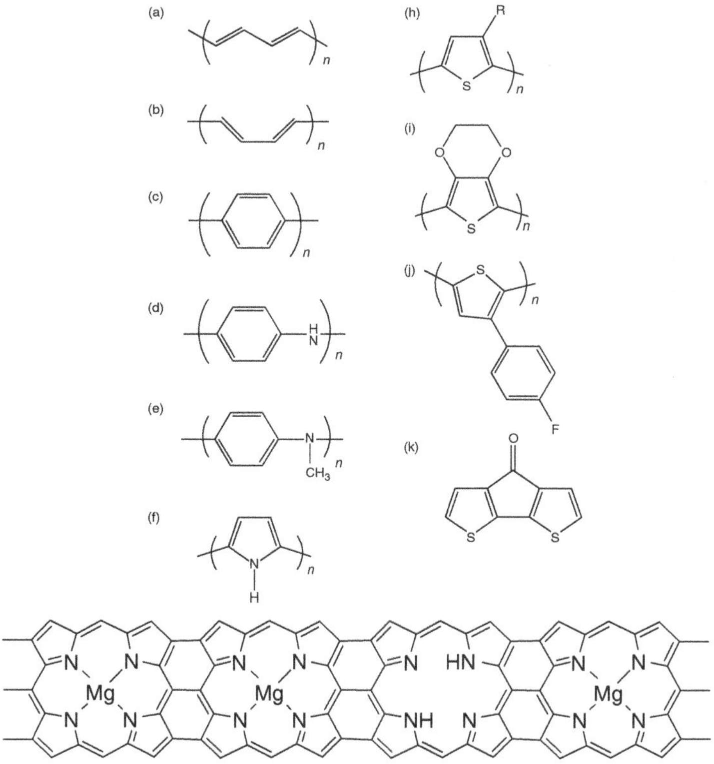

Figure 28.3. Structural formulas of several electron-conducting polymers: (a) *trans*-poly(acetylene), (b) *cis*-poly(acetylene), (c) poly(p-phenylene), (d) polyaniline (PAni), (e) poly(n-methylaniline) (PNMA), (f) polypyrrole (PPy), (g) polythiophene (PTh), (i) poly(3,4-ethylenedioxythiophene) (PEDOT), (j) poly(3-(4-fluorophenyl)thiophene) (PFPT), (k) poly(cyclopenta[2,1-b;3,4-dithiophen-4-one]) (PcDT), and (m) Mg polyporphine.

TABLE 28.1. The Conductivities of Different Conducting Polymers

Polymer	Conductivity (S/cm)
Polyaniline	0.1–5
Polypyrrole	10–50
Poly(3,4-ethylenedioxythiophene)	300 – 500
Polythiophene	300 – 400

TABLE 28.2. Characteristics of ECPs Used in ECSCs: The Molecular Weight (Mw), Dopant Level, Potential Range, Theoretical Specific Capacitance, and Measured Specific Capacitance

Conducting Polymer	Mw (g/mol)	Dopant Level	Potential Range (V)	Specific Capacitance (F/g)
PAni	93	0.5	0.7	250
PPy	67	0.33	0.8	545
PTh	84	0.33	0.8	420
PEDOT	142	0.33	1.2	95

characteristics for ECPs used in ECSCs: the molecular weight (Mw), dopant level, potential range, and specific capacitance.

The dopant level is the average amount of dopants (counterions) per unit monomer.

Figure 28.4 shows a Ragone diagram showing the regions of energy density and power density occupied by ECSCs with ECPs as compared to other rechargeable electrochemical devices (Snook GA, 2011). As we see in the case of ECSCs with ECPs, this region occupies an intermediate position between EDLCs and lithium ion batteries (LIBs). It is higher as regards the energy density values as compared to EDLCs and lower as compared to LIBs; however, ECSCs with ECPs have much higher power density values as compared to LIBs.

Conducting polymers are attractive materials for application in ECSCs, as they possess high charge density and low cost as compared to relatively expensive metal oxides. It is possible to develop devices with low equivalent series resistance (ESR), high power, and high energy. PAni can manifest the charge density of 140 mAh/g that is just slightly lower than that obtained for expensive metal oxides, such as $LiCoO_2$, but is much higher than in the case of carbon EDLCs for which often less than 15 mAh/g is obtained (possibly, ~40 mAh/g for an individual electrode). An AC–AC symmetrical device can reach the power density of 3–5 kW/kg and energy density of 3–5 Wh/kg, while a conducting polymer supercapacitor based on PAni can reach only the slightly lower power density of 2 kW/kg, but double energy density (10 Wh/kg).

EDLCs are highly cyclable; they often yield > 0.5 million cycles, while conducting polymer pseudocapacitors often start degrading at several thousands of cycles

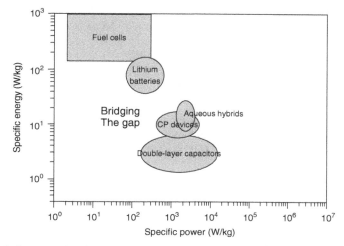

Figure 28.4. Ragone plots for different types of energy-storage devices, where CP is a conducting polymer.

because of a change in their physical structure resulting from the doping/undoping (intercalation/deintercalation) of ions.

A higher energy density can be reached by an increase in the doping degree and therefore the degree of counterion intercalation into the polymer bulk, which, in its turn, cases swelling. This gradually results in the course of cycling in mechanical failures and a decrease in voltage.

Application of a number of ECPs as electrodes for PsCs is based on the high degree of reversibility of redox reactions of their electrochemical doping–undoping. Chen and Wen (2003) measured a cyclic voltammograms for an electrode based on PAni supported on carbon. It was shown that this curve is sufficiently reversible, which is characteristic for electrode processes in supercapacitors. High reversibility of the charging–discharge processes is largely because of their high specific surface area in the electrolyte-swollen state, hundreds of square meter per gram, which was shown using the method of standard contact porosimetry (MSCP). In this state, solvation of ECP counterions occurs resulting in the formation of a large amount of micro- and mesopores. Therefore, solid-state diffusion of counterions in the polymer phase increases drastically and therefore the rate of the doping–undoping processes grows.

The process of reversible electrochemical doping–undoping can be represented by the following reactions:

$$(P)_m - xe + xA^- \leftrightarrow (P)_{x+m}(A^-)_x \tag{28.7}$$

$$(P)_m + ye + yM^+ \leftrightarrow (P)_{y-m}(M^+)_y \tag{28.8}$$

where $(P)_m$ is a polymer with a system of conjugated double bonds, m is the polymerization degree, A^- are anions, and M^+ are cations. Here, the doping reactions are left to right and those in the reverse direction are undoping reactions. Reaction (28.7) is the reaction of oxidative p-doping and reaction (28.8) is the reaction of reductive n-doping. Most ECPs can be only p-doped. However, several ECPs can be both reversibly p-doped and n-doped. These include PAc, PT, and their derivatives.

Three types of electrochemical capacitors (I, II, and III) are known with ECPs as active electrode materials. Type I corresponds to a capacitor based on a symmetrical configuration with identical active materials based on p-doped ECPs (e.g., based on PPy) on both electrodes. Figure 28.5 shows charging–discharge chronovoltammograms characterizing the behavior of ECSCs with two PPy-based electrodes (Volfkovich and Sedyuk, 2002). As we see, they have the shape of straight lines for conventional capacitors.

The type II capacitor has an asymmetrical configuration (this is a hybrid supercapacitor, HSC) with two different p-doped active materials (e.g., PPy and PT) at each of the electrodes. The third and more promising capacitor type is based on the same ECP that can undergo n- and p-type electrochemical doping. A charged type III capacitor consists of an electrode (negative) in the completely n-doped state and another electrode (positive) in the completely p-doped state. After discharge, both ECPs are in the undoped state. A capacitor of this type (III) has the following advantages as compared to capacitors of types I and II: (i) a capacitor of type III can have high voltage (about 3 V in nonaqueous solutions of electrolytes) and can reach full charging in the doping mode. (ii) Both ECPs in the charged state, while being doped, have the maximum electron conductivity. As opposed to this, ECP on one of the electrodes in capacitors of types I and II is in the undoped form and is characterized by low conductivity. Therefore, this leads to a decrease in power density as compared to the capacitor of type III. (iii) Owing to a significant separation of potential regions of n- and p-doping, the whole charge is implemented at high voltage values (in the case of PTs, in the range of $3-2$ V). Therefore, capacitors of type III have the highest energy density values close to the corresponding values for some battery types.

Only a limited number of ECPs can be electrochemically n-doped, usually at high reduction (negative) potentials and therefore the properties of p-doped ECPs

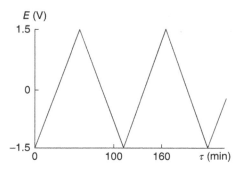

Figure 28.5. Charging–discharge chronovoltammograms characterizing the behavior of ECSCs with two polypyrrole-based electrodes. Electrolyte: 0.5 M LiClO$_4$ in PC.

are better studied: for PPy and PAni. They can be used not only in nonaqueous, but also in aqueous solutions of electrolytes. n-Doped ECPs include PAc. PAc has the working range of potentials suitable for operation both as an anode and as a cathode. Polyparaphenylene has an even wider working range of potentials and can be doped electrochemically. However, PAc and polyparaphenylene are hardly suitable as active materials for capacitors because of the regularities of the processes of their doping and very high impedance values.

High reversibility is characteristic of the processes of PT n- and p-doping and the processes of n-doping of polymers obtained on the basis of thiophene monomer derivatives in position 3 of the phenyl group and other functional groups also in position 3. Therefore, these ECPs are very promising for application in PsCs. Enhanced n-doping was also achieved at higher positive potentials for polymers obtained from bound-together thiophene dimers.

Electrochemical properties of n- and p-doped ECPs based on various PT derivatives used in PsCs of type III are shown in Figure 28.6 (Volfkovich and Sedyuk, 2002). One can see that all the three CV curves have two closed loops each; their peaks are spaced apart by the potentials of 2.5–3 V. One of these loops characterizes the reaction (28.7) of oxidative p-doping and the other one corresponds to the reaction (28.8) of reductive n-doping. The highest capacitance is characteristic of a capacitor with electrodes based on poly-3-(4-fluorophenyl)thiophene.

Rather high characteristics of PsCs with ECPs have by now been obtained, primarily on the basis of various PT derivatives. One should point out the very high discharge efficiency (i.e., efficiency by capacitance) for such electrodes because of reversibility of charging–discharge processes.

Table 28.3 shows the data for the values of average specific anodic and cathodic capacitance (per unit mass of active ECPs) for various derivatives of

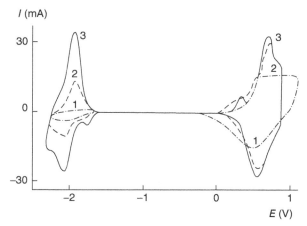

Figure 28.6. CV curves for films of (1) polythiophene, (2) poly-3-phenylthiophene, and (3) poly-3-(4-fluorophenyl)thiophene deposited onto carbon paper; electrolyte: Et_4NBF_4 in acetonitrile. The potential scan rate was 0.025 V/s.

TABLE 28.3. Values of Specific Anodic and Cathodic Capacitance (Per Unit Mass of Active ECP) and Efficiency of Capacitance of Different Derivatives of Fluorine-Substituted Polyphenylthiophene

ECP	Specific Anodic Capacitance (mAh/g)	Specific Cathodic Capacitance (mAh/g)	Average Efficiency Anodic Capacitance (%)	Average Efficiency Cathodic Capacitance (%)
Poly-3,4,5-trifluorophenyltiophene	32.1	28.4	99.53	99.90
Poly-3,5-difluorophenylthiophene	35.2	29.0	98.37	99.92
Poly-3.4-difluorophenylthiophene	21.3	28.5	99.85	99.88
Poly-2.4-difluorophenylthiophene	23.9	23.5	99.76	99.91
Poly-2-fluorophenylthiophene	22.6	26.8	99.80	99.94
Poly-3-fluorophenylthiophene	26.8	26.7	99.88	99.87
Poly-4-fluorophenylthiophene	14.8	30.6	99.88	99.87

fluorine-substituted polyphenylthiophene and the corresponding values of average efficiency of anodic and cathodic capacitance for the first 100 cycles. As follows from this table, poly-3,5-difluorophenylthiophene has the highest values of specific anodic and cathodic capacitance. The average efficiency values for anodic and cathodic capacitance for the first 100 cycles are practically 100%. The maximum voltage values for all these ECPs are close to 3 V.

The characteristics of ECPs can be significantly increased by the formation of composites between ECPs and other materials, for example, carbon (including carbon nanotubes), inorganic oxides and hydroxides, and other metal compounds. Examples of some such composites and their electrochemical characteristics are given in Table 28.4. Such electrodes can be used to manufacture symmetrical devices (type I or type III) of the same positive and negative electrodes. Composite materials

TABLE 28.4. Specific Capacitances of Composite and Treated Materials

Electrode Material	Specific Capacitance (F/g)	Electrolyte
MWNT/PAni	360	1M H_2SO_4
PAni-coated CNF	260	1M H_2SO_4
PPy-funct-SWNTs	200	Aqueous
PPy/Fe_2O_3	420	$LiClO_4$ (aq)
PPy-fast CV deposited	380	1 M KCl (aq)
PEDOT/MoO_3	300	Gel polymer electrolyte
RuO_x/PEDOT	1410	Gel polymer electrolyte
PEDOT/PPy	290	1 M KCl (aq)

CNF is carbon nanofiber, SWNT is single-walled carbon nanotube, MWNT is multiwalled carbon nanotube, and RuO_x is ruthenium oxide.

have enhanced conductivity (especially at higher negative/reduction) and better cyclability, mechanical properties, and specific capacitance and performance.

Later, we briefly characterize the main types of ECPs used in ECSCs, namely, PAni, PPy, and PT derivatives.

28.2.2   Polyaniline

PAni has been very widely studied as an electrode material for supercapacitors. The structural formula of PAni is shown in Figure 28.3d. PAni has many favorable properties for its application in ECSCs. It possesses high electroactivity and high doping degree (0.5, see Table 28.2), good stability and high specific capacitance (400–500 F/g in an acidic medium). In addition, it is characterized by good environmental stability determined by its conductivity (from 0.1 to 5 S/cm, as shown in Table 28.2) and has good processability. The main fault of PAni is that it requires a proton for charging and discharge; therefore, a proton solvent or proton ionic liquid is needed. For PAni, capacitance values were obtained in a wide range of 44–270 mAh/g. This difference is determined by many factors including the synthesis technique, polymer morphology, amount and type of the binder and additives, and the electrode thickness. PAni has a much more variable specific capacitance as compared to other ECPs. Specific capacitance of electrodeposited PAni is higher than that of chemically synthesized PAni. It was reported in one of the works that cyclability of PAni doped by lithium was more than 5000 cycles, during which the specific capacitance dropped from 100 to 70 F/g. In another work, PAni doped by $LiPF_6$ reached specific capacitance of 107 F/g that decreased to 84 F/g after 9000 cycles. The manufacturing of hybrid ECSC with PAni as a positive electrode and a carbon negative electrode is reported. It yielded 380 F/g at the constant current density of 0.5 mA/cm^2 and energy density of 18 Wh/kg and power density of 1.25 kW/kg at the current density of 20 mA/cm^2. Here, 4000 cycles were obtained. This PsC was charged between 1.25 and 1.5 V and was discharged between 1.4 and 1.0 V. PAni can be modified for higher stability forming poly(n-methyl aniline) (Fig. 28.3e). Proton-exchange sites in this polymer are occupied by methyl groups, so that this polymer is stable against degradation and is more redox active. The combination of this ECP with a negative lithium electrode can provide the capacitance of 52 mAh/g.

28.2.3   Polypyrrole

PPy (Fig. 28.3f) has a larger degree of flexibility than other ECPs and it is more suitable for the manufacturing of supercapacitor electrodes. On the other hand, PPy cannot be n-doped like thiophene derivatives and therefore it can be used only as a cathode material. Owing to its higher density, PPy has high capacitance per unit volume (400–500 F/cm^3). But this is an unfavorable aspect, as it results in a limited access of dopant ions to the internal polymer regions. This causes a decrease in capacitance per 1 g, especially for thicker coatings on electrodes. PPy is usually doped by single-charged anions, such as Cl$^-$. PPy was used in both electrodes for the manufacturing of symmetrical PsC of type I and also asymmetrical PsC of type II in combination with poly(3-methylthiophene). Type I HCC showed the discharge capacitance of $8 - 5$ mF/cm^2, which is close to HCC of type II. The voltage range for the device of type I was $0.5 - 1.0$ V and that for device of type II was up to 1.2 V. PPy is also used for the preparation of a solid-state supercapacitor with polymer electrolyte based on polyvinyl alcohol (PVA). It yields stable capacitance of up to 84 F/g for 1000 cycles and energy density of 12 Wh/kg.

28.2.4   Polythiophene Derivatives

Structural formulas of some PT derivatives are shown in Figure 28.3. PT can be both n-doped and p-doped. As follows from Table 28.3, anodic capacitance (under p-doping) of PT derivatives is higher than its cathodic capacitance (under n-doping). Therefore, the cathode in PsCs of type III must be thicker than the anode. It was found that conductivity in the n-doped form is lower than in the p-doped form; conductivity in the n-doped form is rather low. Most of PT derivatives are stable in air and in a moist state both in the p-doped and undoped forms. Symmetrical type III supercapacitors with PT derivatives on both electrodes were manufactured. Herewith, the energy density of 30–40 Wh/kg and power density of 5–10 kW/kg per mass of active materials was reached. Table 28.3 shows the characteristics of PsCs based on poly-3-(3,4-difluorophenyl)thiophene (PDFPT) and poly-3-(4-cyanophenyl)thiophene (PCPT). One can see that rather high energy density values were obtained in this case (Table 28.5).

PsC of type III based on poly-[3-p-fluorophenyl]thiophene (PFPT) has also been studied. The electrolyte used was 2 M Et$_4$NBF$_4$ in acetonitrile. The fraction of capacitance obtained at 200 mA/cm^2 in capacitance obtained at 10 mA/cm^2 was sufficiently high and was 85%. Pulsed discharge for 5 ms was used to obtain the following specific characteristics of PsC per its unit weight: energy density of 2.8 Wh/kg and power density of 2.0 kW/kg. Successful solution of the problem of a less conducting and more stable n-doped form of PT derivatives consisted of the manufacturing of asymmetrical (hybrid) ECSCs with a carbon negative electrode and p-doped polymer as a positive electrode. Thus ECSC manifests at least 10,000 cycles.

TABLE 28.5. Comparison of Characteristics of PsCs Based on Poly-3-(3,4-difluorophenyl)thiophene (PDFPT) [153] and Poly-3-(4-cyanophenyl)thiophene (PCPT)

Polymer	Cell Voltage (V)	Discharge Time (s)	Energy Density, Wh/kg of the Polymer	Power Density, kW/kg of the Polymer
Poly-3	2.7	10	26.5	8
PDFPT	2.8	10	33	11
PCPT	2.9	10	45	11

Specific capacities for several PT derivatives are presented in Table 28.3. A derivative of PT, poly(3,4-ethylenedioxythiophene) (PEDOT), is popular for application with PsC (see Fig. 28.3i). This polymer has high conductivity (see Table 28.1) and can be successfully n-doped. It has a wide potential window of 1.4 V, but its specific capacitance is not high (80–100 F/g). However, on account of all properties, it is assumed that PEDOT is very promising for application in supercapacitors. The main cause for this is its very fast kinetics, high specific surface area, and high conductivity. A fault of PEDOT is that its monomer is not soluble in aqueous electrolyte, so that it is dissolved in toxic organic solvents, such as acetonitrile used for the deposition of this polymer. The best results were obtained when using poly(3-(4-fluorophenyl)thiophene (PFPT). Laboratory hybrid ECSC with PFPT as a positive electrode and a carbon negative electrode was made. Energy density of 48 Wh/kg and power density of 9 kW/kg was obtained for it.

28.2.5 Prototypes

Until now, only devices with small geometric dimensions and mainly with electrolytes with propylene carbonate and acetonitrile solvents have been manufactured on the basis of conducting polymers. Hybrid supercapacitor batteries with poly(3-methylthiophene) (PMeT) as the positive electrode and activated carbon (AC) as the negative electrode were developed and the voltage of 3 V and more was obtained with capacitance of 1.5 kF and power of 2–3 kW. In such devices, carbon-coated aluminum current taps with resistance of 33 mΩ were used. The data on capacitance and ESR for the given device were used to obtain the values of 5 Wh/kg and 2 kW/kg. According to the estimates of the authors, optimization of the ECSC design can allow obtaining the following characteristics: 13 Wh/kg and 5 kW/kg.

An asymmetric device with an anode based on lithium titanate ($Li_4Ti_5O_{12}$ or LTO) and a PFPT cathode was also designed. LTO has excellent characteristics of stability and cycling rate. A supercapacitor device based on the LTO/C system had the following characteristics: 20 Wh/kg, 90% utilization, ability to be charged at the rate of 10 C and discharged at the rate of 100 C; and cycle life of the order of 1000 cycles. However, this system suffers from high self-discharge. The positive electrode (PFPT) has the specific capacitance of 270 F/g and specific charge of 40–45 mAh/g, while the negative electrode (LTO) has the specific charge of 60 mAh/g. One of these devices demonstrated the capacitance loss of 14% after 1500 cycles.

A new PsC wholly based on electrodes of conducting polymers (both the cathode and the anode) was suggested for energy storage. The cathode was PPy and the anode was a styryl-substituted dialkoxy terthiophene (poly(OC(10)DASTT)) on Ni/Cu-coated nonwoven polyester. The discharge performance was 94%; the specific charge was 39.1 mAh/g.

28.2.6 Composite Electrodes: Conducting Polymer/Carbon Support

Composite compounds of ECPs with highly dispersed carbon supports, especially with carbon nanotubes, demonstrate improved mechanical, electric, and electrochemical properties as compared to pure ECPs, which leads to their wide application in sensors, catalysts, and supercapacitors with high specific capacitance and power density. Three methods of manufacturing ECP–CNT composites were developed (Peng C, et al, 2008). Chemical deposition is the simplest and cheapest method of mass production. Electrochemical deposition of ECPs on CNTs results in a nonuniform structure and therefore in the actual absence of the synergistic effect between ECPs and CNTs. Electrochemical codeposition of both components leads to a homogeneous structure and facilitates transport of both ions and electrons. Therefore, this method yields the highest electrochemical capacitance as compared to pure ECPs. As a result of it, the investigation of the ECP/carbon composites obtained by chemical polymerization it was shown that the highest specific capacitance was obtained for composites with carbon supports of MWNT.

28.2.7   Comparison with Other Types of Supercapacitor Materials

Comparison of the values with specific capacitance of ECPs with other supercapacitor materials is shown in Table 28.6.

Obviously, replacement of carbon materials in ECSCs by conducting polymers and metal oxides results in an increase in specific capacitance, but herewith cyclability decreases. Another important fact is that ruthenium oxide can provide the highest specific capacitance among all other supercapacitor electrode metals. However, ruthenium oxide is too expensive to be widely used in supercapacitors. PAni is the most promising material as regards specific capacitance of conducting polymers. However, specific capacitance of PAni varies considerably depending on the method of its preparation. Besides, using PAni as a supercapacitor material is complicated by the requirements for sufficient activity of protons in the solution of electrolyte to provide sufficient cyclability. A wide range of specific capacitance values are shown in Figure 28.2 for manganese oxide, which is explained by a change in the phase composition, morphology, and crystalline form of oxide that depend on the method of synthesis. Nevertheless, this material turns out to be the most important rival of polymer as a supercapacitor material. Specific capacitance of manganese dioxide is comparable to the corresponding values in the case of PAni and, for example, for a

TABLE 28.6.   Typical Specific Capacitance for Different Supercapacitor Materials

Carbon	C (F/g)	CP	C (F/g)	CP/Composite	C (F/g)	Metal Oxides	C (F/g)	RuO_x	C (F/g)
Functionalized carbon	200	PAni	400–600	PEDOT-RuO_2	420	MnO_2	100–500	RuO_2 (sol–gel)	670–950
AC	100	PPy	200–300	PPy-Fe_2O_3	410	NiO	300	RuO_2/CB	500
		PEDOT	100	PEDOT PPy	300	SnO_2/ Fe_2O_3	80	RuO_2 (CXG)	250
				PPy/CNT	200				

CXG is carbon xerogel and CB is carbon black. Funct-C refers to functionalized carbons, AC refers to activated carbon, CNT is carbon nanotube, CPs refers to conducting polymers, and CPcomp refers to conducting polymer composites.

composite mixture of PEDOT and iron oxide. An advantage of the latter component is high capacitance and flexibility of the film, which is convenient for the preparation of devices. Other composites of conducting polymers (e.g., with carbon materials) often do not possess high specific capacitance, but have other advantages, such as increased charging and discharge rates.

28.3 REDOX CAPACITORS BASED ON ORGANIC MONOMERS

The main fault of electrodes based on ECPs is their limited chemical reproducibility and thus insufficient cyclability in aqueous electrolytes. Therefore, replacement of ECPs in PsCs by a redox couple based on organic monomers covalently bound to the carbon support of a highly dispersed carbon electrode is considered (Hashmi SA, et al., 2004). Naturally, a synergistic effect between the EDL capacitance charging and the concurrent redox (pseudocapacitive) reaction occurs in this case. The latter, alongside with EDL capacitance, contributes to the overall capacitance of such a composite electrode.

The following asymmetrical PsC capacitor was studied as a typical example of a redox capacitor. 1,2-dihydroxybenzene (DHB; also referred to as o-benzoquinone and catechol) obtained by modification of a carbon electrode was used in it as a positive electrode. The negative electrode used was an anthraquinone (AQ)-modified carbon electrode (Algharaibeh and Pickup, 2011).

A Spectracarb 2225 carbon fabric (Engineered Fibres Technologies) was used as a basis for the synthesis of an AQ-modified carbon cloth (C-AQ) electrode. This cloth was immersed into the salt of anthraquinone-1-diazonium chloride 0.5 $ZnCl_2$ in acetone. Then, water and 50 wt% hypophosphorous acid were added. Then, the thus-modified cloth was washed by deionized water and dried at 110°C for 20 min.

DHB-modified carbon cloth electrodes (C-DHBs) were manufactured by the modification of a Spectracarb 2225 carbon cloth. This cloth was immersed into 0.25 M HCl and then 4-aminocatechol and water (8 ml) were added. This was followed by the cooling in an ice bath and slow addition of 3 M $NaNO_2$. After that, the modified cloth was washed by deionized water and dried in air.

The separator used was a Nafion membrane. Each electrode was separated by a carbon paper disk (Toray™ TGP-H-090) from the Ti plate current tap to minimize contact resistance. The assembled cell was immersed into 1 M H_2SO_4(aq) to impregnate electrodes by electrolyte.

CV curves of individual electrodes were measured in the same cell with a Ag/AgCl reference electrode immersed in the H_2SO_4 (aq) electrolyte and with a different AC working electrode as a counterelectrode. Specific capacitances, energy densities, and power densities were calculated using the electrode masses, which were summed up for both electrodes for measurements in a two-electrode mode. A DHB-modified carbon cloth electrode shows an average specific capacitance of 201 F/g between 0.2 and 0.8 V, while a nonmodified ACC electrode manifests on the average 141 F/g in the same range of potentials. An increase in specific capacitance by 43% is because of reversible pseudocapacitive redox reactions at the potentials of 0.41 and 0.65 V.

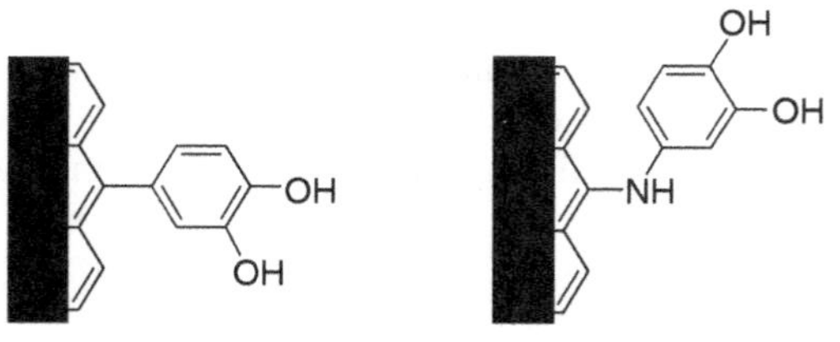

Structure 1   Structure 2

Figure 28.7. Structures 1 and 2 implemented on a positive (C-DHB) electrode.

The presence of two peaks in CV curves for DHB is a result of using 4-aminocatechol as a precursor, as it has two different ways of bonding with the carbon cloth, either through a C–C bond (Structure 1) or a C–N bond (Structure 2) (Fig. 28.7).

The negative C–AQ electrode also manifests a significant increase in capacitance to 367 F/g at the potential of about -0.05 V as compared to a nonmodified ACC electrode. Figure 28.8 and b presents dependencies of energy and power densities on current for two ECSCs: a nonmodified C–C type capacitor and modified type C-AQ/C-DHB ECSC. As we see, PsC of type C-AQ/C-DHB provides approximately a twice higher value of energy density as compared to a nonmodified carbon capacitor. However, the values of power density for both ECSC types are close and depend on their compression degree.

ACs were also modified by treatment by 2-nitro-1-naphthol to manufacture composite supercapacitor electrodes, in which EDL capacitance and pseudocapacitance of the following redox reactions of organic compounds is used: o-aminonaphthol $\leftrightarrow$ o-naphthaquinoneimine (Leitner et al., 2004):

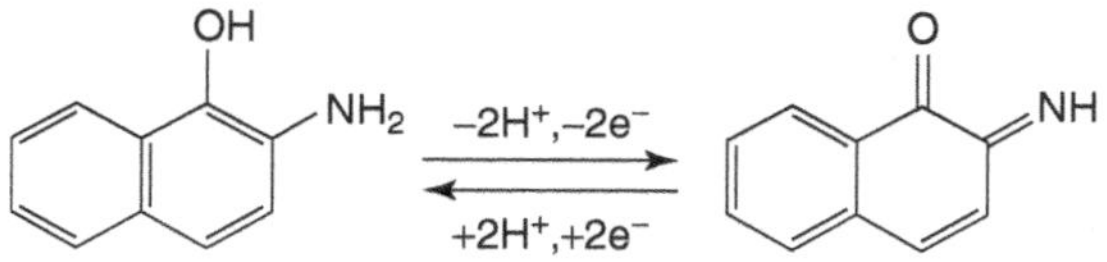

Specific charge of 35 mAh/g was obtained on such electrodes. If such an electrode can be combined with another battery-type electrode (e.g., with such a proton carrier as Mo_2N or WO_3), where the redox potential is sufficiently far apart and the kinetics are rather fast, then such a system can be used in supercapacitors. Other examples of redox-type PsCs are also mentioned in literature, for example, modified AC electrodes, in which the following variant of a redox reaction is used: o-naphthaquinone $\leftrightarrow$ o-naphthahydroquinone:

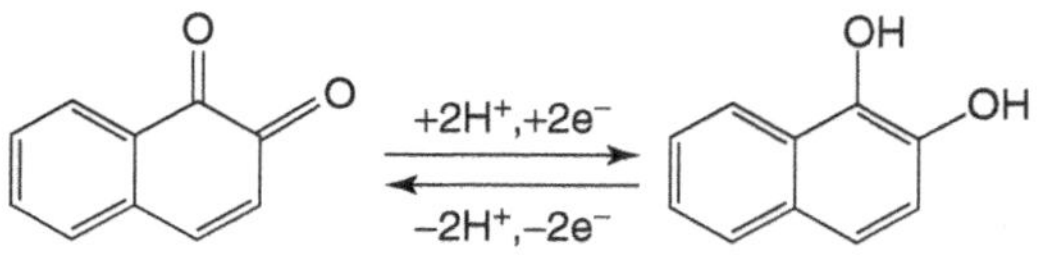

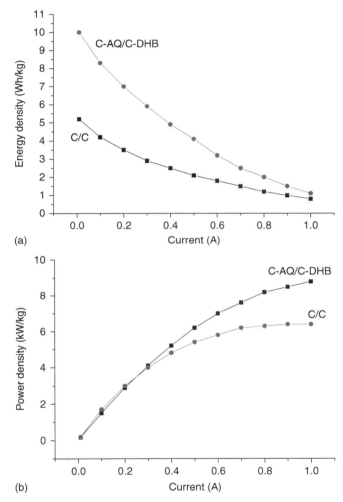

Figure 28.8. Dependencies of (a) energy densities and (b) power densities on current for two ECSCs: a nonmodified C–C type capacitor and modified type C-AQ/C-DHB ECSC.

28.4   LITHIUM-CATION-EXCHANGE CAPACITORS

In most recent years, original pseudocapacitors based on the exchange of lithium cations between the anode and cathode have been developed. Such PsCs approach LIBs by their mechanism of operation. However, there are significant differences between them. LIBs function on the basis of faradaic reactions in the bulk of the active material. This accumulation mechanism allows obtaining high energy density (160 Wh per kilogram of the weight of LIB with graphite anodes) as compared to supercapacitors. However, storage of lithium in the bulk of the material implies that lithium must pass from within an active cathode particle to within an active anode particle under charging; inverse movement of lithium occurs under discharge. This means occurrence of intercalation–deintercalation processes in both electrodes. These processes are kinetically controlled because of a very slow solid-state diffusion rate. Therefore, LIBs possess very low power density (100–1000 W per kilogram of dead weight), so that recharging requires much time.

In the lately developed supercapacitors with exchange of lithium cations between the anode and cathode, both are based on graphene with a high specific surface area and high porosity. Herewith, a large fraction of the graphene surface is in contact with liquid electrolyte thus providing fast and direct surface adsorption of lithium ions and/or surface interaction of functional groups with lithium therefore removing the need for intercalation. When the cell is assembled, lithium from the initial lithium particles or lithium foil is intercalated into the anode and it is ionized in the course of the first discharge supplying a great amount of lithium cations. The thus-generated lithium cations move through liquid electrolyte to the nanostructured cathode passing through the pores and reaching directly the surface of the internal cathode part while avoiding the necessity of solid-state intercalation. Under further discharge the flux of lithium ions is released over a larger part of the cathode surface and moves to the anodic region. Large surface area of the nanostructured anode makes possible fast deposition of a large amount of lithium ions restoring the difference of electrochemical potentials between the lithium-containing anode and the cathode. A particularly suitable nanostructured electrode material is nanographene (NGP) representing multilayer graphene plates. Materials produced by Nanoteck Instruments, Inc. and Angstron Materials, Inc., Ohio, United States, contained 85% graphenes, 5% of a conducting additive (carbon black), and 10% of a binder. It was found that these graphene materials have certain surface functional groups ($> C{=}O$ and $> COOLi$). Charging and discharge result in the following reversible redox reactions (Jang et al., 2011):

or $> C{=}O_{graphene} + Li^+ + e^- \leftrightarrow > C{-}O_{Li\,graphene}$

Therefore, these PsCs can be classified as redox capacitors. The performed studies allowed developing a new generation of energy-storage devices. These lithium ion-exchange cells with fully utilized surface can support the energy density of 160 Wh per kilogram of their proper mass, which exceeds 30-fold the indicators of EDLCs. The power density of 100 kW per kilogram of dead weight exceeds 10-fold the power density of usual EDLCs (10 kW per kilogram of dead weight) and exceeds 100-fold the power density of LIBs (1 kW per kilogram of dead weight).

While several hours (e.g., 7 h) are required for conventional LIBs with a graphite anode and a $LiFePO_4$ cathode for the completion of the necessary intercalation processes in one of the electrodes and deintercalation in the other electrode, the time of ion migration in lithium-cation-exchange capacitors is several minutes. Further studies are required for a more clear determination of the main mechanisms of energy storage in such PsCs.

REFERENCES

Algharaibeh Z, Pickup PG. Electrochem Commun 2011;13:147.

Chen W, Wen T. J Power Sources 2003;117:273.

Hashmi SA et al. J Power Sources 2004;137:145.

Jang BZ et al. Nano Lett 2011;11:3785.

Kudo T et al. Solid State Ion 2002;152–153:833.

Leitner KW et al. Electrochim Acta 2004;50:199.

Li Z et al. J Power Sources 2011;196:8160.

Peng C, Zhang S, Jewell D, Chen GZ. Prog Nat Sci 2008a;18:777.

Vorotyntsev MA et al. Electrochim Acta 2011;56:3436–3442.

Yue-feng S et al. New Carbon Mater 2007;22:53.

MONOGRAPH AND REVIEWS

Conway BE. *Electrochemical Supercapacitors*. New York: Kluwer; 1999.

Peng C et al. Prog Nat Sci 2008b;18:777.

Snook GA, Kao P, Best AS. J Power Sources 2011;196:1–12.

Volfkovich YM, Sedyuk TM. Russ Electrochem 2002;38:1043.

29

HYBRID (ASYMMETRIC) SUPERCAPACITORS (HSCs)

Electrodes of various types are used in hybrid (asymmetric) supercapacitors (HSCs). For example, one of the electrodes is highly dispersed carbon, that is, a double-layer electrode, and the other electrode is a battery one or one of the electrodes is carbon and the other one is a pseudocapacitor, for example, based on electron-conducting polymer (ECP). The main advantage of HSCs as compared EDLCs is an increase in energy density because of the wider potential window. The main fault of HSCs, meanwhile, as compared to electric double-layer capacitors (EDLCs), is a decrease in cyclability following the limitations posed by the nondouble-layer electrode.

29.1 HSCs OF MeO_x/C TYPES

One of the examples of such HSCs is the system of $^+$(NiOOH↔NiO$_2$)/KOH/C$^-$, in which the positive alkaline battery electrode is used. This HSC was developed by the Elit and Esma companies (Russia) (Beliakov and Brintsev, 1997 and Burke A, 2000). Application of ACs as the basis for the negative electrode in hybrid systems allowed achieving the double potential variation range as compared to EDLC, that is, to a system with two AC-based electrodes, and enhancing energy density by four to five times. All processes occurring in the bulk of the positive electrode and on its surface are similar to the processes occurring in alkaline batteries. Evolution of free oxygen is possible at the 70–80% discharge degree of the NiO_x electrode, which results in

Electrochemical Power Sources: Batteries, Fuel Cells, and Supercapacitors, First Edition.
Vladimir S. Bagotsky, Alexander M. Skundin, and Yurij M. Volfkovich.
© 2015 John Wiley & Sons, Inc. Published 2015 by John Wiley & Sons, Inc.

the destruction of nonstoichiometric nickel oxides ($Ni_xO_{x+0.8\ldots 1}$) under overcharging. This process is accelerated with an increase in the temperature. Therefore, NiO_x electrodes of a capacitor are not to be fully charged. Its discharge results in the protonation of the crystalline lattice of the positive electrode. In the range of high discharge currents, capacitance of a hybrid C/NiO_x capacitor is controlled by capacitance of the NiO_x electrode that depends on the electrode process rate on this electrode. The working potential range of the capacitor is 0.8–1.7 V. The main characteristics of HCs produced by Elit are: the energy density is 4–5.1 Wh/kg, the power density is 1.3–2.3 kW/kg. The energy density of this HSC is higher than that of EDLCs and lower than that of batteries. The maximum energy density of HSCs produced by Esma was 11 Wh/kg. The HSC lifecycle was determined by the positive electrode, on which the degradation processes occur. A total of 1000 charging–discharge cycles were obtained for such HSC.

A new HSC type containing a negative polarizable electrode based on activated carbon and a positive weakly polarizable electrode based on the known system of $PbSO_4/H_2SO_4/PbO_2$ has been developed in Inkar CJSC and in A.N. Frumkin Institute of Physical Chemistry and Electrochemistry (Russia) (Volfkovich and Shmatko, 2003).

The overall reaction in the capacitor is

$$PbO_2 + H_2SO_4 + (H^+)_{ad}/e \leftrightarrow PbSO_4 + 2H_2O + (HSO_4^-)_{ad}/-e \qquad (29.1)$$

Figure 29.1 shows the schematic drawing of this HSC.

The electrolyte is in the porous space of the two electrodes and the separator and is absent in the free state. The activated carbon electrodes (ACEs) used were made of ACC Viskumak with $S = 1000$ m^2/g. The maximum voltage of discharge

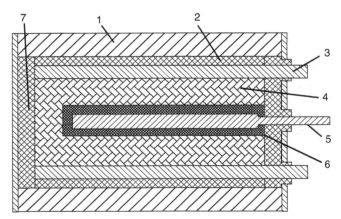

Figure 29.1. Schematic drawing of the HSC device of the C–PbO$_2$ type. (1) casing, (2) insulator, (3) current tap of the negative electrode, (4) negative electrode, (5) positive electrode, (6) porous separator, and (7) sealant.

initiation of HSC $U_{max} \sim 2.1$ V and the minimum discharge voltage is 0.7 V. The minimum potential of the carbon electrode at the start of discharge is -0.2 V, while the maximum potential at the end of discharge is 1.0 V. The specific capacitance of this cloth in the range of potentials of 0.1–0.9 V in the initial state is 140–160 F/g and it is largely determined by electric double-layer (EDL) capacitance. However, as follows from Section 27.10.1 and from Volfkovich YM, et al. (2012), the limiting specific capacitance reaches 1110 F/g under deep cathodic charging (this corresponds to 1560 C/g) because of the faradaic reaction of hydrogen intercalation into carbon. Owing to this, the used mode of operation of this HSC provides deep cathodic charging of the activate carbon cloth (ACC) electrode, which allows obtaining very high pseudocapacitance values and therefore high overall energy density of the whole HSC. The above very high U_{max} value on aqueous electrolytes is determined by the high charging potential value of the PbO$_2$/PbSO$_4$ electrode, $E_{max}^+ \sim 1.9$ V, because of the high overpotential of hydrogen evolution of lead and on lead compounds and also by the low negative cathodic charging potential $E_{min}^- \sim -0.2$ V because of the high overpotential of hydrogen evolution on carbon. Figure 29.2 shows charging–discharge cycles for HSC at $I = 10$ mA/cm^2. 1 is voltage U on HSC terminals; 2 and 3 are potentials of the positive and negative electrodes versus the hydrogen electrode in the same solution (Volfkovich and Serdyuk, 2002).

In practice, the ratio of active materials of both the electrodes was often chosen such that the capacitor should operate in the mode, in which capacitance was determined by the polarizable carbon electrode. Owing to the relative smallness of the volume of electrolyte in HSC, a change in its volume and concentration occurs in the course of charging and discharge. It was found that a 5–20% decrease in the liquid phase volume occurs in the course of charging, and a 15–50% decrease in the concentration of electrolyte is observed in the course of discharge. Herewith, an increase

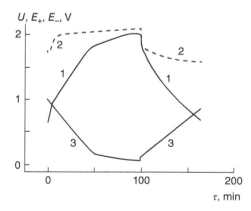

Figure 29.2. Charging–discharge curves for HSC at $I = 10$ mA/cm^2. (1) Voltage U on HSC terminals; (2) and (3) potentials of the positive and negative electrodes versus the hydrogen electrode in the same solution.

in the solid-state volume occurs because of liquid–solid-phase reactions on the $PbSO_4/PbO_2$ electrode. This is probably related to the significant change in internal resistance of HSC in the course of charging and discharge observed in practice. Resistance is at its minimum at the medium charging degrees of HSCs and it increases at a decrease and at an increase in the charging degree. An important property of HSCs is the effect of the lead electrode on the ACC electrode, owing to which capacitance of the latter increases significantly. Various analytical methods were used to establish that lead transported from the positive HSC electrode at the amount of $0.5–2$ mg/cm^2 is deposited. This is one of the causes for the abovementioned increase in ACE capacitance following its hydrophilization. Besides, in this case, leakage currents decrease considerably. Other methods of chemical and electrochemical modification of the ACC inner surface were also used, which also resulted in an increase in the capacitance of this electrode. Application of these electrodes allowed obtaining the HSC energy density values of up to $20–25$ Wh/kg, which is much higher than energy density of symmetric carbon DSCs (Chapter 27). Another important advantage of this HSC as compared to EDLC is that no HSC contains any positive carbon electrodes that are known to gradually corrode already at $E > 0.9$ V, which considerably limits the voltage window and energy density of HSCs. An important property of this HSC is that it does not deteriorate under overcharging. Oxygen is transported at high positive potentials through large pores of the separator from the PbO_2 electrode to ACE, where its electroreduction occurs and the latter catalyzes O_2 reduction. Thus, even considerable overcharging leads to no significant increase in internal pressure and therefore this HSC is manufactured as fully sealed, that is, it contains no valves for gas release. The main characteristics of HSCs are listed in Table 29.1.

HSCs are used as tractive power sources in electric buses, wheelchairs, and trolleys and in lanterns. Besides, it is planned to be used in portable devices (radio telephones, camera recorders, players), and so on. One of the advantages of HSCs as compared

TABLE 29.1. The Main Characteristics of HSCs, Developed by "Inkar" Company and A.N. Frumkin Institute of Physical Chemistry and Electrochemistry, Moscow

No	Parameter	Value of Parameter
1	Specific capacitance of carbon electrode (F/g)	600
2	Specific weight energy (Wh/kg)	20–25
3	Specific volume energy (Wh/dm^3)	60–75
4	Maximum discharge voltage (V)	2.1
5	Minimum discharge voltage (V)	0.5
6	Internal resistance (mOhm)	3–5
7	Lifetime (cycles number)	10,000
8	Charge time (min)	20–30
9	Temperature interval (°C)	$-40 \div +60$

to many batteries is the possibility of fast charging (20–30 min) instead of several hours for a lead/acid battery.

HSCs with a $RuO_2 \cdot xH_2O$ positive electrode and a negative electrode based on activated carbon were also studied. HSCs of this type have the capacitance of 770 F/g and high energy density (26.7 Wh/kg).

The recently developed HSCs of the $(+)MnO_2–C/C(-)$ type are quite promising. AC carbon supports or carbon black were used in the positive electrode and they were also used in the negative electrode as an active polarizable material (Gao P, et al., 2011). The maximum voltage of such HSCs in the aqueous electrolyte of 0.5 M Na_2SO_4 was 2 V. Herewith, the range of working potentials (vs the calomel electrode) for the negative electrode is from −1.1 to 0 V and that for the positive electrode is 0–0.9 V. The following maximum values were obtained: the energy density of 18.2 Wh/kg and power density of 10.1 kW/kg.

Higher values of energy density in HSCs were obtained in the system of $(+)MnO_2–C/C(-)$ where graphenes were used. Deng et al. (2011) carried out a comparison between an asymmetric system (HSC) of the graphene–MnO_2//graphene type, symmetric systems of graphene–MnO_2//graphene–MnO_2, and symmetric systems of graphene//grapheme.

According to Deng et al. (2011), the highest values of energy density (25 Wh/kg) and power density (3 kW/kg) were obtained for HSC of the graphene–MnO_2//graphene type. Here, 10,000 cycles were obtained.

Of interest are HSCs, in which the active anode material and material of support for the cathode is graphene and the cathode material is lithium iron phosphate $LiFePO_4$ (or some other substance intercalating lithium) that is used as the cathodic material in lithium ion batteries (LIBs). Such HSCs contain nonaqueous electrolytes, for example, 1 M $LiPF_6$/EC+DM. These HSCs have considerably higher power density values but lower energy density values as compared to LIBs.

29.2 HSCs OF ECP/C TYPE

HSCs of the polyaniline (PAni)/C type were developed, in which mesoporous carbon CMK-3 was used as the negative electrode and the positive electrode was a PAni/CMK-3 composite. An aqueous solution of 1 M H_2SO_4 was used as electrolyte. As a result, the HSC discharge capacitance of 87.4 F/g was obtained at the current density of 5 mA/cm^2 and cell voltage of 1.4 V; the HSC energy density was up to 23.8 Wh/kg and the power density was 206 W/kg. Besides, this HSC is characterized by high charging–discharge efficiency.

According to Ragone diagrams for HSC PAni/CMK-3 and, for comparison, for EDLC CMK-3/CMK-3, HSC energy density is twice higher than that of the corresponding symmetric EDLC (Cai et al., 2010). About 90% of the initial capacitance remained after 1000 cycles. A decrease in capacitance only by 10% offers fair prospects for practical application of this HSC.

REFERENCES

Beliakov AI, Brintsev AM. Development and application of combined capacitors: Double electric layer - pseudocapacity. *Proceedings of 7th International Seminar on Double Layer Capacitors and Similar Energy Storage Devices*; Deerfield Beach, Florida; 1997. V 7.

Cai JJ, Kong LB, Zhang J, Luo YC, Kang L. Chin Chem Lett 2010;21:1509.

Deng L, Zhu G, Wang J, Kang L, Liu Z-H, Yang Z, Wang Z. J Power Sources 2011;196:10782.

Gao P, Lu A-H, Li W-C. J Power Sources 2011;196:4095.

Volfkovich YM, Serdyuk TM. Russ Electrochem 2002;38:1043.

Volfkovich YM, Shmatko PA, inventors; Inkar CJSC, assignee. Electric double layer capacitor, US patent 6,628,504. 2003.

Volfkovich YM, Mikhailin AA, Bograchev DA, Sosenkin VE, Bagotsky VS. Chapter 7. *Recent Trend in Electrochemical Science and Technology*. INTECH open access publisher; 2012. www.intechopen.com. p 159.

REVIEW

Burke A. J Power Sources 2000;91:37.

30

COMPARISON OF CHARACTERISTICS OF SUPERCAPACITORS AND OTHER ELECTROCHEMICAL DEVICES. CHARACTERISTICS OF COMMERCIAL SUPERCAPACITORS

It is most convenient to use Ragone diagrams for the comparison of the main characteristics of electrochemical devices. Figure 30.1 shows a Ragone diagram for diverse commercial electrochemical devices: conventional electrolytic capacitors, electrochemical supercapacitors (ECSCs), batteries, and fuel cells (Pandolfo and Hollenkamp, 2006). One should point out that it is mainly electric double-layer capacitors (EDLCs) of all the available ECSC types that are produced at present.

The principal average characteristics of commercial capacitors, supercapacitors, and electrolytic capacitors are compared in Table 30.1.

Comparative characteristics are listed in Table 30.2 for some particular commercial ECSCs and capacitors.

As seen in Figure 30.1, Tables 30.1 and 30.2, ECSCs occupy an intermediate position between electrolytic capacitors and batteries by energy density and power density. As compared to most battery types, ECSCs have lower energy density values, but these values are close in some cases. For example, the energy density values in hybrid supercapacitors of the C/PbO_2 type are close to and, in some cases, even exceed the corresponding characteristics of lead–acid batteries (see Chapter 29).

In the case of ECSCs, the range of power density and energy density values is much wider than in other devices, which explains the wider scope of application of ECSCs. The principal advantages of ECSCs as compared to batteries are the following:

(1) High power characteristics.

Electrochemical Power Sources: Batteries, Fuel Cells, and Supercapacitors, First Edition.
Vladimir S. Bagotsky, Alexander M. Skundin, and Yurij M. Volfkovich
© 2015 John Wiley & Sons, Inc. Published 2015 by John Wiley & Sons, Inc.

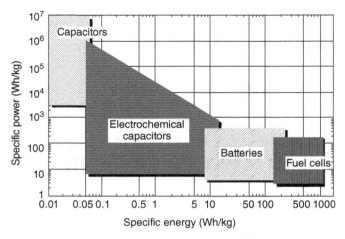

Figure 30.1. Ragone diagram for electrolytic capacitors, ECSCs, batteries, and fuel cells.

TABLE 30.1. Comparison of the Main Characteristics of Batteries, Supercapacitors, and Lectrolytic Capacitors

Parameter	Batteries	ECSC	Electrolytic Capacitors
Specific energy (Wh/kg)	10–100	1–10	<0.1
Charging time	$1 < t < 5\,h$	1–30 s	$10^{-3} < t < 10^{-6}\,s$
Discharging time	$t > 0.3\,h$	1–30 s	$10^{-3} < t < 10^{-6}\,s$
Lifetime	1000	10^6	10^6
Specific power (W/kg)	<1000	10,000	10^{-6}
Charge/discharge efficiency	0.7–0.85	0.85–0.98	>0.95

(2) High cyclability comparable to that of conventional capacitors (about hundreds of thousands and more than 1 million cycles for high-power ECSCs). At the same time, high-power ECSCs can readily give up and accumulate energy exceeding by hundreds and thousands of times the energy of various conventional capacitors with similar mass and volume.

(3) EDLCs operate reliably under extreme temperatures from −50 to +60°C, as they are not limited by the kinetics of electrode reactions, but are in fact controlled by the law of electrophysics, and not electrochemistry.

(4) Another important advantage some types of ECSCs as compared to batteries is a very high (up to 90–98%) energy efficiency value (the ratio of discharge energy to charging energy), which allows using ECSCs in various devices for energy accumulation, storage, and output in electric mains and for peak load smoothing.

(5) A very important advantage of ECSCs is that they can be charged and discharged in a very wide range of times from thousandths of a second to

TABLE 30.2. Characteristics of a Number of Supercapacitors and Pulse Batteries (Burke A., 2000)

	Voltage (V)	Capacitance (Ah)	Weight (kg)	Resistance (mΩ)	Wh/kg	P_{ef} (95% discharge) (W/kg)	P_{pi}, W/kg
Supercapacitors							
Maxwell							
2700 F	3	2.25	0.85	0.5	4.0	593	5294
1000 F	3	0.83	0.39	1.5	3.1	430	3846
Panasonic							
800 F	3	0.67	0.32	2.0	3.1	392	3505
2000 F	3	1.67	0.57	3.5	4.4	127	1128
Superfard (250 F)	50	3.4	16	20	5.4	219	1953
SAFT							
Gen2 (144 F)	3	0.12	0.030	24	6.0	350	3125
Gen3 (132 F)	3	0.11	0.025	13	6.8	775	6923
Power Stor	3	0.083	0.015	10	8.33	1680	15000
Batteries							
Panasonic NiHD	7.2	6.5	1.1	18	42	124	655
	12.0	98	17.2	8.7	68	46	240
Ovonic NiHD	13.2	88	17.0	10.6	70	46	245
	12.0	60	12.2	8.5	65	80	420
	7.2	3.1	0.522	60	43	79	414
Sanyo Li ion	3.6	1.3	0.039	150	121	105	553
Hawker Pb acid	2.1	36	2.67	0.83	27.0	95	498
	12	13	4.89	15	29.0	93	490
Optima Pb acid	6	15	3.2	4.4	28	121	635
Horizon Pb acid	2.1	85	3.63	0.5	46	115	607

hours, that is, in the range of seven orders of magnitude (see Fig. 27.4). This extends dramatically their scope of application, including the possibility of their application for smoothing peak loads in electric mains.

(6) Many types of ECSCs have yet another advantage, that is, their environmental safety. The fact is that, for example, billions of lithium batteries are thrown to garbage dumps or buried in the earth and pollute the environment after their lifecycle is over. Thus, such toxic elements as lithium, fluorine, sulfur, nickel, and so on are released. As opposed to this, the widespread types of ECSCs with carbon electrodes and aqueous electrolytes are quite environment-friendly, that is, practically safe.

(7) All ECSC types are completely leakproof and therefore lossless.

(8) Many ECSC types, especially those based on carbons, are much cheaper as compared to batteries; besides, not only precious metals, but also nonferrous metals are not used in these ECSCs, as opposed to batteries and fuel cells.

TABLE 30.3. Characteristics of Best-Developed and Commercial Supercapacitors (Burke A, Miller MJ, 2011)

Company	Country	Characteristics	Energy Density (Wh/kg)	Power Density (Wh/kg)	Status
Carbon particulate composites					
Panasonic	Japan	3 V, 800–2000 F	3–4	200–400	Commercial
EPCOS	Germany	2.7 V, 3400 F	4.3	760	Commercial
NESS	United States	2.7 V, 4640 F	4.2	930	Commercial
SAFT/Alcatel	France/United States	3 V, 130 F	3	500	Packaged prototypes
Cap xx	Australia	3 V, 120 F	6	300	Packaged prototypes
NEC	Japan	5–11 V, 1–2 F	0.5	5–10	Commercial
Elit	Russia	450 V, 0.5 F	1.0	100,000 cycles	Commercial
BatScap	France	2.7 V, 2700 V	4.2	18,230	Commercial
Carbon fiber composites					
Maxwell	United States	3 V, 1000–2700 F	3–5	400–600	Commercial
Superfarad	Sweden/Ukraine	40 V, 250 F	5	200–300	Packaged prototypes
Power Systems	United States	2.7 V, 1800 F	8.0	4320	Commercial

Aerogel carbon					
PowerStor	United States	3 V, 7.50 F	0.4	250	Commercial
Conducting polymer					
Los Alamos National	United States	2.8 V, 0.8 F	1.2	2000	Laboratory prototype
Mixed metal oxides					
Pinnacle Research Institute	United States	15 V, 125 F, 100 V, 1 F	0.5–0.6	200	Packaged prototypes
US Army, Fort Monmouth	United States	5 V, 1 F	1.5	4000	Unpacked lab prototype
Hybrid					
Evans	United States	28 V, 0.02 F	0.1	30,000	Packaged prototype
ESMA	Russia	1.7 V cells/17 V modules /20 Ah (50,000 F)	8–10	80–100 (95% efficiency) cycle life 10,000–20,000)	Commercial
Fuji Heavy Industry	Japan	3.8 V, 2000 F	12.1	9,223	Commercial

Table 30.3 contains the characteristics of the best-developed commercial supercapacitors.

As follows from these tables, symmetrical double-layer supercapacitors based on C/C-type carbon electrodes and, in limited quantities, some hybrid C/NiOOH-type ECSCs are produced at present. In the case of pseudocapacitors based on electron-conducting polymers, there are as yet only laboratory prototypes.

REFERENCE

Burke A, Miller M. J Power Sources 2011;196:514.

REVIEWS

Burke A. J Power Sources 2000;91:37.
Pandolfo AG, Hollenkamp AF. J Power Sources 2006;157:11.

31

PROSPECTS OF ELECTROCHEMICAL SUPERCAPACITORS

Although the size of the world market of supercapacitors is at present only $0.4 billion, it will reach $3.8 billion by 2015 as the analysis of NanoMarkets/Smart Grid shows.

There are several directions in which electrolytic capacitors and supercapacitors (ECSCs) will be developed in the nearest future and in midterm according to our forecasts.

1. At present, single-wall carbon nanotubes (SWCNTs) possess the highest kinetic parameters of all the numerous types of carbon electrode materials. Pulse ECSCs with very high power density approaching 100 kW/kg at sufficiently high energy density of 20–25 Wh/kg were obtained in laboratories on their basis. But a significant fault of SWCNTs is their high price. Therefore, one can assume that a large-scale, rather cheap technology of SWCNT production will be developed in midterm that will result in the commercial production of SWCNT-based ECSCs.

2. It is expected that the technology of graphene synthesis will be improved and their porous hierarchical structure will be optimized so as to considerably increase the carbon/electrolyte interface surface area while retaining sufficient electrode strength. Graphenes are characterized by fast kinetics of charge–discharge processes as opposed to the majority of activated carbons

Electrochemical Power Sources: Batteries, Fuel Cells, and Supercapacitors, First Edition.
Vladimir S. Bagotsky, Alexander M. Skundin, and Yurij M. Volfkovich.
© 2015 John Wiley & Sons, Inc. Published 2015 by John Wiley & Sons, Inc.

and carbon blacks so that they are to be used primarily in power (pulse) type ECSCs. Graphenes approach SWCNTs by their kinetic capacitive properties but are much cheaper. This improvement of the technology of graphene synthesis is to be expected in the short term before a sufficient cheap technology of SWCNT synthesis is developed in midterm (see Section 27.10.5).

3. In the nearest future, one should expect wide commercialization of hybrid supercapacitors of the C/PbO_2, C/NiO_2, C/MnO_2, C/ECP, and other supercapacitor types with elevated energy density of the order of 40–50 Wh/kg, that is, with the energy density not lower than that of conventional batteries, but with a much higher cyclability.

4. A very promising ECSC type is lithium cation–exchange pseudocapacitors (see Section 28.3.4) with graphene-based electrodes. Electrodes exchange lithium cations. By their operation mechanism, these PsCs approach lithium ion batteries (LIBs), but no lithium intercalation–deintercalation processes are used in them that would limit the rate of charging–discharge processes and therefore power density of LIBs , but fast processes of adsorption/desorption processes surface lithium ions contained in functional groups are applied and therefore the necessity of intercalation is eliminated. A new generation of energy-storage devices has been developed as a result of the laboratory studies performed. These PsCs can maintain the energy density of 160 Wh/kg of the mass of active materials, which exceeds by dozens of times the parameters of conventional double-layer capacitors. Power density of 100 kW per kg of their weight is tenfold that of conventional EDLCs. In the measurable future, one can expect on the basis of the data of studies development of commercial ECSCs with the energy density of 70–90 Wh/kg and power density of 60 kW/kg. Thus, these PsCs approach LIBs by their energy characteristics, but possess much higher power density and cyclability.

5. Because of the functioning of ECSC electrodes in a very wide range of charging–discharge times (from fractions of a second to hours), dramatic expansion of the scope of application of diverse ECSCs is contemplated for different types of consumers and devices from high-frequency electronic circuitry to electric transport and other applications of the energy-type ECSCs. Among the new fields of ECSC applications, one should particularly point out energy storage and output for electric mains for the smoothing of peak loads with different charging and discharge times, recovery of the braking energy of internal combustion engines (ICEs) (locomotives, cars, airplanes, ships), power backup in railway transport, underground and other types of public transport, large buffer power storage units in electric power industry, telecommunications systems, uninterrupted power supply of critical objects, alternative electric power industry, replacement of electric batteries in electric power substations, supply of peak power demand, efficient renewable energy storage, uninterrupted power supply in the case of accidents. Significant consumption potential of buffer storage systems is preserved in the case of wind power and solar energy engineering.

One should particularly point out the promising applications of ECSCs in such an economically important industry as the smoothing of peak loads in electric mains, as ECSCs are practically beyond competition in this field as compared to batteries owing to their higher energy efficiency, higher cyclability, wider range of charging–discharge times, and lower cost.

32

ELECTROCHEMICAL ASPECTS OF SOLAR ENERGY CONVERSION

32.1 PHOTOELECTROCHEMICAL PHENOMENA

The effects of luminous (or other electromagnetic) radiation on the properties of electrodes and on electrochemical reactions are the subject of photoelectrochemistry.

Luminous radiation (light) can produce changes in the open-circuit potentials and in the polarization characteristics of electrodes; at constant potential the current may change (anodic or cathodic photocurrents appear), while at constant current, the electrode potential may change (photopotentials appear). It is an important special feature that electrochemical reactions may become possible that at the same potentials in the dark are thermodynamically prohibited, that is, associated with an increase in Gibbs energy; under illumination, such reactions are possible because of the energy supplied from outside.

The first observations of photoelectrochemical phenomena were made in 1839 by Antoine Becquerel (1788–1879). He used symmetric galvanic cells consisting of two identical metal electrodes in a dilute acid. When illuminating one of the electrodes he observed current flow in the closed electric circuit (Becquerel, 1839).

Studies of photoelectrochemical phenomena are of great theoretical value. With light as an additional energy factor, in particular, studies of the elementary act of electrochemical reactions are expedited. Photoelectrochemical phenomena are of a great practical value as well. One of the most important research activities nowadays is the development of electrochemical devices for a direct conversion of luminous (solar)

Electrochemical Power Sources: Batteries, Fuel Cells, and Supercapacitors, First Edition.
Vladimir S. Bagotsky, Alexander M. Skundin, and Yurij M. Volfkovich
© 2015 John Wiley & Sons, Inc. Published 2015 by John Wiley & Sons, Inc.

into electrical energy and for the photoelectrochemical production of hydrogen. Other applications of photoelectrochemistry are the photosynthesis of organic compounds and the photoreduction of carbon dioxide CO_2.

In photoelectrochemistry the incident radiation is characterized by the energy of the photons (light quanta), which is given by hv (where h is the Planck constant and v is the frequency of the light wave), and by the light intensity or photon flux striking the object. For visible light the photon energies are between 1.63 eV (for the red limit, with a wavelength of $\lambda = 760$ nm) and 3.1 eV (for the violet limit, with $\lambda = 400$ nm). The energies are higher in the ultraviolet, for example, 6.2 eV at $\lambda = 200$ nm. Electrochemical systems are less strongly influenced by photons having lower energies (in the infrared part of the spectrum).

32.2 PHOTOELECTROCHEMICAL DEVICES

Photoelectrochemical phenomena can be used for the design of practical devices in which luminous (solar) energy is directly converted to electrical or chemical energy (evolution of hydrogen used in fuel cells or other devices).

Galvanic cells in which the anodic reaction occurring at an illuminated electrode is fully compensated by the reverse (cathodic) reaction occurring at a dark electrode, that is, there is no overall current-producing reaction, and no overall chemical change whatever occurs in the system. Cells of this type, where the electrodes are operated under different conditions but where the same electrode reaction occurs in different directions at the two electrodes, are called regenerative.

Devices for the production of chemical energy differ from the ones described, in that different electrode reactions occur at the cathode and anode. As a result of the overall current-producing and current-consuming reactions, energy-rich products are generated (photoelectrolysis); their chemical energy can be utilized. Of greatest interest is the photoelectrolytic production of hydrogen. The first cells of this type were built in 1972 by Akira Fujishima and Kenichi Honda in Japan. Their photoanode was titanium dioxide TiO_2 (an n-type semiconductor) and their cathode was platinized platinum on which hydrogen was evolved. A porous separator was used in the cell in order to keep the anodic and cathodic reaction products apart.

32.3 PHOTOEXCITATION OF METALS (ELECTRON PHOTOEMISSION INTO SOLUTIONS)

Events of electron photoemission from a metal into an aqueous solution was first documented in 1963 by Geoffrey C. Barker and Arthur W. Gardner on the basis of indirect experimental evidence. The formation of solvated electrons in nonaqueous solutions (e.g., following the dissolution of metallic sodium in liquid ammonia) had long been known, but it was only in the beginning of the 1950s that their existence in aqueous solutions was first thought possible. It is probably for this reason that even nowadays in aqueous solutions we more often find the term *solvated* than *hydrated* electrons.

When photons are absorbed in a metal the ensemble of electrons are excited and some of the electrons are promoted to higher energy levels. The excited state is preserved in the metal for only a short time, and the system rapidly returns to its original state. When the photon energy hv is higher than the metal's electron work function, $\lambda^{(M,E)}$, in the solution at a given potential, individual excited electrons can be emitted into the solution. Usually the quantum yields of this process are low, for example, around 10^{-4}, and depend on the depth of the layer in which the photons are absorbed.

Photoemission produces "dry" (nonsolvated) electrons leaving the metal surface with a high initial velocity. An excess photon energy of 0.05 eV is enough to impart a velocity of about 10^5 m/s to the electron. Already in the first layers of solution the electrons are strongly decelerated and "thermalized," that is, their kinetic energy is reduced to values typical for thermal motion at the given temperature.

Several picoseconds after photon adsorption, the electrons end up in a localized "solvated" state at distances of $1-5$ nm from the electrode surface. The electron's hydration energy in aqueous solutions is about $1.5-1.6$ eV.

The further fate of the solvated electrons depends on solution composition. When the solution contains no substances with which the solvated electrons could react quickly, they diffuse back and are recaptured by the electrode. A steady state is attained after about 1 ns; at this time the rate of oxidation has become equal to the rate of emission.

32.4   BEHAVIOR OF ILLUMINATED SEMICONDUCTORS

Semiconductor electrodes exhibit electron photoemission into the solution, like metal electrodes, but in addition they show further photoelectrochemical effects following excitation of the electrode under illumination. In 1955, W.H. Brattain and Ch.G.B. Garrett published a paper in which they established the connection between the photoelectrochemical properties of single-crystal semiconductors and their electronic structure.

The difference between metals and semiconductors becomes apparent when the further fate of these excited charges is considered. In metals an excited electron will very quickly (within a time of the order of 10^{-14} s) return to its original level, and the photon's original energy is converted to thermal energy. Photoexcitation of metals has no other consequences.

In semiconductors, which have a band gap, recombination of the excited carriers—return of the electrons from the conduction band to vacancies ("holes" in the valence band)—is greatly delayed, and the lifetime of the excited state is much longer than in metals.

Moreover, in n-type semiconductors excess electrons in the conduction band will be driven away from the surface into the semiconductor by the electrostatic field, while positive holes in the valence band will be pushed against the surface. The electrons and holes in the pairs produced are thus separated in space. This leads to an additional stabilization of the excited state, to the creation of some steady concentration of excess electrons in the conduction band inside the semiconductor, and to the creation of excess holes in the valence band at the semiconductor surface.

Excitation of semiconductors is now widely used for the development and manufacturing of devices for the direct conversion of solar energy to electric energy (solar cells). The field of direct conversion of the solar radiation energy to electric energy is called *photogalvanics* and the corresponding devices *photogalvanic devices*.

Two types of solar cells are discussed in the following sections: (i) semiconductor solar batteries (SC-SB) (Section 32.5) and (ii) dye-sensitized solar cells (DSSC) (Section 32.6).

32.5 SEMICONDUCTOR SOLAR BATTERIES (SC-SB)[*]

For semiconductor solar batteries mostly (but not always) highly purified silicon is used. The purification of silicon and its further processing are expensive highly energy-consuming processes and are also connected with environmental problems. This leads to high costs of solar batteries. The high costs prevented for many years a widespread application of semiconductor solar batteries. Partially these costs are compensated by the electric energy produced by the solar battery. Silicon can be used for the conversion of visible sunlight (up the violet part of the spectrum), and ultraviolet (UV) radiation leads only to heat evolution and an increase in the battery temperature.

32.5.1 Mechanism of Energy Conversion

For preparing a single solar cell a highly purified silicon wafer is doped by adding corresponding additives. One side is doped as an *n*-type semiconductor, and the other side is doped as *p*-type conductor. Thus, inside the cell a *p*–*n*-junction is created. The *n*-type side is directed toward the sun. When a photon is adsorbed by silicon its energy is given to an electron in the valence band that is tightly bound between neighboring atoms. If this energy is high enough to overcome the silicon band gap, the photon pushes the electron from the lower energy valence band into the higher energy conduction band, where it is free to move around the lattice (photoexcitation of the semiconductor). The missing electron in the valence band is equivalent to a hole (electron-hole), which also can move around the lattice. Thus, the photon creates a mobile electron and hole pair.

Inside the silicon semiconductor, electrons diffuse from the *n*-type side (with a high electron concentration) in the direction of the *p*-type side. Holes diffuse from the p-side in the opposite direction. This results in a region where all charge carriers are depleted (space charge region). At the *p*–*n* junction both electrons and holes accumulate: electrons on one side of the junction, and holes at the other side. Thus, a double layer of electric charges is formed, which leads to a potential difference of about 0.6–0.7 V. This is the solar cell's open-circuit voltage (OCV). When both sides of the

---

[*]Semiconductor solar cells are low voltage power sources (<1 V) and therefore are used in series and parallel connected batteries. For this reason, the term *semiconductor solar battery* is more appropriate than the often used term *semiconductor solar cell*.

semiconductor are connected by an external load, electrons move to the *p*-type side, thus forming a continuous electric current through the load. As in electrochemical batteries, the cell voltage decreases with increasing discharge current. The maximal discharge power is reached at a voltage of about 0.5 V.

32.5.2 Efficiency of Solar Energy Conversion

The maximal theoretical efficiency of solar energy conversion (for normal intensity of solar radiation) in silicon solar batteries with one *p*-*n*-junction is about 30% (Shockley–Queisser limit, described in 1961). For multijunction tandem solar batteries and for batteries with an optical light condenser the theoretical limit is higher. The practical efficiency is lower and depends on several factors, including the crystalline modification of silicon and the thickness of the semiconductor, where photons are adsorbed. For high thickness silicon batteries the efficiency reaches about 20%. For low thickness batteries made from amorphous silicon it is 5–10%. The maximum practical conversion efficiency silicon has is at a temperature of about 25°C. With rising temperatures the efficiency diminishes. The battery is often covered by a thin layer of silicium nitride that reflects UV light and prevents a temperature rise (antireflection layer).

32.5.3 Development of Semiconductor Solar Batteries

The first silicon solar battery was developed in 1954 at the American company Bell. Due to its high production costs, which could not compete with production costs for electric energy produced in conventional thermal power plants, this new device at first drew little attention from the scientific community. In 1954 began the era of artificial earth satelites. The first satelites were equipped with electrochemical batteries, which allowed only for a limited operational time. Soon it was realized that semiconductor solar batteries are the only alternative for the power supply of satelites (and later spaceships) with an extended operational lifetime, and extended research and development (R&D) work was started in this field. In 1958 the first sattelite with a silicon solar battery Vanguard 1 was launched. The conversion efficiency of its solar battery was 10%. The battery remained operable for about 8 years.

32.5.4 Semiconductor Solar Batteries with Nonsilicon Materials

Other materials that can be used for semiconductor solar batteries are gallium arsenide GaS, cadmium telluride TeCd, and copper-indium selenide CuInSe.

Gallium arsenide solar cells have an efficiency up to 40% (higher than that of the theoretical efficiency limit of silicon solar cells). In contrast to their silicon counterparts they are temperature stable. Because of their very high production costs they are used only for space applications.

32.5.5 The Use of Solar Batteries for Production of Grid Energy

As a result of the huge amount of R&D work on semiconductor solar batteries' cells for space applications, the production of these batteries was improved and simplified, and steadily their costs went down and they became competitive with other, conventional means of power production. At the current prices the cost of a stationary solar battery installation can be recapturated by the electric energy it produces within 2–5 years. The oil crisis in the 1970s initiated attempts to find alternative energy sources, including the energy of solar radiation. Gradually solar batteries began to be used in many countries for large-scale production of electric grid power. As an example the Charaka Solar Park 600 MW power system occupying a 2000 ha site in the district of Patan (India) can be mentioned. Now the electric power of all stationary solar batteries worldwide is over 60,000 MW and their annual production of electric energy is estimated as approaching a trillion kilowatt-hours.

32.6 DYE-SENSITIZED SOLAR CELLS (DSSC)

In contrast to semiconductor solar batteries (which contain no electrolytes), DSSC are galvanic cells, containing two electrodes and an ion-conducting electrolyte. As material for one of the electrodes titanium dioxide TiO_2 is used. This material is an n-type semiconductor and in comparison with highly purified silicon is much cheaper. In particular it is used as pigment in white dye-ware.

32.6.1 First DSSC Prototypes and Working Principle

DSSCs were developed in 1991 in the group of Dr. Michael Graetzel at the École Polytechnique Fédérale in Lausanne (Switzerland). In 1992 they were patented in the United States. For this discovery, M. Graetzel was awarded with the prestigious 2010 Millennium Technology Grand Prize in 2010.

In the first DSSC prototypes the two electrodes were mounted on transparent glass supports. The supports were covered with a thin (also transparent) layer of fluorine-doped tin dioxide SnO_2 (FTO). This layer has a good electron conductivity. At one of the electrodes (the photoanode) a comparatively thick porous layer of nanocrystalline TiO_2 particles was deposited above the FTO layer. The semiconductor TiO_2 can be excited only by UV radiation, not by visible light. For this reason dyes were deposited on the TiO_2 layer, where they are held by covalent bonds. The dyes adsorb (according to their color) photons from the visible sun spectrum. After adsorbing a photon the dye becomes excited, generates an electron, and transfers it to the conducting level of TiO_2, thus sensitizing the semiconductor and leaving behind a dye cation.

The second electrode (cathode) often called *counter electrode* also has an electron-conducting sublayer of FTO on which a thin layer of platinum was deposited. The electrolyte was a solution of iodine in a potassium iodide KI solution, forming I_3^- ions, and thus, an I^-/I_3^- redox couple.

After the photoexciting of TiO_2, between the two electrodes, an OCV of 0.7 V is established. When the FTO layers of both electrodes are connected by an external load, electrons flow from the anode's TiO_2 layer to the cathode. At this electrode the electrons reduce I_3^- ions to iodide ions I^-. These ions diffuse to the anode where they are reoxidized to I_3^-, thus replenishing the amount of the I_3^- ions. The net result of this shuttle action is the transport of electrons trough the electrolyte layer from the cathode to the anode, thus closing the electric circuit for a continuous electric current, initiated by the solar radiation. The energy conversion efficiency of the first DSSC prototypes was very low—about 2–3%. Another important parameter for solar cells is the maximal current they can generate per unit surface area in a circuit with a zero resistance external load, called *short circuit current*. For DSSCs this parameter is about $20\,mA/cm^2$ (a value close to that for silicon semiconductor solar batteries).

Advantages of DSSCs

This new type of solar cells has many advantages: in comparison with semiconductor solar batteries, they are much less expensive as their silicon counterparts, since only cheap materials are used. Their production process is relatively simple and is environmental friendly. The main disadvantage of the first DSSC prototypes was the use of a liquid electrolyte solution that required a thorough hermetization of the cell. Moreover the liquid state of the electrolyte made the cell sensitive to temperature changes. At very low temperatures the electrode froze and ice crystals could destroy other parts of the cell. At high temperatures the expanding electrolyte could weaken the cell's hermetization.

32.6.2 Further Improvements of DSSCs

The advantages of DSSCs immediately drew the attention of the scientific community. After the first publications of this invention (1991/1992), in many countries extended R&D work in this field was started. These efforts continue up to the present time (2012). Every year hundreds of papers connected with improvements of DSSCs are published in different scientific journals. These investigations are devoted to different aspects of DSSC design and operation.

Replacement of the Liquid Electrolyte

Several methods for a solidification of the electrolyte were proposed, such as addition of gel-forming agents or a large volume of nanosized inert particles. These methods gave fairly good results and it became possible to produce "dry versions" of DSSCs.

Replacement of Noble Metals

In the first DSSC prototypes the catalyst for the process of I_3^- reduction was a thin platinum layer. In these prototypes dyes based on organic ruthenium complexes were used. Ruthenium compounds are not only expensive but also the world resources of ruthenium are very limited. The problem of replacing the platinum catalyst was not easy to solve. As a result of many investigations it was shown that combined catalysts containing some carbonaceous materials (e.g., carbon nanotubes) had an activity and stability comparable to those of platinum catalysts. Gradually, new dyes with acceptable properties were also found.

Flexible DSSC Varieties

The replacement of the electrode's glass supports by transparent plastic materials led to the possibility to produce flexible DSSC varieties that in many cases were more convenient for practical applications. Plastic supports (in comparison to brittle glass supports) resulted in an increased stability from mechanical stresses.

Increase of the Conversion Efficiency

As already mentioned, the efficiency for converting light radiation energy to electric energy of the first DSSC prototypes in 1991 was very low (2–3%). For this reason, a great amount of efforts was directed to increase this very important operational parameter. This was a very painstaking job since the conversion efficiency depends on a multitude of factors:

(a) the nature of the dye and the thickness of the dye layer;
(b) the nature of the semiconductor (TiO_2 can be replaced by other materials with similar properties, for example, zinc oxide ZnO);
(c) the crystalline and geometric structure of the semiconductor layer (which can be influenced by different additives to the layer);
(d) the activity of the cathode's catalytic layer;
(e) the nature of the electrolyte's redox couple, and so on.

Therefore the rise of conversion energy was comparatively slow. In 2006 on a small area cell ($0.2\ cm^2$) an efficiency of 11% was achieved by (Chiba et al., 2006).

32.6.3 Possible Applications for DSSCs

The first laboratory-made DSSC prototypes were small sized and apart from low efficiency had many problems including those of stability. For this reason the extended production of these cells was delayed approximately for a decade. Only at the beginning of the twenty-first century improved versions of DSSCs with an efficiency of about 10% became commercially available.

DSSCs are competitive with thin-film silicon solar batteries, which have an efficiency 12–15%, but cannot be used in applications, requiring higher values of efficiency (e.g., in spacecrafts).

DSSC with their low price to power ratio can be successfully used in large-scale terrestrial installations for production of electric grid energy.

Following particulars of their conversion mechanism DSSCs work not only under direct solar radiation, but also from diffuse low-intensity light even in rooms with artificial lighting and without daylight. This is in contrast to silicon solar batteries, which are almost useless in cloudy days at a lower limit of radiation.

Thus, in addition to cheap large units for the production of electric grid power, DSSCs can find large in-door applications for powering different electronic equipments like calculators, external memory devices, and liquid crystal displays, and so on. This power source would also be convenient for grid-independent charging units, used in many households for recharging storage batteries.

REFERENCES

Barker C, Gardner A. J Electrochem Soc 1966;113:1182.

Becquerel A. Compt Rend Acad Sci 1839;6:145.

Brattain WH, Garrett CGB. J Bell Syst Techn 1955;34:129.

Chiba Y, Islam A, Watanabe Y, et al. J Appl Phys 2006;45:638.

Fujishima A, Honda K. Nature 1972;238:37.

Graetzel M, Liska P. US patent 508,4365. 1992.

Shockley W, Queisser HJ. J Appl Phys 1961;32:510.

REVIEWS AND MONOGRAPHS

Brabec C. *Organic Photovoltaics – Materials, Device Physics, and Manufacturing Technologies.* Wiley-VCH Verlag GmbH: Weinheim; 2008.

Kalyanasundaram K, editor. *Dye-Sensitized Solar Cells.* Boca Raton, FL: CRC Press; 2010.

Luque A, Hegedus S. *Handbook of Photovoltaic Science and Technology.* Hoboken, NJ: John Wiley & Sons, Inc.; 2003.

Markvart T, Castañer L. *Solar Cells – Materials, Manufacture and Operation.* Oxford: Elsevier; 2006.

Pleskov YV. *Solar Energy Conversion: a Photoelectrochemical Approach.* Berlin: Springer; 1990.

AUTHOR INDEX

Electrochemical Power Sources: Batteries, Fuel Cells, and Supercapacitors, First Edition.
Vladimir S. Bagotsky, Alexander M. Skundin, and Yurij M. Volfkovich
© 2015 John Wiley & Sons, Inc. Published 2015 by John Wiley & Sons, Inc.

SUBJECT INDEX

Electrochemical Power Sources: Batteries, Fuel Cells, and Supercapacitors, First Edition.
Vladimir S. Bagotsky, Alexander M. Skundin, and Yurij M. Volfkovich
© 2015 John Wiley & Sons, Inc. Published 2015 by John Wiley & Sons, Inc.

THE ELECTROCHEMICAL SOCIETY SERIES

Corrosion Handbook
Edited by Herbert H. Uhlig

Modern Electroplating, Third Edition
Edited by Frederick A. Lowenheim

Modern Electroplating, Fifth Edition
Edited by Mordechay Schlesinger and Milan Paunovic

The Electron Microprobe
Edited by T. D. McKinley, K. F. J. Heinrich, and D. B. Wittry

Chemical Physics of Ionic Solutions
Edited by B. E. Conway and R. G. Barradas

High-Temperature Materials and Technology
Edited by Ivor E. Campbell and Edwin M. Sherwood

Alkaline Storage Batteries
S. Uno Falk and Alvin J. Salkind

The Primary Battery (in Two Volumes)
Volume I *Edited by* George W. Heise and N. Corey Cahoon
Volume II *Edited by* N. Corey Cahoon and George W. Heise

Zinc-Silver Oxide Batteries
Edited by Arthur Fleischer and J. J. Lander

Lead-Acid Batteries
Hans Bode
Translated by R. J. Brodd and Karl V. Kordesch

Thin Films-Interdiffusion and Reactions
Edited by J. M. Poate, M. N. Tu, and J. W. Mayer

Lithium Battery Technology
Edited by H. V. Venkatasetty

Quality and Reliability Methods for Primary Batteries
P. Bro and S. C. Levy

Techniques for Characterization of Electrodes and Electrochemical Processes
Edited by Ravi Varma and J. R. Selman

Electrochemical Oxygen Technology
Kim Kinoshita

Synthetic Diamond: Emerging CVD Science and Technology
Edited by Karl E. Spear and John P. Dismukes

Corrosion of Stainless Steels
A. John Sedriks

Semiconductor Wafer Bonding: Science and Technology
Q.-Y. Tong and U. Göscle

Uhlig's Corrosion Handbook, Second Edition
Edited by R. Winston Revie

Atmospheric Corrosion
Christofer Leygraf and Thomas Graedel

Electrochemical Systems, Third Edition
John Newman and Karen E. Thomas-Alyea

Fundamentals of Electrochemistry, Second Edition
V. S. Bagotsky

Fundamentals of Electrochemical Deposition, Second Edition
Milan Paunovic and Mordechay Schlesinger

Electrochemical Impedance Spectroscopy
Mark E. Orazem and Bernard Tribollet

Fuel Cells: Problems and Solutions, Second Edition
Vladimir S. Bagotsky

Lithium Batteries: Advanced Technologies and Applications
Edited by Bruno Scrosati, K. M. Abraham, Walter van Schalkwijk, and Jusef Hassoun

Electrochemical Power Sources: Batteries, Fuel Cells, and Supercapacitors
Vladimir S. Bagotsky, Alexander M. Skundin, and Yurij M, Volfkovich